D0844691

Hydraulic Fracturing

G. C. Howard

Special Research Associate

Pan American Petroleum Corporation

C. R. Fast

Research Group Supervisor

Pan American Petroleum Corporation

Henry L. Doherty Memorial Fund of AIME

Society of Petroleum Engineers of AIME

New York 1970 Dallas

DEDICATION

We wish to dedicate this book to our wives, Anne and Virginia, without whose cooperation this book would not have been possible.

ISBN 0-89520-201-8

Contents

SPE Monograph Committees

1965

J. M. C. Gaffron, Chairman, Freeport Oil Co., New Orleans
Buck Joe Miller, Petroleum Consultant, Dallas
Henry J. Ramey, Jr., Stanford U., Stanford, Calif.
J. G. Richardson, Humble Oil & Refining Co., Houston
Shofner Smith, Phillips Petroleum Co., Bartlesville, Okla.
Bruce Vernor, Atlantic Richfield Co., New York City
M. R. J. Wyllie, Gulf Research & Development Co., Pittsburgh

1966

J. M. C. Gaffron, Chairman, Freeport Oil Co., New Orleans
Buck Joe Miller, Petroleum Consultant, Dallas
Don G. Russell, Shell Oil Co., New York City
M. R. J. Wyllie, Gulf Research & Development Co., Pittsburgh
Robert S. Cooke, Union Oil Co. of California, Midland, Tex.
Henry J. Ramey, Jr., Stanford U., Stanford, Calif.
Charles Blomberg, Chevron Oil Co., New Orleans
F. F. Craig, Jr., Pan American Petroleum Corp., Tulsa

1967

Henry J. Ramey, Jr., Chairman, Stanford U., Stanford, Calif.
M. R. J. Wyllie, Gulf Research & Development Co., Pittsburgh
Robert S. Cooke, Union Oil Co. of California, Midland, Tex.
Don G. Russell, Shell Oil Co., New York City
J. R. Elenbaas, Michigan Consolidated Gas Co., Detroit
B. F. Burke, Humble Oil & Refining Co., Long Beach, Calif.
B. L. Sledge, Humble Oil & Refining Co., New Orleans
C. Max Stout, Dowell, Tulsa

1968

Don G. Russell, Chairman, Shell Oil Co., New York City
Robert S. Cooke, Union Oil Co. of California, Midland, Tex.
J. R. Elenbaas, Michigan Consolidated Gas Co., Detroit
B. F. Burke, Humble Oil & Refining Co., Long Beach, Calif.
William E. Brigham, Continental Oil Co., Ponca City, Okla.
J. G. Richardson, Humble Oil & Refining Co., Houston
Lowell Murphy, Eppler, Guerin & Turner, Dallas
Roland F. Krueger, Union Oil Co. of California, Brea, Calif.

1969

B. F. Burke, Chairman, Humble Oil & Refining Co., Long Beach, Calif.
J. R. Elenbaas, Michigan Consolidated Gas Co., Detroit
William E. Brigham, Continental Oil Co., Ponca City, Okla.
J. G. Richardson, Humble Oil & Refining Co., Houston
James H. Henderson, Gulf Research & Development Co., Houston
Oren C. Baptist, U. S. Bureau of Mines, San Francisco
Richard L. Perrine, UCLA, Los Angeles
Robert A. Wattenbarger, Scientific Software Corp., Englewood, Colo.

1970

J. G. Richardson, Chairman, Humble Oil & Refining Co., Houston
William E. Brigham, Continental Oil Co., Ponca City, Okla.
Oren C. Baptist, U. S. Bureau of Mines, San Francisco
James H. Henderson, Gulf Research & Development Co., Houston
W. T. Bell, Schlumberger Well Services, Houston
Roland F. Krueger, Union Oil Co. of California, Brea, Calif.
George H. Bruce, Esso Production Research Co., Houston
Lincoln F. Elkins, Sohio Petroleum Co., Oklahoma City

Preface

Most books on technical subjects, although they represent the work and views of their authors, more correctly represent a summation of the work and thinking of many people. This Monograph is a book that, perhaps more than most, brings together the papers of many investigations and the thinking of many engineers involved in the study of hydraulic formation fracturing and its application to well stimulation.

It is interesting to note that at one time or another every major research laboratory associated with the oil industry has devoted considerable effort and money to the study of the fracturing process. This fact is dramatically demonstrated by the hundreds of technical papers and articles that we reviewed in preparing this book. These works have been used directly in writing this Monograph, and without the contributions of these authors, this book would not have been possible.

In writing a book of this type, the authors are inevitably indebted to more people than they are aware of. The works that have been read by the authors in the past and the discussions that have resulted in the molding of ideas and opinions often, unintentionally, are not acknowledged. However, several of our associates have contributed to this book. W. G. Bearden, J. B. Clark, R. F. Farris, D. H. Flickinger, M. A. Mallinger and F. H. Rixe are co-workers who have helped considerably in the preparation of this Monograph in various ways. Acknowledgment is given to Bill Bearden, Joe Clark and Floyd Farris for their early work on hydraulic fracturing, to Don Flickinger for his knowledge of and assistance with the mechanics and hydraulics of the process, to Arnold Mallinger for his contribution to the mechanics of the treatment, and to Fred Rixe for his work on the theory of fracturing.

E. H. Gras of the Halliburton Co. devoted considerable time and effort assembling information on hydraulic fracturing mechanical equipment. A. W. Coulter and H. E. David, Dowell Div. of Dow Chemical Co., devoted a similar amount of effort developing information on the many types of fracturing fluids and additives that are available to the industry. The help of these men and the cooperation of their companies made this Monograph more nearly complete.

We wish to thank the management of Pan American Petroleum Corp. for their cooperation; and in particular we wish to thank Lloyd Elkins and George Roberts, Jr., who have directed our work on hydraulic fracturing. Preparation of this Monograph would not have been possible without access to company files and technical reports.

GEORGE C. HOWARD
C. ROBERT FAST

Tulsa, Oklahoma
December, 1967

Chapter 1

Introduction

1.1 Scope

This Monograph is designed to be a thesis on hydraulic fracturing* covering the state of the art from the theory and technique of hydraulic fracturing to the application of nuclear energy as a means of cracking the reservoir rock and forming rubble. Considerable space in this book has been devoted to a study of the wells that are applicable to fracturing. The information is presented to provide the practicing engineer with procedures that will facilitate the selection of wells for this stimulation process.

Effectiveness of hydraulically created fractures is measured both by the orientation and areal extent of the fracture system and by the post-treatment production. Those factors that control or affect fracturing are reviewed and analyzed, and those that are controllable are discussed in detail. Methods of calculating the fracture extent are presented.

The hydraulic fracturing of wells has grown in complexity, with each service company adding to its service lists many combinations of fracturing fluids, fluid-loss control agents, and propping agents. To help the engineer in his planning and in his use of these aids, we have included here a comprehensive study of their relative merits. We have considered such factors as the fluid loss to be expected during fracturing, friction losses in the tubular goods, compatibility of the fracturing fluid with the reservoir, and well conditions. We also discuss the need to select a propping agent capable of maintaining a fracture of the desired flow capacity.

The chapter dealing with the mechanics of hydraulic fracturing covers the methods used by various operators to achieve the desired fracture placement and orientation and also discusses the preplanning that must be done if this type of stimulation treatment is to be successful. The problems of interzonal isolation and associated hydraulics also are discussed.

The chapter covering the evolution of the mechanical equipment used presents a history of fracturing. Such items as high volume, high pressure pumps; sand-liquid blenders; and down-hole opposed packer zonal-isolation tools were not in existence when fracturing was born. The fact that these mechanical devices were designed and built and that they worked is a story in itself. Discussion of these items provides the reader with an insight into the tools and procedures that make fracturing work.

The pressure of economics dictated the development of the optimized fracturing job — that is, a job designed to accomplish the desired results with a minimum expenditure of time and materiel. The method evolved considers such factors as size of tubular goods, pump rate, formation characteristics, fracturing fluid properties, cost of all the materiel used, the present worth of the oil, and the increase in ultimate recovery that might result from hydraulic fracturing.

The effect of hydraulic fracturing on well productivity and ultimate recovery is discussed from the standpoint of both theory and observation. This discussion provides the student of fracturing with a review of what can be expected from well stimulation by hydraulic fracturing. The effect of time on production decline is discussed, as are the factors causing this decline and the probability of successful retreatments.

Nuclear fracturing, an untried process in early 1967, is now definitely to be considered as a technique for the future. The calculated effect of such a treatment on oil or gas production, based on information obtained from the underground detonation of various sizes of nuclear devices, is also included in this book.

1.2 Objective

This Monograph proposes to serve a dual purpose: (1) to provide a comprehensive review of the state of the art, and (2) to provide the practicing engineer with a source of information that will aid him in

*Hydraulic fracturing, a method for increasing well productivity by fracturing the producing formation and thus increasing the well drainage area, was originally conceived and patented by R. F. Farris, U.S. Patent reissued Nov. 10, 1953. Re. 23733.

judging the relative merits of various hydraulic fracturing treating procedures and the results to be expected from such a stimulation method. In other words, this Monograph is a *basic* reference book and a *working* text for the practicing engineer.

1.3 Historical Background

The fact that injection pressure decreases when water, acid, cement or oil is pumped into a formation at a high rate and at a high initial pressure has been the subject of a number of studies.

Acidizing of Oil and Gas Wells

The Pure Oil Co. in cooperation with Dow Chemical Co. performed the first acidizing of an oil well (Feb. 11, 1932) on Pure's Fox No. 6 well in Midland County, Mich. A 15 percent (by weight) hydrochloric acid with an arsenic inhibitor was used. By 1934, acidizing was an accepted practice for well stimulation in areas where the producing formation was limestone.

From 1945 to 1963, the technological improvements in acidizing were basically limited to the development of acid fracturing techniques and materials, and to the use of surface active agents. No change in acid composition was noted during this period. As a result of the development of high-pressure, high-rate pumping equipment, oil and gas wells were acidized at fracture inducing rates and pressures.[1,2]

Fitzgerald. In his comments on J. B. Clark's paper on the Hydrafrac process[3], P. E. Fitzgerald reflected the thinking of many engineers when he stated that pressure parting or formation lifting played an important part in the treatment of many wells where fluids were injected.

The pressure parting phenomenon had long been recognized in well acidizing operations. For example, at pressures below those required to lift the overburden, the formation would take substantially no fluid, but when a pressure high enough to part or fracture the formation was reached, the rate of fluid injection could be raised with little or no increase in the injection pressure.

After the acid entered the formation the chemical reaction dissolved the formation, thus further enlarging the established fracture. Since the characteristics of the rock are not uniform, more rock was dissolved by the acid in some places than in others, so that when the treatment was concluded the pressure-parted fracture could not completely close, and remained open to serve as a flow channel to the well.

Water Injection

Dickey and Andersen. From their study of water input wells, Dickey and Andersen[4] concluded that when the pressure at the bottom of an input well was raised above a certain value, the well took much more water than it normally would take. This critical pressure at the sand face ranges between 1.0 and 1.7 psi per foot of depth in the northwestern Pennsylvania and eastern Illinois oil and gas fields (260 to 2,075 ft).

The critical pressure also was observed to vary with depth and was, therefore, some function of the weight of the overlying rock. Assuming the specific gravity of water-saturated sedimentary rocks to be about 2.2, the pressure of the overburden would be approximately 1.0 psi per foot of depth.

Dickey and Andersen concluded that these breakthroughs, or breaks in injection pressure, were the result of a rupture or fracture of the formation.

Grebe. In the same vein, Grebe[5] reported in 1943 that a sand formation in a well 810 ft deep was broken down with a brine solution at a surface pressure of 720 psi. Backflow tests indicated that there was an exact balance at 360 psi (surface pressure) at which water would go in or out of the formation with very little pressure change. The weight of the earth above the point of formation fracture was determined by adding 360 psi to the head of 810 ft of brine. The average density of the earth proved to be about 2.2 (0.9548 psi/ft).

Grebe cited another case: a well 3,000 ft deep in which a formation breakdown was observed at a surface pressure of 1,500 psi while the well was being acidized. The formation lifting factor was calculated to be 0.968 psi/ft.

Yuster and Calhoun. In their study of pressure parting of formations in waterflood operations, Yuster and Calhoun[6] observed that overburden lifting does not mean that the entire overburden from a given input point to the surrounding producers is lifted by the water and actually floated on it. (While such a situation is theoretically possible, it would be the very rare exception rather than the rule.) Lifting of the overburden was defined as the parting of the rock or matrix at any bedding plane by the injection of fluid at pressures in excess of those tending to hold the formation together.

The implication that the downward force in a lifting process is the complete weight of the overburden does not necessarily hold true at all times. The force depends upon the physical condition of the overlying strata and is controlled by such factors as plasticity, compressibility, elasticity, and attitude of the strata. An analogy may illustrate this point. If the overburden were made up of a series of pillows topped by a series of books, it would be possible to part the formation locally by compressing the pillow or pillows upward and/or downward without lifting the entire overburden. The point or elevation of the wellbore where the formation has the lowest tensile strength will break first, while other fractures may occur if the pressure is great enough.

If a fracture initiated by excessive pressures has impermeable boundaries, it will continue to advance into the formation until it reaches a sink or barrier.

Yuster and Calhoun[6] further stated that if one or both boundaries of the horizontal fracture formed by excessive pressure were permeable, the fracture would extend into the formation until the friction of the fluid flowing into it caused just enough drop in pressure to create a balance between the pressure in the liquid and the combined counter-force of the tensile strength of the formation and the downward pressure of the overburden.

In a second article on waterflooding, Yuster and Calhoun[7] concluded that the parting of formations in waterflood operations is indicated by a sudden increase in the rate of input without a corresponding increase in pressure. A graph of water input rate vs pressure might even show, at the point where parting or fracturing of the formation occurs, a definite decrease in injection pressure.

The formation parting factor for injection wells in the Bradford and Allegheny fields in Pennsylvania varied from 0.8 to 1.4 psi/ft. The low parting factors were confined generally to wells on the crest of the structure while the higher parting factors were associated with wells low on the structure. A plot of these factors vs depth indicated a trend of decreasing factors with increasing depth.

As a side-light to their study of waterflooding, Yuster and Calhoun noted that for successful acidizing and squeeze-cementing operations, pressures must be high enough to part the formation.

Squeeze Cementing

Torrey. Early recognition of the fracturing phenomenon in squeeze cementing was reported by P. D. Torrey.[8] He presented geological and engineering information to show that the fluid pressures involved in squeeze cementing part the rocks, generally along bedding planes or other lines of sedimentary weakness. The fracture formed provides channels or passageways in which the cement slurry can lodge beyond the wall of the hole. This phenomenon has been confirmed from cores of sand formations in sidetracked holes adjacent to sections of hole that have been squeezed previously. The cores reveal that the cement slurry sets as relatively thin laminations between the individual sand beds.

Rock samples (Fig. 1.1) recovered in shallow well squeeze-cementing tests[9] illustrate this parting of the formation.

Teplitz and Hassebroek. In their investigation of squeeze cementing, Teplitz and Hassebroek[10] observed that to inject the cement slurry injection pressures must be great enough to lift the overlying formations. This allows the cement slurry to flow into the parting between the formations in a pancake shape and form a barrier to vertical movement of fluids in the formation surrounding the casing. Teplitz and Hassebroek reasoned that it would be advisable to stop the injection of cement shortly after the pressure against the formation has exceeded that of the overburden, and after a reasonable quantity of cement has been forced out, since prolonged injection of cement at high pressure can only fracture the zone to a degree detrimental to the well. The sand core pictured in Fig. 1.2 demonstrates the results of such harmful action. This core was obtained from the producing horizon in a sidetracked well after the original hole had been subjected to a number of squeeze jobs. According to the best estimates, the lateral displacement of this core from the squeeze zone in the original well was approximately 21 ft. It should be noted that the prominent cement vein runs perpendicular to the plane of bedding in the sand.

A microscopic examination of the core revealed a thin filter cake, containing barites, between the cement and the face of the sand; but no trace of cement or mud

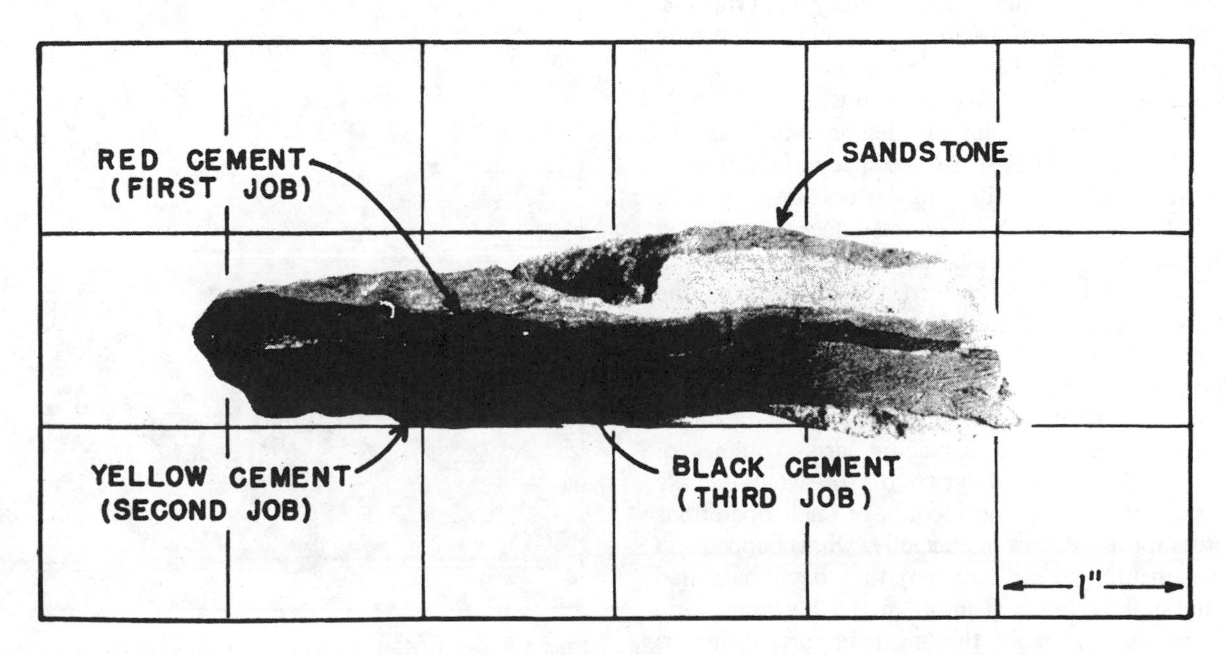

Fig. 1.1 Relative location of cement pancakes on three successive squeeze cementing operations.

Fig. 1.2 Sand core resulting from continued injection of cement.[10]

particles was visible within the sand itself. This is in agreement with the results of laboratory experiments, which indicate that even at pressures greatly exceeding those involved in cement squeezing, cement slurries can be injected into only the coarsest sands.

Howard and Fast. A similar phenomenon was observed by Howard and Fast[9] in two special field tests in which a "rat hole" was drilled in the bottom of a well. The rat-hole portion of the well was squeeze cemented and later cored. The results of the test revealed a cement-filled vertical fracture in the bottom of the 9,530-ft well (Fig. 1.3) and a horizontal fracture filled with cement in the 2,635-ft well (Fig. 1.4).

Walker. In commenting on the Teplitz-Hassebroek paper, A. W. Walker added considerable insight to the observed phenomenon of displacing large volumes of cement into a restricted interval of the wellbore. He stated that in reviewing the records of such operations in an attempt to determine exactly what happens in squeeze cementing, he discovered that one phase had been only slightly touched upon in the literature, and that phase was probably the most important factor: vertical fissuring or fracturing.

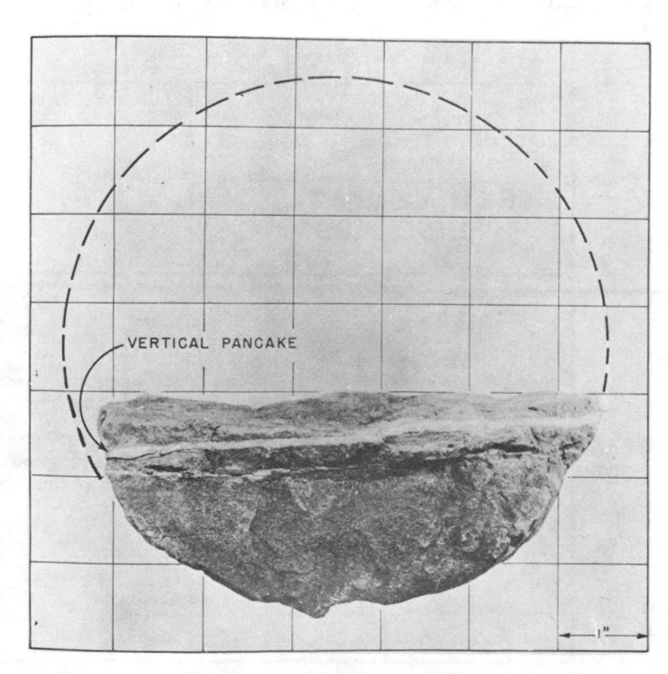

Fig. 1.3 Core from 9,530-ft well, showing vertical cement pancake.[9]

Fig. 1.2 is a photograph of a cement-filled, vertical fracture. Although well records available for study indicated that vertical fractures probably existed, this was the first time positive evidence was available. Analytical studies of the forces involved during squeeze cementing indicated that vertical fracturing was to be expected. It is difficult, according to Walker, to conceive how a formation can be parted horizontally and lifted unless the hydraulic pressures exceed those that would correspond to the weight of the overburden. Yet, in more than half of the cases studied in which cement was squeezed into the formation, the pressures at the formation were considerably less than the corresponding theoretical weight of the overburden. In fact, it is entirely possible to fracture formations vertically at lower pressures and thus inject large quantities of cement.

Reistle. To control the influx of water or gas into a wellbore Reistle[12] proposed placing a disk of cement or other impervious material about the well. To place the disk, the formation was weakened at the desired place by some suitable method such as underreaming, and then a "cement squeeze job" was performed at the weakened point.

Well Stimulation with High Explosives

In the years between 1890 and 1950 the oil industry in the U. S. and Europe used liquid and later solidified nitroglycerin to stimulate wells.[12] Although the hazards associated with the use of liquid explosives limited their use, these materials were immediately and spectacularly successful for oilwell shooting. The object of shooting a well was to fracture or rubble the oil-bearing formation to increase both the initial flow and the ultimate recovery of oil. This same fracturing principle was soon applied with equal effectiveness to water and gas wells.

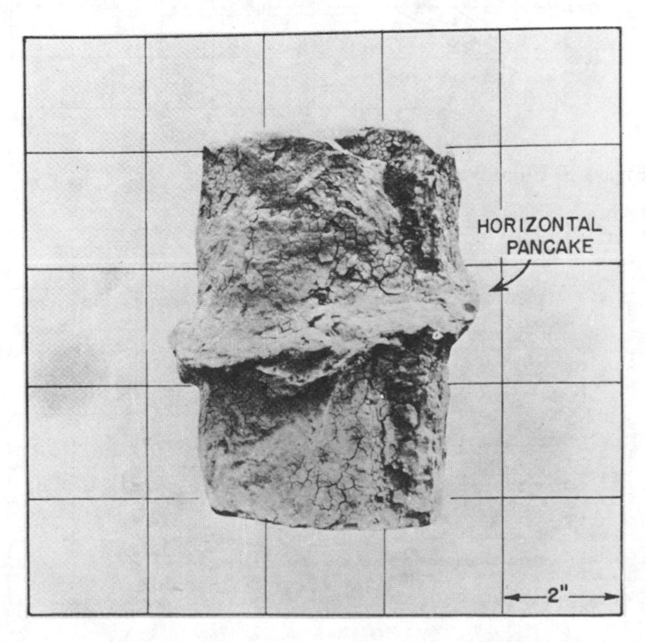

Fig. 1.4 Core from 2,635-ft well, illustrating horizontal pancake.[9]

In 1949, Grant *et al.*[13] conducted a series of shallow-well tests to study the effect of well shooting on the injection rate vs injection pressure relationship of an input well. A plot of typical data obtained from these tests is shown in Fig. 1.5. Note that the higher injection capacity of the well after shooting as compared with that after "breakthrough" or breakdown of the formation was probably due to the failure to extend and prop open the hydraulically formed fracture.

Extensive shattering of the wellbore made major postshooting clean-up operations necessary. The enlarged wellbores and frequently damaged well casing prevented subsequent selective treatment of the producing interval.

The advent of commercial hydraulic fracturing in 1949, coupled with the danger of damaging the well being treated and the possibility of severe injury or death to the person handling or "loading" the well with a high explosive charge, has, for all practical purposes, eliminated oil and gas well shooting.

1.4 Field Study of Hydraulic Parting of Formations

Texas-Louisiana Gulf Coast Area

To understand "formation breakdowns", or fracturing, it is necessary to study acidizing, oil and cement squeezing, and water injection. Farris[14] approached the study by carefully investigating well files to establish a relationship between observed well performance and

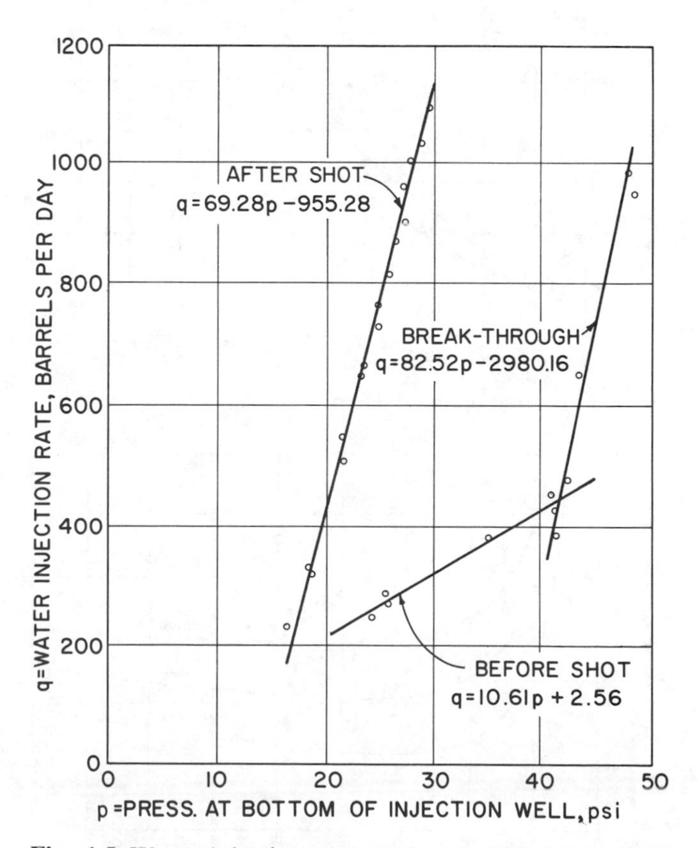

Fig. 1.5 Water injection test, Hole 14. (shot with 17.75 lb solidified nitroglycerin loaded 2.96 lb/ft in 2½-in. × 6-ft shell).[13]

treating pressures. He studied the records of 115 cement squeeze jobs in the Gulf Coast area, and calculated the down-hole pressure (at the elevation of casing perforations) required to "break down" the formation. These data are shown on Fig. 1.6.

The trend shown in Fig. 1.6 could not be represented by straight lines passing through the origin, but rather by a maximum, average and minimum curve. Since formation breakdown pressures are equal to the pressure required to fracture the rock (i.e., to overcome the tensile or compressive strength) plus the effective overburden pressures, it was apparent that the pressure required to rupture the rock would dominate at the shallow depths and deflect the curve away from the origin. In general, therefore, these data indicated that formations were fractured when they were "broken down" prior to squeezing cement and that the breakdown pressure was greater than effective overburden pressure at the depths investigated.

Fig. 1.7 shows the pressure at the perforations at the time of the breakdown divided by the depth to the perforations in each case. These data suggest that formation lifting factors are large at shallow depths, reach a minimum at about 7,000 ft, then gradually increase at the greater depths. The arithmetic average of all points gives a formation lifting factor of 0.85 psi/sq in./ft.

Fig. 1.8 is a plot of the effective overburden pressure (pressure at the perforations while pumping mud after the first breakdown) vs the depth of the perforations. It is the strength of the rock that prevents the curves from passing through the origin. The "average" curve, for example, is displaced from the origin by approximately 750 psi.

Fig. 1.9 represents the effective overburden pressure gradient as determined by dividing the pressure at the perforations while pumping mud after formation breakdown by the depth to the perforations. These data also showed that the effective overburden pressure gradient tended to be greater at the shallow depths. The arithmetic average of all these points showed a pressure gradient of 0.725 psi/ft.

Fig. 1.10 presents a comparison of the various estimates of overburden pressures. What most observers call overburden pressure compares reasonably well with a curve drawn through points that describe maximum formation breakdown pressure. The average breakdown pressure of formations determined from field data was observed to be less than most theoretical estimates, except at very shallow depths, and effective overburden pressure was observed to be less than what most ob-

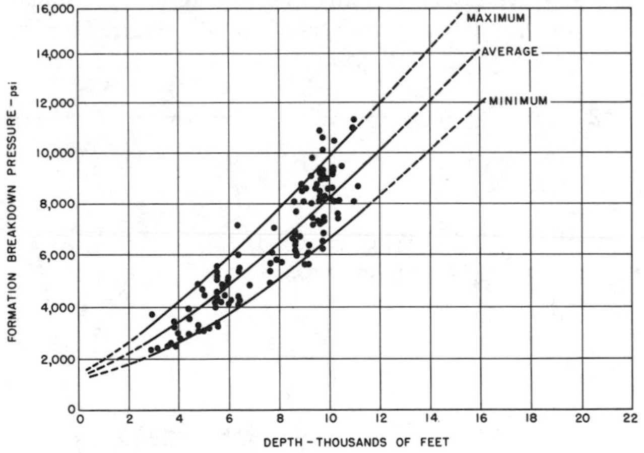

Fig. 1.6 Formation breakdown pressures.[14]

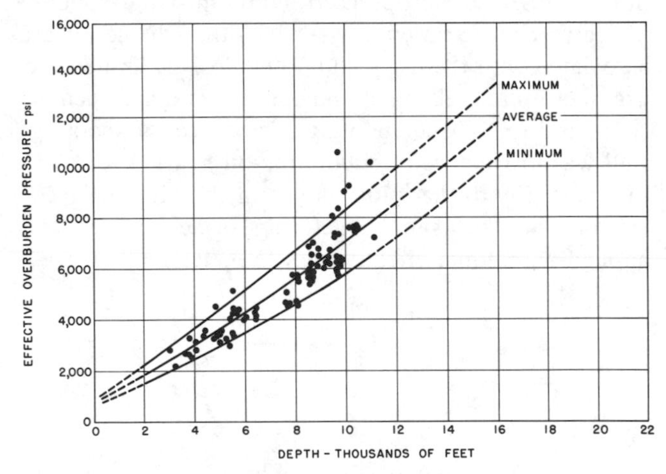

Fig. 1.8 Effective overburden pressure.[14]

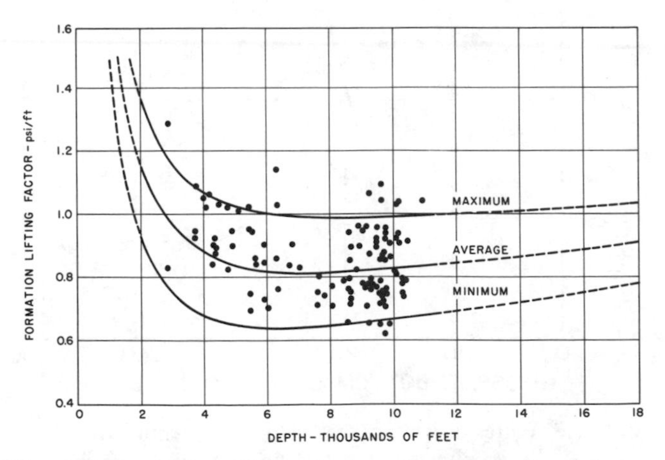

Fig. 1.7 Formation lifting factors based on breakdown pressures.[14]

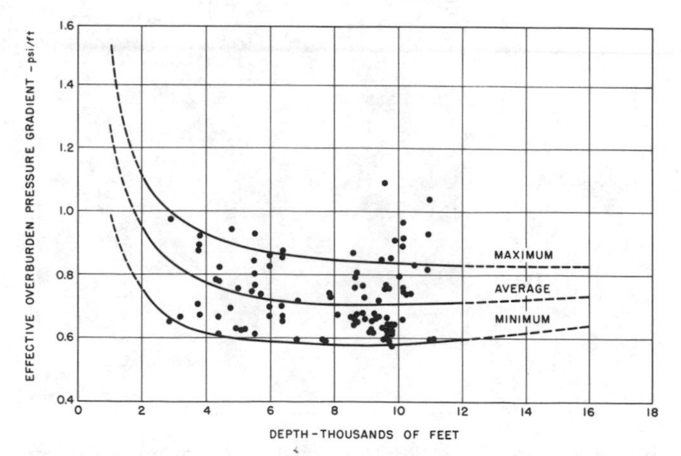

Fig. 1.9 Effective overburden pressure gradients.[14]

servers consider to be overburden pressure.

Well records on cement squeeze jobs showed that many wells had been squeeze cemented several times at or near the same depths. When plotting the information on Figs. 1.6 through 1.10, it was observed that formation parting factors (lb/sq in./ft) tended to become larger as the number of squeeze jobs performed at a given elevation was increased. To illustrate this, formation lifting factors of wells that had been squeeze cemented more than once are plotted on Fig. 1.11. Since the wells used in the construction of this graph were not all the same depth, there was a wide spread of points.

Oklahoma, Kansas, North Texas and New Mexico Areas

The study of formation breakdown pressures was expanded to include acidizing jobs and squeeze-cementing operations in the Oklahoma, Kansas, North Texas and New Mexico areas. These data are shown on Fig. 1.12.

Fig. 1.13 shows the effective overburden pressure on these Mid-Continent wells as indicated by the pump pressure required for injection after the formations were broken down.

Figs. 1.14 and 1.15 present a comparison of average formation breakdown pressures and average effective overburden pressures for the Texas-Louisiana Gulf Coast and the Oklahoma, Kansas, North Texas and New Mexico area wells.

Fig. 1.16 compares the theoretical overburden pressure (rock density method) with the effective overburden pressure calculated from field data; it also compares the corresponding overburden pressure gradients. The reasonably close agreement between the theoretical overburden pressure and the average field data at the shallow depths probably explains why many observers have drawn erroneous conclusions regarding overburden pressure at the greater depths.

Fig. 1.17 was constructed by duplicating Fig. 1.9 and adding the formation lifting factors determined by Yuster-Calhoun, Dickey-Andersen, and Grebe. Note that the values derived by these observers fall within the extrapolated portion of the curves based on well-file formation breakdown data. The data taken at shallow depths by these observers agree with the information that is not limited to shallow depths. It is apparent that the so-called overburden pressures and formation lifting factors derived by these observers are applicable and correct only at the depths at which they were observed.

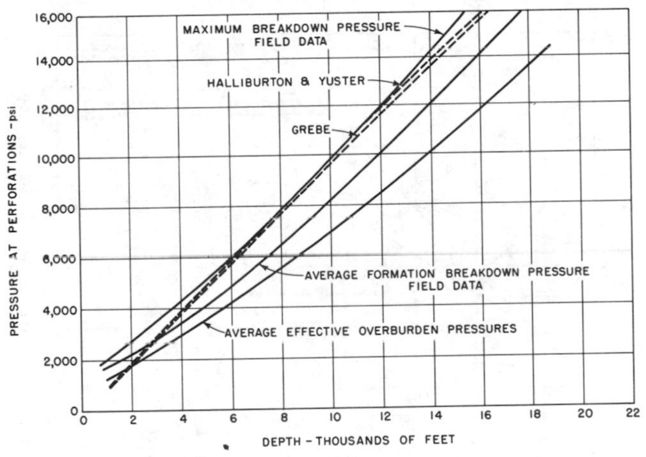

Fig. 1.10 Comparison of various estimates of overburden pressures.[14]

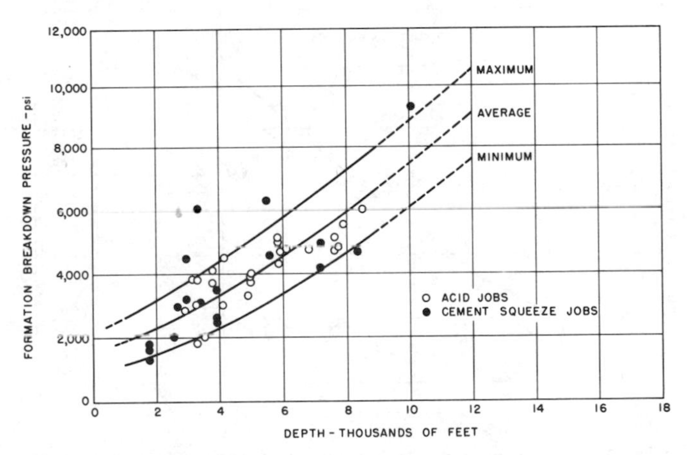

Fig. 1.12 Formation breakdown pressures.[41]

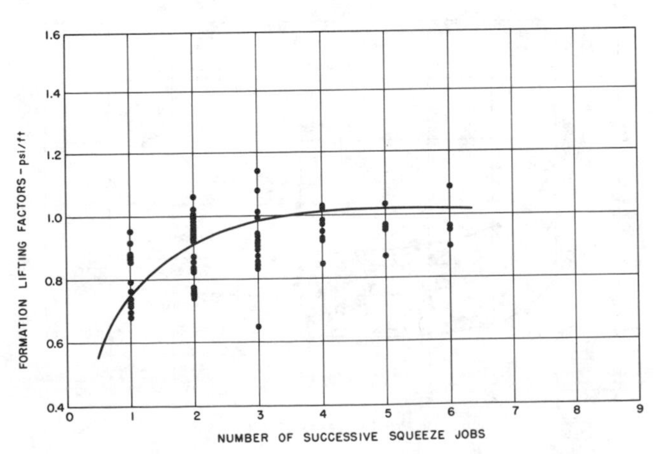

Fig. 1.11 Trend of formation lifting factors with number of squeeze jobs.[14]

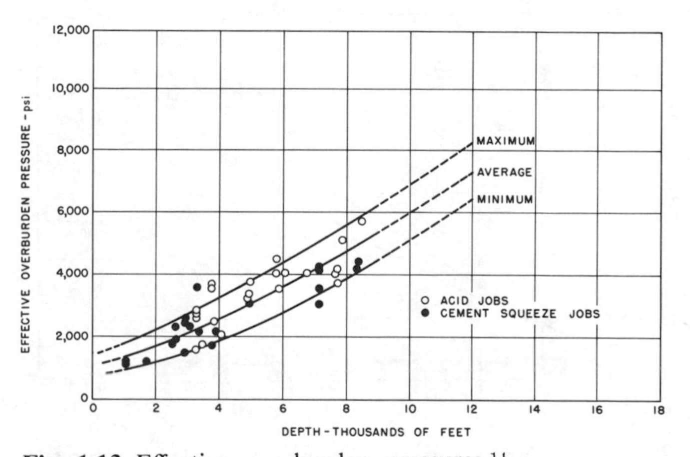

Fig. 1.13 Effective overburden pressures.[14]

It must be kept in mind that effective overburden pressure is the breakdown pressure that controls well manipulations. This pressure varies with folding and faulting (tectonic) stresses in the earth at a given depth and can be defined only between certain limits. The magnitude of effective overburden pressure and the limits of its variation at any given depth in oil fields have been described with reasonable accuracy in Fig. 1.8.

1.5 The First Hydraulic Fracturing Operation

Hydraulic fracturing was introduced first in the Hugoton gas field in western Kansas in 1947. This treatment was conducted on the Klepper No. 1 well located in Grant County. The well was completed with four gas-productive limestone pay zones from 2,340 to 2,580 ft. The bottom-hole pressure was approximately 420 psi. Klepper No. 1 had been completed originally with a down-hole acid treatment and was chosen for hydraulic fracturing because it had a low deliverability and would offer a direct comparison of acidizing and fracturing.

The mechanical pumping equipment used consisted of a centrifugal pump for mixing the gasoline-base napalm gel fracturing fluid and a positive displacement pump for pumping the gel into the well. Fig. 1.18 shows the layout of the fracturing equipment at the well site. Due to the fire hazard, all units, including the mixing tanks, were placed 150 ft apart, which greatly complicated the first operation.

This particular Hydrafrac treatment actually involved four separate treatments — one on each of the three perforated zones and one on the bottom open-hole section — that were conducted through tubing equipped with a cup-type straddle packer. The treatment of each zone consisted of 1,000 gal of napalm-thickened gasoline followed by 2,000 gal of gasoline containing 1 percent of an amine gel breaker.

There was a sharp break in treating pressure during the initial stages of the treatment of each of the three perforated zones. The bottom open-hole section apparently received all of the initial acid treatment, since there was no formation breakdown.

Flow meter tests, conducted after the well was cleaned up by blowing it down, revealed that the relative producing characteristics of the four zones were altered by the fracturing treatments. However, the deliverability of the well was not changed appreciably.

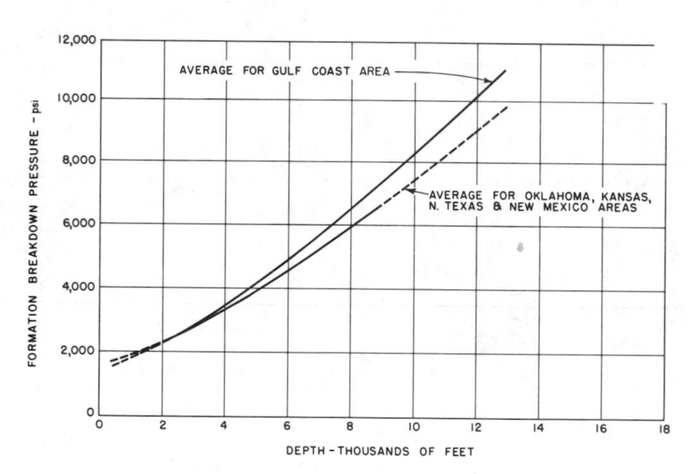

Fig. 1.14 Comparison of average formation breakdown pressures.[14]

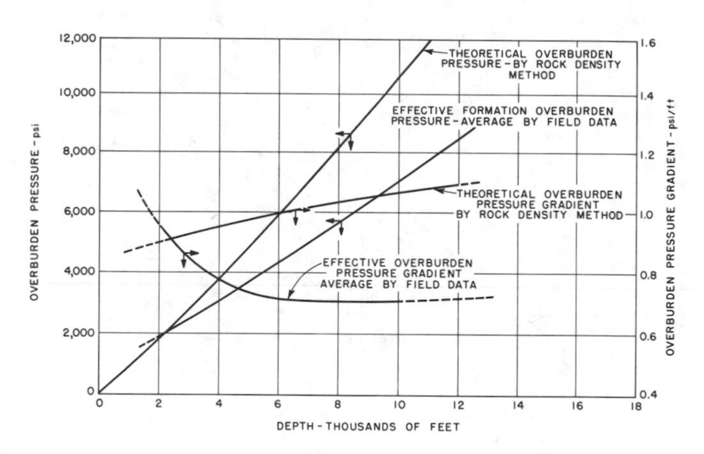

Fig. 1.16 Comparison of theoretical with effective overburden pressure.[14]

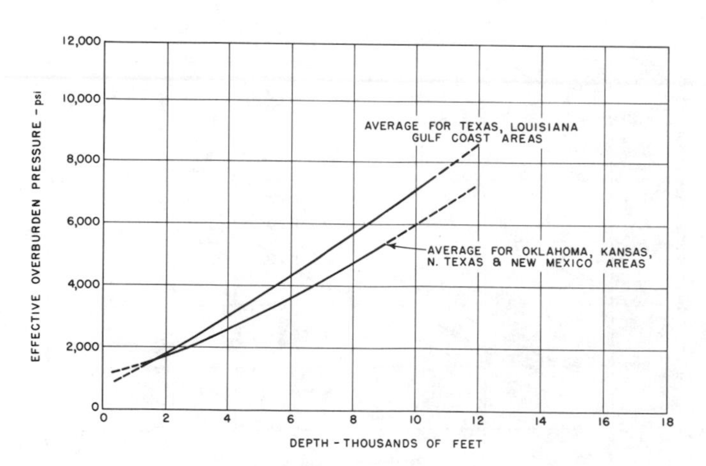

Fig. 1.15 Comparison of effective overburden pressures.[14]

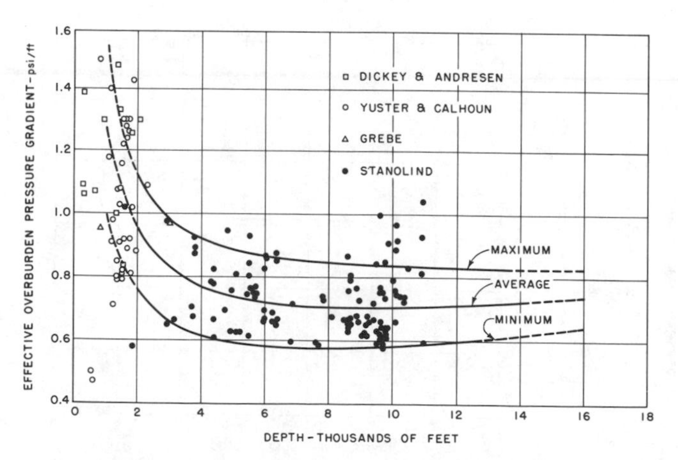

Fig. 1.17 Comparison of various estimates of effective overburden pressure gradients.

It was concluded that the Hydrafrac process could not compete economically with acidizing in this limestone reservoir. However, the assumption has been disproved, since by 1966 the primary method of stimulation in this field was hydraulic fracturing. The use of large volumes of low-cost water-base fluid pumped at very high rates has proved to be an effective and economical procedure for fracturing and stimulating wells. Very little acid stimulation was used in the second and third restimulation workovers in the Hugoton field. Therefore, the place of inception of hydraulic fracturing, where initial response was so small that the economics of the process were questionable, has since proved to be the area in which treatments using the largest volumes and highest injection rates are the most successful.

1.6 Uses of Hydraulic Fracturing

Hydraulic fracturing has been used to accomplish four basic jobs: (1) overcome wellbore damage, (2) create deep-penetrating reservoir fractures to improve the productivity of a well, (3) aid in secondary recovery operations, and (4) assist in the injection or disposal of brine and industrial waste material.

Overcoming Wellbore Damage

Modern drilling operations are geared to using fluids that will assist in effecting the most efficient drilling system. Many of these fluids and the solids in them invade the matrix and cause damage to formation permeability. This phenomenon frequently causes a marked reduction in the ability of the producing formation to flow oil or gas into the bore.

Fracturing, as practiced early in the history of this stimulation procedure, consisted of making very short radius fractures. In most instances, penetration into the surrounding rock was limited to 10 or 20 ft. Production increases ten- to fifty-fold over pretreatment rates can be explained only by the fracture's breaking through a damaged zone immediately adjacent to the well. The success of multi-fracturing (using successive small batches of fracturing fluid to create numerous short radius fractures at various elevations in the well) further illustrates the advantage to be gained from breaking through a damaged wellbore.

Making Deep-Penetrating Reservoir Fractures

A further study of the effect of fracture systems on well production revealed that only by utilizing deep-penetrating, high-flow-capacity fracture systems can a low-resistance flow path to the well be made in large portions of the reservoir. The deeply invading fracture provides the average reservoir with the means of producing large volumes of oil. In addition, it provides large drainage areas into which formations of very low permeability can slowly feed oil, thus utilizing all available reservoir energy to the maximum. It was the recognition of this latter phenomenon that opened to commercial exploitation many areas that previously had been considered non-productive.

The deep-penetrating reservoir fracture, with its ability to rejuvenate and extend the economic life of a well and to open new areas to commercial production, is responsible in large measure for the more than 7 billion bbl of increased oil reserves.

Aiding Secondary Recovery Operations

In the field of secondary recovery of oil, fracturing has played two important roles: (1) it has increased the capacity of the water injection well to accept fluid at a predetermined pressure, and (2) it has produced high capacity flow channels into the producing well, thus increasing the efficiency of the gas or waterflooding project.

By selectively placing the fracture, the effects of gravity drainage can be utilized at the desired point in a well.

The thermal combustion recovery process, in many instances, has required help from the fracturing process to insure that the proper quantities of air are injected. Some processes even rely on preformed fractures to establish a duo-directional burning front.

In secondary recovery operations, fracturing is the means by which the point of fluid injection is selectively placed, and it furnishes a way for producing wells to capture displaced oil.

Disposing of Oilfield Brines

The large volumes of salt water produced by some oil wells threatened to limit oil production severely. However, it was found that with the aid of fracturing, a low-pressure, high-fluid-injection capacity well could be established anywhere.

The adaptation of the hydraulic fracturing process to the disposal of industrial waste has gained wide acceptance by industry. The Atomic Energy Commission

Fig. 1.18 Klepper Gas Unit No. 1, Hugoton field, Kans. The first well to be hydraulically fractured to increase well productivity.

has experimented extensively with the use of this process to dispose of radioactive material.

1.7 Summary

The concept of fracturing or formation breakdown, recognized by the oil industry many years ago, played a very important role in well stimulation, acidizing, water injection and cementing. While little understood, it provided a means for fracturing a formation by displacing a fluid into a matrix at rates and volumes greater than the rock itself could accept.

In all of these early well completion-workover operations only the problem of injection was considered, and the effect of such actions on well productivity was generally overlooked. Some investigators were concerned over the possibility that water would invade a well as a result of fractures caused by squeeze-cementing operations.

Most investigators agree that the theoretical overburden pressure of the earth is equal to 1.0 psi per foot of depth. The effective overburden pressure, or that pressure observed to be required to inject fluid into a fractured well, varies from 0.75 to 1.11 psi/ft at 2,000 ft, to 0.57 to 0.85 psi/ft at 8,000 ft.

Hydraulic fracturing has found wide usage as a well stimulation procedure for overcoming damaged matrix permeability surrounding a wellbore, and for creating deep-penetrating fractures that provide high capacity channels for flow from deep within the producing formation to the well. Water injection and waste disposal are also important uses for this process.

Klepper No. 1, the first well to be hydraulically fractured, heralded a new well stimulation procedure that used hydraulic pressure accompanied by the injection of large volumes of fracture forming material at high rates to break the rock and form flow channels to the well. The procedure has become the most widely used well stimulation method ever developed.

References

1. Hendrickson, A. R. and Wieland, D. R.: "Personal Interview with R. C. Phillips, Union Oil Co. of California", Dowell Div. of Dow Chemical Co., Tulsa, Okla. (July 5 and 14, 1966).
2. Grebe, John J. and Stosser, S. M.: "Increasing Crude Production 20,000,000 Barrels from Established Fields", *World Petroleum* (Aug., 1935) **6**, No. 8, 473.
3. Clark, J. B.: "A Hydraulic Process for Increasing the Productivity of Wells", *Trans.*, AIME (1949) **186**, 1-8.
4. Dickey, P. A. and Andersen, K. H.: "Behavior of Water Input Wells — Part 4", *Oil Weekly* (Dec. 10, 1945).
5. Grebe, J. J.: "Tools and Aims of Research", *Chem. and Eng. News* (Dec. 10, 1943) **21**, No. 23, 2004.
6. Yuster, S. T. and Calhoun, J. C., Jr.: "Pressure Parting of Formations in Water Flood Operations — Part I", *Oil Weekly* (March 12, 1945).
7. Yuster, S. T. and Calhoun, J. C., Jr.: "Pressure Parting of Formations in Water Flood Operations — Part II", *Oil Weekly* (March 19, 1945).
8. Torrey, P. D.: "Progress in Squeeze Cementing Application and Technique", *Oil Weekly* (July 29, 1940).
9. Howard, G. C. and Fast, C. R.: "Squeeze Cementing Operations", *Trans.*, AIME (1950) **189**, 53-64.
10. Teplitz, A. J. and Hassebroek, W. E.: "An Investigation of Oil Well Cementing", *Drill. and Prod. Prac.*, API (1946) 76.
11. Reistle, C. E.: U. S. Patent No. 2,368,424 (Jan. 30, 1945).
12. *Blaster's Handbook* (du Pont), Sesquicentennial ed., 49 and 443.
13. Grant, B. F., Duvall, W. I., Obert, L., Rough, R. L. and Atchison, T. C.: "Research on Shooting Oil and Gas Wells", *Drill. and Prod. Prac.*, API (1950) 303.
14. Farris, R. F.: unpublished report, Pan American Petroleum Corp., Tulsa, Okla. (1946).

Chapter 2

Theories of Hydraulic Fracturing

2.1 Introduction

Hydraulic fracturing may be defined as the process of creating a fracture or fracture system in a porous medium by injecting a fluid under pressure through a wellbore in order to overcome native stresses and to cause material failure of the porous medium. Briefly, it is the creation and preservation of a fracture in a reservoir rock. To fracture a reservoir rock, energy must be generated by injecting a fluid down a well and into the formation.

To understand the fracturing process, one must consider how losses in energy hamper the efficiency of the process. If it were possible to deliver the fracturing force directly to the formation, as can be done in a laboratory, there would be little concern other than for the actual mechanics of failure of the reservoir rock. However, to fracture the actual formation, energy that is delivered at the wellhead by pumping equipment must be transmitted underground. This process of energy transmission is hampered by energy losses due to (1) frictional pressure drop in the injection string, (2) viscous pressure drop due to flow through the fracture itself, and (3) pressure drop due to flow (leakoff) from the fracture into the formation. A considerable portion of this Monograph is devoted to discussing the various sources of energy loss in the fracturing process. In this chapter, however, we shall consider theories of fracturing as applied to rock formations without particular regard to the mechanics of energy transmission.

This chapter presents the principal theories of material failure and relates them to hydraulic fracturing of rock formations. This method of well stimulation has been used more than half a million times, and over a period of time, the fracturing materials and techniques have been greatly improved. In fact, they have improved much faster than has the understanding of the mechanics of in-situ fracturing of the actual rock formations.

Several schools of thought exist about the mechanics of hydraulic fracturing. The differences in opinion may be related to fundamental assumptions such as (1) the concept of failure, (2) elasticity vs plasticity, (3) the state of stress in general, (4) the effect of fluid penetration into the formation, and (5) homogeneity and isotropy vs heterogeneity.[1] Another cause of disagreement is the meagerness of such direct data as the subsurface condition under which the type of fracture was formed at the well, and the orientation of the fracture in the interwell area.

Card,[2] in his review of fracturing theories, states that the general problem of failure of materials must be considered as two problems: theories of strength, associated with the theory of deformation of sizable bodies; and theories concerned with the mechanism of fracture itself. These two problems have been considered separately in the past because the study of strength and its practical application is devoted almost entirely to macroscopic phenomena, whereas fracture theories have been concerned more with problems on a microscopic scale. In recent work, the tendency has been to use an energy criterion to relate theories of energy and theories of fracture, since a body of any size may be submitted to strength analyses and since a crack or fracture extending through a rock may become large compared with its original, microscopic size. The real difference between these two types of theories is that the strength theories describe the ultimate conditions leading to failure, whereas the fracture theories attempt to reconstruct the reasons for the beginning of failure and to define the conditions under which the fracture may be extended.

2.2 Concept of Failure

Attempts have been made to apply several theories of the mechanics of failure to rock that is subjected to internal fluid pressure. These are: (1) maximum-stress theory, (2) maximum-shearing theory, (3) maximum-strain theory, (4) maximum-strain-energy theory, (5) Mohr's theory, (6) Coulomb-Navier theory, (7) Griffith's theory, (8) octahedral-shearing-stress theory, (9) Walsh-Brace theory, (10) single-plane-of-weakness theory, and (11) variable-cohesive-strength theory.

There is frequent disagreement among authors concerning the orientation of fractures. However, most authors work from the assumption that when the pressure is increased in the hole, rupture will occur in the plane that is perpendicular to the direction of the least compressive stress. Hubbert demonstrated this premise in experiments with a gelatin model.[3] By using a nonpenetrating plaster of paris "fracturing fluid", and by creating pre-established states of stress, he was able to anticipate the planes of fracture.

2.3 Elasticity vs Plasticity of Formation Rock

Most investigators assume that rock surrounding the wellbore is in the elastic state of stress. For example, Miles and Topping[4] assumed elastic rock properties and concluded from their study that a plastic deformation of the rock through geologic time tends to mitigate the stress concentrations around the well.

Teplitz and Hassebroek[5] suggested that during squeeze operations cement could be lost through enlargement of the wellbore caused by squeeze cementing pressures. Wellbore enlargement was made possible by plastic deformation of shales.

Both McGuire *et al.*[6] and Hubbert[3] suggested that plastic deformation of rock could be expected, particularly in rock salt, clays, shales, and unconsolidated sands.

Scott *et al.*[7] discussed the possibility that rock in a plastic state of stress immediately surrounding the wellbore is itself surrounded by undisturbed rock in the elastic state of stress. This thought has been expressed mathematically by both Fenner[9] and Westergaard.[8]

The stresses that cause failure at any point in the well may be calculated on the assumption that the well performs as a thick-walled cylinder of infinite wall thickness. The problem, therefore, is to find the stresses at any point in the material in terms of the internal and external pressure and radii of the internal and external surfaces. These values of the stresses depend upon whether the material is assumed to be in an elastic or in a plastic state.

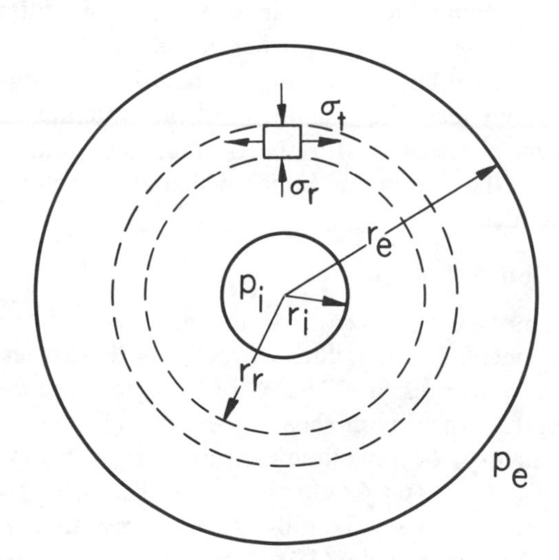

Fig. 2.1 Nomenclature for thick-walled cylinders.[7]

Material in Elastic State

When the material is assumed to be elastic, the stresses in a cylinder are given by Lamé's formulas.[14] In the following formulas the axial stress is assumed to be zero. This assumption facilitates correlating the formulas with results of experiments in which cylinders were ruptured while free to expand in the axial direction.

Lamé's Theory.[14] Lamé's theory was expressed mathematically by Scott *et al.*[7] as follows:

$$\sigma_t = \frac{p_i r_i^2 - p_e r_e^2 + \dfrac{r_i^2 r_e^2}{r_r^2}(p_i - p_e)}{r_e^2 - r_i^2}, \quad \ldots \ldots \quad (2.1)$$

$$\sigma_r = \frac{p_e r_e^2 - p_i r_i^2 + \dfrac{r_i^2 r_e^2}{r_r^2}(p_i - p_e)}{r_e^2 - r_i^2}, \quad \ldots \ldots \quad (2.2)$$

where, as shown in Fig. 2.1, σ_t is the circumferential or tangential stress (a positive value is a tensile stress), and σ_r is the radial stress (a positive value is a compressive stress).

By making r_r in Eqs. 2.1 and 2.2 equal to r_i, the maximum stresses are at the inner surface and are

$$(\sigma_t)_{\max} = \frac{p_i(r_i^2 + r_e^2) - 2p_e r_e^2}{r_e^2 - r_i^2} \quad \ldots \ldots \quad (2.3)$$

$$(\sigma_r)_{\max} = p_i \quad \ldots \ldots \ldots \ldots \quad (2.4)$$

If the internal pressure is p_i and the external pressure (p_e) equals zero,

$$\sigma_t = p_i \frac{r_i^2}{r_e^2 - r_i^2}\left(\frac{r_e^2}{r_r^2} + 1\right), \quad \ldots \ldots \quad (2.5)$$

$$\sigma_r = p_i \frac{r_i^2}{r_e^2 - r_i^2}\left(\frac{r_i^2}{r_r^2} - 1\right) \quad \ldots \ldots \quad (2.6)$$

These equations show that the maximum values of σ_t and σ_r occur at the inner surface when r_r equals r_i. The equations also show that σ_t is always greater than σ_r and is always a tensile stress.

To determine when failure occurs in the thick-walled cylinder, many criteria have been proposed that are founded on Lamé's expressions of stresses and pressure. In the discussion of these criteria, the terms "inelastic action" and "rupture" are used synonymously, since it has been observed that when samples of rock formations at atmospheric temperature are subjected to loads, inelastic action and rupture occur almost simultaneously. This observation may be substantiated by plotting the stress-strain relationship of samples of rock subjected to direct tensile loads. The increase in stress between the elastic limit and the rupture point will be small. A comparison of the relationship of stress to strain for sandstone and ductile steel is illustrated in Fig. 2.2.

Maximum-Principal-Stress Theory. Timoshenko,[10] Nadai,[11] and Seely[12] define maximum stress as the criterion for strength. They assume that for ductile materials yielding starts when the maximum stress becomes equal to the yield-point stress of the material in simple tension, or when the minimum stress becomes equal to the yield-point stress of the material in simple compression. This theory disregards the normal or shearing stresses that occur in other planes. Homogeneous and isotropic materials that may be weak in compression can be subjected to large hydrostatic pressures without yielding.

According to the maximum-principal-stress theory. inelastic action occurs when the maximum principal stress at any point reaches a value equal to the tensile strength of the material. Therefore, inelastic action should occur first when σ_t equals the value of stress at the elastic limit of the material in pure tension.

By letting $r_r = r_i$ in Eq. 2.5 (Scott *et al.*[7]),

$$(\sigma_t)_{\max} = p_i \frac{r_e^2 + r_i^2}{r_e^2 - r_i^2} \quad . \; . \; . \; . \; . \; . \; . \; . \; . \quad (2.7)$$

When the value of the external radius is large with respect to the internal radius, $\sigma_t = p_i$.

The maximum-principal-stress theory may then be summarized as follows:

The maximum internal pressure to which a cylinder of infinite thickness may be subjected before inelastic action occurs is equal to the ultimate strength of the material in tension.

Maximum-Shearing-Stress (Guest's) Theory.[26] The maximum-shearing-stress theory assumes that yielding starts when the maximum shearing stress becomes equal to the maximum shearing stress at the yield point in a simple tensile test. This theory is applied best to ductile materials where the tensile yield stress and the compressive yield stress are nearly equal.[9] Failure in brittle materials is not predicted very well by this theory.[10]

According to the maximum-shearing-stress theory, inelastic action in a material begins only when the maximum shearing stress reaches the ultimate shearing strength of the material. The expression for the maximum shearing stress τ at any point at the distance r_r from the center of a thick-walled cylinder is

$$\tau_{\max} = \frac{r_i^2 r_e^2 (p_i - p_e)}{r_r^2 (r_e^2 - r_i^2)} \quad . \; . \; . \; . \; . \; . \; . \; . \; . \quad (2.8)$$

The maximum shearing stress occurs on a plane making an angle of 45° with the directions of the tangential and radial stress (see Fig. 2.3).

The maximum value of τ occurs at the inner surface of a cylinder where $r_r = r_i$.

If a cylinder is subjected to internal pressure only, the maximum shearing stress is found by making $r_r = r_i$ and $p_e = 0$.

$$\tau_{\max} = p_i \frac{r_e^2}{r_e^2 - r_i^2} \quad . \; . \; . \; . \; . \; . \; . \; . \; . \; . \quad (2.9)$$

As r_e becomes large with respect to r_i, the value of τ approaches p_i.

The maximum-shearing-stress theory says, then, that the maximum pressure that may be imposed in a thick-walled cylinder of infinite thickness before inelastic action occurs is equal to the shear strength of the material.

Maximum-Strain Theory. The maximum-strain theory, usually attributed to Saint Venant, states that permanent deformation in a material subject to any combination of stresses begins only when the maximum principal strain reaches a value equal to the strain value that occurs when inelastic action begins in a bar of the material in simple tension. This limiting strain ε_e is the strain that occurs at the tensile proportional limit and is equal to S_e/E, where S_e is the proportional limit, and E is the modulus of elasticity.

If the external pressure is zero, then

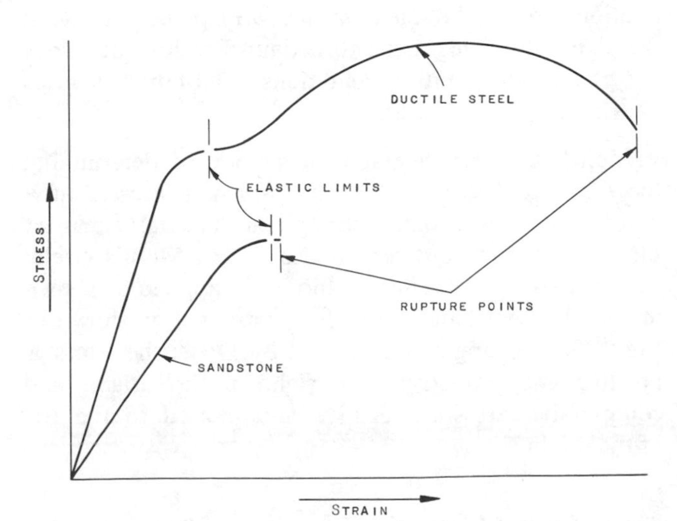

Fig. 2.2 Stress vs strain for steel and sandstone.[7]

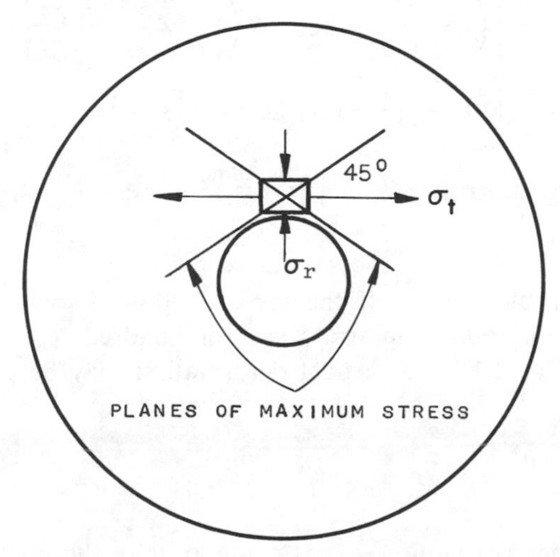

Fig. 2.3 Planes of maximum shearing stress.[7]

$$\sigma_\epsilon \text{ (maximum)} = p_i\left(\frac{r_e^2 + r_i^2}{r_e^2 - r_i^2} + \nu\right) \quad . \quad . \quad . \quad . \quad (2.10)$$

This theory has limitations because, in a case where there are tensile stresses in two directions on a plate, it does not predict the correct yield point. The results of experiments using specimens under uniform hydrostatic pressure also contradict this theory.[10]

The maximum-strain theory may be summarized as follows (see Eq. 2.10): The maximum pressure that can be imposed in a cylinder of infinite thickness before inelastic action occurs is equal to the tensile elastic limit divided by one plus Poisson's ratio.

Maximum-Strain-Energy Theory. This theory, first proposed by Beltrami in 1885, proposes that the yield point be determined on the basis of the quantity of strain energy stored per unit volume of the material.[11] This theory breaks down because under high hydrostatic pressure large amounts of elastic energy may be stored without causing either fracture or permanent deformation.[10] Direct measurement of the stored energy within the material is difficult.

The maximum-strain-energy theory states that the maximum internal pressure that may be imposed in a thick-walled cylinder before inelastic action occurs can be determined by the following equation.[27]

$$(\sigma_t)_{\max} = \frac{p_i\sqrt{10 + \dfrac{6r_i^4}{r_e^4}}}{2\left(1 - \dfrac{r_i^2}{r_e^2}\right)} \quad . \quad . \quad . \quad . \quad . \quad . \quad . \quad . \quad . \quad . \quad (2.11)$$

In this equation Poisson's ratio is assumed to be 0.25.

Unidirectional Horizontal Compression in Strata Penetrated by a Vertical Hole. This problem was first described by Kirsch.[13] Let σ_i be the initial uniform stress perpendicular to the axis of the hole. Assume further that the rock is free to become slightly thicker according to Poisson's ratio. The stresses about the borehole can then be calculated by using the following equations.

$$\sigma_r = \frac{\sigma_i}{2}\left[\left(1 - \frac{r_i^2}{r_r^2}\right) + \cos 2\theta\left(1 - 4\frac{r_i^2}{r_r^2} + \frac{r_i^4}{r_r^4}\right)\right], (2.12)$$

$$\sigma_t = \frac{\sigma_i}{2}\left[\left(1 + \frac{r_i^2}{r_r^2}\right) - \cos 2\theta\left(1 + 3\frac{r_i^2}{r_r^2}\right)\right], \quad . \quad (2.13)$$

$$\tau = -\frac{\sigma_i}{2}\left(1 + 2\frac{r_i^2}{r_r^2} - 3\frac{r_i^4}{r_r^4}\right)\sin 2\theta \quad . \quad . \quad . \quad . \quad (2.14)$$

A problem arises if the rock is allowed to become thicker or thinner in strata several hundred feet thick. From Eq. 2.15 the vertical deformation may be calculated.

$$\epsilon = \frac{\nu}{-E} 2\sigma_i h \quad . \quad . \quad . \quad . \quad . \quad . \quad . \quad . \quad . \quad . \quad . \quad (2.15)$$

If the assumption can be made that the material surrounding the well provides enough support to keep the formation from distorting in this manner, the vertical stresses imposed are just great enough to restore this distortion to zero.

$$\Delta\sigma_v = \pm 2\nu\sigma_i \quad . \quad . \quad . \quad . \quad . \quad . \quad . \quad . \quad . \quad . \quad (2.16)$$

This equation shows the maximum additional vertical stress concentration.

Horizontal Rock Strata in Uniform Compression from All Directions in Its Plane, and Penetrated by a Vertical Hole. This case was developed by Lamé and Clapeyron[14] and Timoshenko[10] for a thick-walled cylinder with an external pressure equal to the initial compressive stress in the formation. If we assume a cylinder of infinite wall thickness, the following equations will describe the stresses.

$$\sigma_r = \sigma_{ie}\left(\frac{r_i^2}{r_r^2} - 1\right) \quad , \quad . \quad . \quad . \quad . \quad . \quad . \quad . \quad . \quad . \quad (2.17)$$

$$\sigma_t = -\sigma_{ie}\left(\frac{r_i^2}{r_r^2} + 1\right) \quad , \quad . \quad . \quad . \quad . \quad . \quad . \quad . \quad . \quad (2.18)$$

$$\tau_{\max} = \pm\sigma_{ie}\frac{r_i^2}{r_r^2} \quad , \quad . \quad . \quad . \quad . \quad . \quad . \quad . \quad . \quad . \quad (2.19)$$

$$\sigma_r = -\sigma_{ii}\frac{r_i^2}{r_r^2} \quad , \quad . \quad . \quad . \quad . \quad . \quad . \quad . \quad . \quad . \quad (2.20)$$

$$\sigma_t = \sigma_{ii}\frac{r_i^2}{r_r^2} \quad , \quad . \quad . \quad . \quad . \quad . \quad . \quad . \quad . \quad . \quad (2.21)$$

$$\tau_{\max} = \pm\sigma_{ii}\frac{r_i^2}{r_r^2} \quad . \quad . \quad . \quad . \quad . \quad . \quad . \quad . \quad . \quad . \quad (2.22)$$

Mohr's Theory of Fracture. Mohr's theory assumes that, at failure across a plane, the normal and shear stresses across the plane are connected by some functional relation,[1]

$$\tau = f(\sigma), \quad . \quad . \quad . \quad . \quad . \quad . \quad . \quad . \quad (2.23)$$

which is characteristic of the material. A plot of this relationship may be made on the (σ,τ) plane. Changing the sign of τ changes the direction of failure but does not change the limiting conditions. The curve is symmetrical about the σ-axis.

Mohr's theory is a graphical method of determining the limits of failure.[10-12] If the principal stresses at a point are known, the shearing and normal stress at these points can be determined by using Mohr's circle. Let the principal stresses σ_1 and σ_2 be applied as shown in Fig. 2.4. A Mohr's circle is constructed as shown in Fig. 2.5. The origin is indicated by O; tensile stress is positive and plotted to the right of the origin, and compressive stress is negative and plotted to the left of the origin.

$$\tau = AB = AC \sin 2\theta = \frac{\sigma_1 - \sigma_2}{2}\sin 2\theta \quad , \quad . \quad . \quad (2.24)$$

$$\sigma_y = OB = OC + CB = \frac{\sigma_1 + \sigma_2}{2} + \frac{\sigma_1 - \sigma_2}{2} \cos 2\theta \qquad (2.25)$$

This theory can be applied in a converse situation where the normal and shearing stresses are known and the principal stresses are to be determined. If sufficient data are obtained to construct three or more Mohr's circles, then an envelope can be drawn tangent to the circles and symmetrical with the σ-axis (see Figs. 2.6A and 2.6B). Stresses that fall within the envelope MNN_1, M_1 are below the point of failure, but outside the envelope the stresses will cause failure.

The circle of center C (Fig. 2.6A), which just touches the curve, illustrates a limiting case. Failure, in this case, will occur under conditions corresponding to the points P and P′ — that is, over the planes whose normals are inclined at angles one-half the size of the angle PCD to the direction of greatest principal stress.

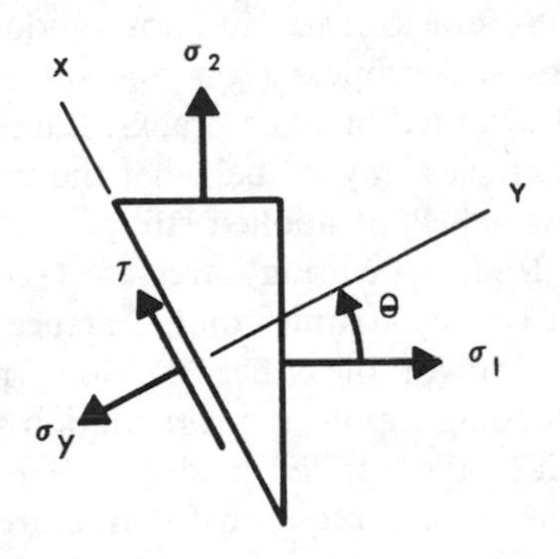

Fig. 2.4 Stress diagram.[25]

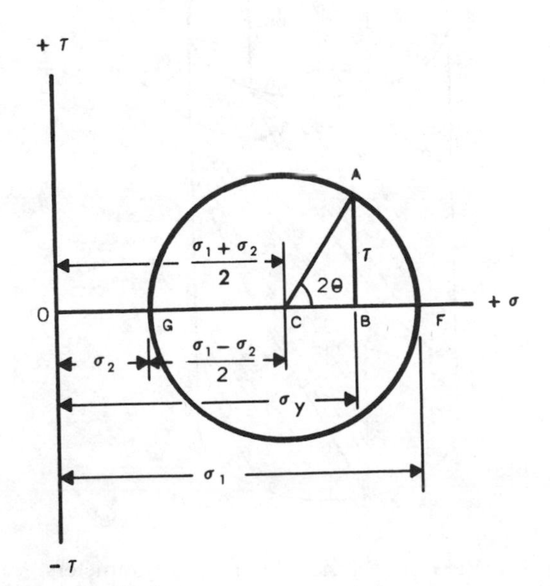

Fig. 2.5 Mohr's circle.[25]

The curve MN will be the envelope of all circles corresponding to all the conditions at which fracture takes place, and is known as Mohr's envelope. Three circles that touch it can be found from simple experiments. These circles are those with centers C_1, O, C_2 in Fig. 2.6B. These correspond to tension, simple shear, and compression. Since it is difficult to perform shear and tensile tests on rocks, the triaxial test is preferred. By varying the hydrostatic pressure, any number of circles, all to the left of the τ-axis, can be found. Since the resistance to fracture generally increases with hydrostatic pressure, the Mohr envelope is usually open to the left. When the Mohr envelope consists of two straight lines, it becomes the Coulomb-Navier theory, which is discussed later.

One disadvantage of the Mohr theory is that it does not predict the fracture of brittle material in tension. Also, it recognizes only the maximum and minimum principal stresses and neglects the effect of the mean principal stress, with the result that it yields answers that are not always consistent with experimental results.[11]

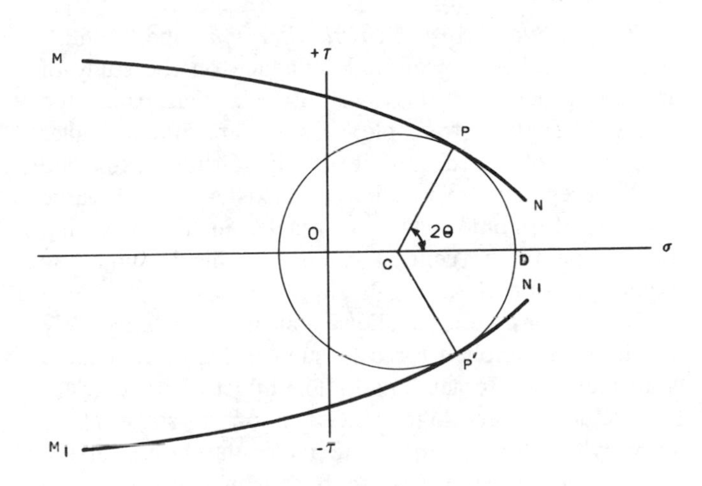

Fig. 2.6A Mohr's envelope.[2,10]

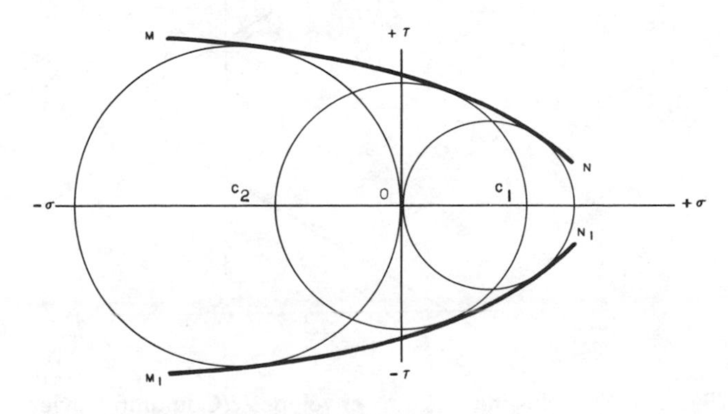

Fig. 2.6B Mohr's envelope.[2,10]

Coulomb-Navier Theory. This theory is a special case of Mohr's theory in which the envelope is a pair of straight lines symmetrical with the σ-axis (see Fig. 2.7). Stress combinations (σ,τ) that fall within the envelope will not cause failure, but if they fall outside the envelope, failure will result. The rest of the theory development is the same as for the previously described Mohr's circle. The results of triaxial tests on most rocks are represented quite well by the straight-line Mohr envelope. Again, in three dimensions, only the Mohr circles in the planes of the greatest and smallest principal stress are of consequence, and the fracture always takes place in the plane parallel to the intermediate principal-stress axis. Results of experiments do not always confirm this. One advantage of the Coulomb-Navier theory is that it does set up a criterion for failure stresses and direction of fracture. These quantities involve the mean stress and the maximum shear stress. Because of its simplicity, the Coulomb-Navier theory is popular in rock mechanics, although it does not differentiate failure by yielding from brittle fracture even though it does allow for different values of tensile and compressive strengths.[2]

The Griffith Theory of Brittle Strength. The Griffith[16] studies have had a profound influence on the study of fracture processes.[2] It is well known that calculated tensile strengths of simple crystals are much higher than those observed experimentally. Griffith explained this discrepancy by proving the existence of a large number of minute flaws or cracks in the material, demonstrating the general validity of his formulas by experiments on glass.

Griffith was able to show that a crack effectively multiplies an applied force to such a degree that there is an increase in crack length. The effect of the crack is to produce a very high concentration of stress at its apex, which focuses upon the molecular bonds of the rock material at its vertex, causing further fracturing.[15, 17–19] This stress concentration can be calculated from the maximum tensile stress in a flat plate containing an elliptical hole of major axis 2C and subjected to an average stress σ in a direction perpendicular to the major axis.

$$\sigma_{\max} = \overline{\sigma}\left(\frac{c}{r_{ee}}\right)^{1/2}, \quad \ldots \ldots \quad (2.26)$$

when $r_{ee} \rightarrow 0, \sigma_{\max} \rightarrow \infty$.

Orowan,[24] who developed Griffith's theory, gave the criterion of failure as

$$(\sigma_1 - \sigma_3)^2 - 8T_o(\sigma_1 + \sigma_3) = 0 \text{ if } \sigma_1 + 3\sigma_3 > 0, \quad \ldots \ldots \quad (2.27)$$

$$\sigma_3 + T_o = 0 \text{ if } \sigma_1 + 3\sigma_3 < 0, \quad \ldots \ldots \quad (2.28)$$

when $\sigma_1 > \sigma_3$.

Walsh-Brace Theory. The Walsh-Brace theory[21] assumes that failure is tensile and that the body is composed of long cracks that are not randomly oriented and that are superimposed on an isotropic array of randomly distributed smaller cracks. The long as well as the short crack array is such that the cracks close at relatively low values of applied stress, thus transmitting both normal and shearing stresses (see Fig. 2.8). Walsh and Brace assume that fracture may occur through the growth of either the long or the small cracks, depending upon the orientation of the long crack system to the applied load $(\sigma_3 - \sigma_1)$. The fracture stress $(\sigma_3 - \sigma_1)$ required for fractures originating

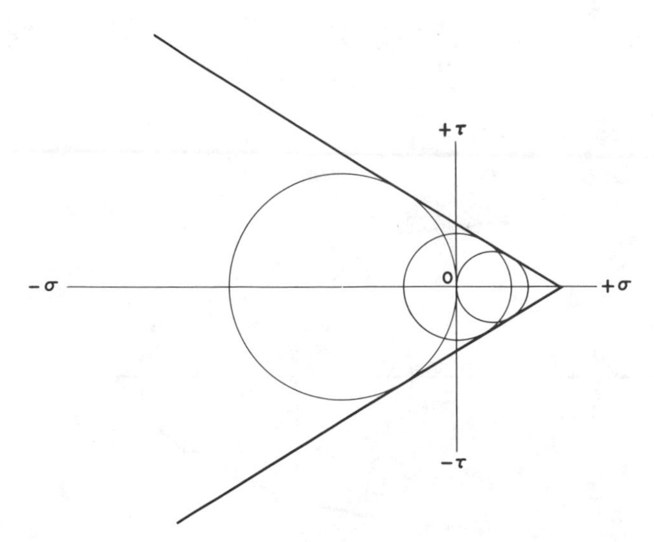

Fig. 2.7 Straight-line Mohr envelope. (Coulomb-Navier Theory[2,10])

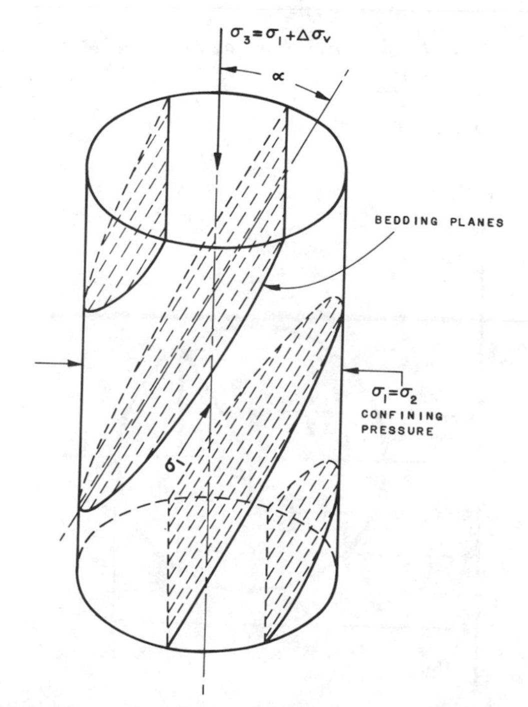

Fig. 2.8 View of rock sample showing parameters.[21]

at the small, randomly oriented cracks, is given for any confining pressure, σ_1, by

$$(\sigma_1 - \sigma_3)_s = C_{os} + \frac{2 f_s \sigma_1}{(1+f_s^2)^{1/2} - \nu_s} \quad . \ . \ . \ . \ (2.29)$$

If fracture occurs as a result of the growth of the long crack system, which is oriented at an angle α to σ_3, then the fracture stress $(\sigma_1 - \sigma_3)$ at any confining pressure, σ_1, is given by

$$(\sigma_1 - \sigma_3)_L = \frac{C_{oLc}[(1+f_L^2)^{1/2} - f_L] + 2 f_L \sigma_1}{2 \sin \alpha \cos \alpha \, (1 - f_L \tan \alpha)}, \quad . \ (2.30)$$

where C_{oLc} is the atmospheric compressive strength for the most critical orientation of α, say 30° (see Fig. 2.9).

McLamore and Gray,[20] in discussing Eqs. 2.29 and 2.30, showed that to use this theory C_{oL}, C_{os}, f_L, and f_s must be determined. They proposed that the C_{os} is found by determining the atmospheric compressive strength for samples with orientations of 0° and 90°. The value of C_{oLc} is found by determining the minimum value of the atmospheric compressive strength as the orientation of the sample is varied. This orientation usually occurs at about 30°.

The friction coefficients, f_s and f_L, can be determined by running a series of compression tests at various confining pressures and fixed orientations — for example, 0° and 90° for f_s and 30° for f_L. The slope of the plot for compressive strength vs confining pressure for any orientation is equal to

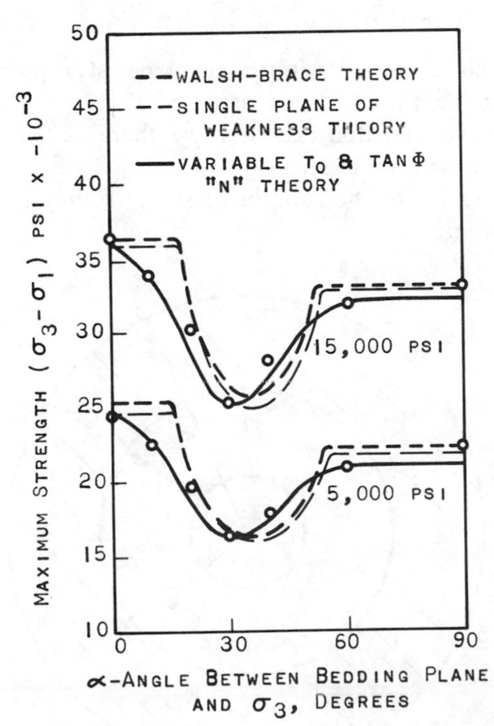

Fig. 2.9 Comparison of various failure theories with Green River Shale-2 data for confining pressures of 5,000 psi and 15,000 psi.[20]

$$\frac{f_{L,s}}{[(1+f_{L,s}^2)^{1/2} - f_{L,s}]}, \quad . \ . \ . \ . \ . \ . \ . \ . \ . \ (2.31)$$

where the subscripts refer to the particular orientation and crack system.

The values of f_s and C_{os} should be determined for both the 0° and the 90° orientations, and the corresponding fracture strength should be calculated for both cases as a function of confining pressure, since the fracture strengths at these two orientations, which represent failure due to the short crack system, are not necessarily identical.

Once the parameters f_s, f_L, C_{os}, and C_{oL} have been determined, the theory may be evaluated by calculating the values of $(\sigma_1 - \sigma_3)_s$ and $(\sigma_1 - \sigma_3)_L$, using Eqs. 2.29 and 2.30 for a given confining pressure and orientation, and using the smaller of the two values as the fracture strength.

Single-Plane-of-Weakness Theory. The single-plane-of-weakness theory proposed by Jaeger[22] assumes that the rock fails in shear. This theory is a generalization of the well known Mohr-Coulomb linear-envelope-failure theory and describes an isotropic body that contains a single plane or a system of parallel planes of weakness. The failure of the matrix material is given by

$$\tau = T_{ma} - \sigma \tan \Phi \quad . \ . \ . \ . \ . \ . \ . \ . \ . \ (2.32)$$

McLamore and Gray[20] describe failure in the plane of weakness by

$$\tau = T_{ma}' - \sigma \tan \Phi' \quad . \ . \ . \ . \ . \ . \ . \ . \ (2.33)$$

Using the well known Mohr circle relationship that relates τ and σ to σ_1, σ_3, and the failure angle θ, the final form of the single-plane-of-weakness theory can be derived from Eqs. 2.32 and 2.33.

For failure within the matrix, the equation is

$$(\sigma_3 - \sigma_1) = \frac{2T_{ma} - 2\sigma_1 \tan \Phi}{(\tan \Phi - \sin 2\theta - \cos 2\theta \tan \Phi)} \quad . (2.34)$$

The fracture strength of the material in the plane of weakness is given by

$$(\sigma_3 - \sigma_1) = \frac{2T_{ma}' \cos \Phi' - 2\sigma_1 \sin \Phi'}{\sin \Phi' - \sin (2\alpha + \Phi')}, \quad . \ . \ (2.35)$$

where α is the angle between σ_3 and the plane of weakness, and where, in both cases, σ_1 represents the confining pressure.

The theory is evaluated by running tests at 0°, 90°, and 30° orientations for various confining pressures, plotting linear Mohr-Coulomb envelopes and determining the values of Φ, Φ′, θ, T_{ma}, and T_{ma}'. Once these values have been determined, the fracture strength is calculated for a certain pressure and orientation using the two previous equations, with the lowest value taken as the strength of the material. The same argument made in the previous section relative to the matrix strength near the 0° and 90° orientations is true here also, and fracture strengths should be calculated for both of these orientations to determine the strength of the matrix on either side of the area of anisotropic strength behavior.

Variable-Cohesive-Strength Theory. Jaeger[22] also proposed the variable-cohesive-strength theory to describe a body that fails in shear and has a variable cohesive strength T_{ma}, and a constant value of internal friction, tan Φ. The governing equation describing failure for this case in terms of tan Φ, σ_1, and T_{ma} is

$$(\sigma_3 - \sigma_1) = \frac{2T_{ma} - 2\,\sigma_1 \tan\Phi}{(\tan\Phi - \sqrt{\tan^2\Phi + 1})}, \quad . \quad . \quad . \quad . \quad (2.36)$$

where $\sigma_3 - \sigma_1$ is the fracture stress.

$$T_{ma} = A - B\,[\cos 2\,(\xi - \alpha)]\,, \quad . \quad . \quad . \quad . \quad . (2.37)$$

and tan $\Phi \cong$ constant.

In Eq. 2.37, A and B are constants and ξ represents the orientation of α that has a minimum value of T_{ma}. Usually, $\xi = 30°$.

McLamore and Gray[20] noted that the variation of T_{ma} can be described by the following relationship:

$$T_{ma} = A_{1,2} - B_{1,2}\,[\cos 2(\xi - \alpha)]^n, \quad . \quad . \quad (2.38)$$

where A_1 and B_1 are constants that describe the variance over the range of $0° \leq \xi \leq \alpha$, and A_2 and B_2 describe the variance over the range of $\alpha < \xi \leq 90°$. The factor n is an "anisotropy-type" factor and has the value of 1 or 3 for "planar" anisotropy (cleavage and possibly schistosity) and the value of either 5 or 6 (or greater), for the "linear" type of anisotropy associated with bedding planes.

Octahedral-Shearing-Stress Theory. This theory, derived by Nadai,[11] is explained as a limiting state of mechanical strength. Basically, Mohr's theory states that the intermediate principal stress had no influence on failure and that failure occurred in the plane of the intermediate principal stress. The octahedral-shearing-stress theory assumes that the octahedral shearing stress at the limit of yielding is a function of the octahedral normal stress. Further, the octahedral normal stress is the mean of the three principal stresses, which implies that the intermediate principal stress does have an influence on failure. This contradicts Mohr's theory, in which the intermediate principal stress does not influence failure.

$$\tau_{oct} = f(\sigma_{oct})\,, \quad . \quad . \quad . \quad . \quad . \quad . \quad . \quad . \quad . \quad . \quad . (2.39)$$

$$\sigma_{oct} = \frac{\sigma_1 + \sigma_2 + \sigma_3}{3} \quad . \quad . \quad . \quad . \quad . \quad . \quad . \quad . \quad . (2.40)$$

A derivation of the octahedral shear stress is found in *Theory of Flow and Fracture of Solids* by A. Nadai.[11] This theory measures the intensity of stress that is responsible for bringing a solid substance into the plastic state.[11]

Material in Plastic State

If the material in a thick-walled cylinder is assumed to yield, a complete redistribution of stresses results. The material may yield only partially so that there is a plastic region surrounded by a region stressed below the limit of plasticity (see Fig. 2.10). In the outer portion of the cylinder the equations must satisfy the conditions of stress and strain of a perfectly elastic body.

If the cylinder is short and the longitudinal stress (σ_v) is assumed to be zero, the internal pressure that initiates a plastic ring along the inner surface may be expressed by

$$p_i = \sigma_o \frac{\dfrac{r_e^2}{r_i^2} - 1}{\sqrt{3\dfrac{r_e^4}{r_i^4} + 1}} \quad . \quad . \quad . \quad . \quad . \quad . \quad . \quad . \quad . \quad . \quad . \quad . \quad (2.41)$$

Fenner,[9] in his study of a well drilled into a plastic medium, allowed the formation to flow into the well. The particles moved toward the center of the well, and the zone of flow expanded until the boundary stresses became such that no further increase of the flow zone was possible. The radial and tangential stresses both increased, extending outwards until the sum of these stresses became constant at the outer boundary of the flow zone.

Westergaard[8] reasoned that the stresses around a wellbore, determined according to Lamé calculations for an elastic medium, would be so great that the combination of stresses could not be maintained at any great depth. This reasoning suggested the existence of a plastic ring surrounding the wellbore. Westergaard stated that the relation between the tangential and radial stresses is

$$\sigma_t - (K + 1)\,\sigma_r = \sigma_c \quad . \quad . \quad . \quad . \quad . \quad . \quad . \quad . \quad . \quad (2.42)$$

In this case, the Mohr's envelope of rupture is a straight line. Westergaard assumed that when the left side of this equation becomes less than σ_c, the rock is in an elastic state of stress.

The extent of the zone of plasticity b may be calculated from

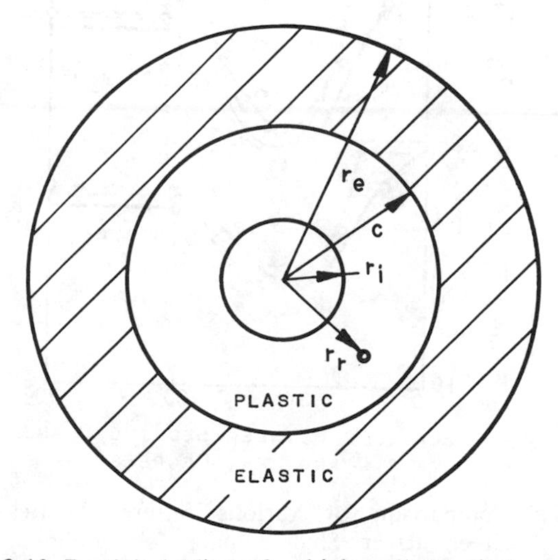

Fig. 2.10 Partial yielding of a thick-walled cylinder.[7]

$$(h/r)\ K = \frac{2\ (K\sigma_{iv} + \sigma_c)}{(K\ \lambda\ 2)(K\sigma_i + \sigma_c)} \quad . \quad . \quad . \quad . \quad . \quad (2.43)$$

The extent of the flow zone is similar when calculated by either Fenner's or Westergaard's method. In this particular case, the extent is approximately 80 times the radius of the wellbore.

2.4 Effect of Environmental Conditions on Properties of Rock

Before any attempt was made to apply the theories of behavior of thick-walled cylinders of homogeneous, isotropic, impermeable material to the behavior of rock formations adjacent to bores, consideration was given to the relationship of the earth's overburden to horizontal natural stresses in the normal range of well depths.

There is some diversity of opinion regarding the distribution of stresses (such as the stress acting to collapse a well or acting against the pressure applied in a well) resulting from the earth's overburden. It can be shown that the stress in the earth's crust exerted in a horizontal direction theoretically would be equal, at the minimum, to the axial stress (effective overburden) multiplied by the expression $\frac{\nu}{1-\nu}$, where ν is Poisson's ratio for rock, and where effective overburden is defined as the acting or net-bearing pressure per unit area at any given depth. It is the pressure that results from geologic changes such as folding and faulting that controls the orientation of a fracture. The other limit or maximum horizontal confining stress would be equal to the effective overburden.

Let σ_v denote the vertical stress, and σ_x and σ_y the two other principal stresses. Let w equal the weight above any one point of the overlying mass per unit area. Then in a formation, in situ, $\sigma_v = w$.

Assume that at any depth the horizontal stresses are equal in all directions and are denoted by σ_h. Then

$$\sigma_x = \sigma_y = \sigma_H \quad . \quad . \quad . \quad . \quad . \quad . \quad . \quad . \quad . \quad . \quad (2.44)$$

(σ_H will be referred to as "confining stress".)

The value of σ_H depends upon which of the following assumptions is made.

Case I — Elastic stresses have disappeared over long geologic periods of time.

Case II — While the overburden has increased with deposition, the rock formations have not moved horizontally and have remained perfectly elastic.

In Case I, the stresses are hydrostatic, i.e., equal in all directions.

$$\sigma_H = \sigma_v = P_{oB} \quad . \quad . \quad . \quad . \quad . \quad . \quad . \quad . \quad . \quad . \quad . \quad (2.45)$$

In Case II, since no horizontal movement has occurred, the horizontal strain ε_H equals zero. According to Hooke's law, the horizontal strain is expressed as

$$\epsilon_H = \frac{1}{E}\left[\sigma_x - \nu\ (\sigma_y + \sigma_v)\right] \quad . \quad . \quad . \quad . \quad . \quad . \quad (2.46)$$

Since ϵ_H is zero and $\sigma_H = \sigma_x = \sigma_y$,

$$\sigma_x = \nu\ (\sigma_y + \sigma_x) = \sigma_h = \nu\ (\sigma_h + \sigma_v) = 0,$$

or

$$\sigma_h = \frac{\nu}{1-\nu}\sigma_v = \frac{\nu}{1-\nu}(w) \quad . \quad . \quad . \quad . \quad . \quad . \quad (2.47)$$

CASE I: CONFINING STRESS EQUAL TO EFFECTIVE OVERBURDEN

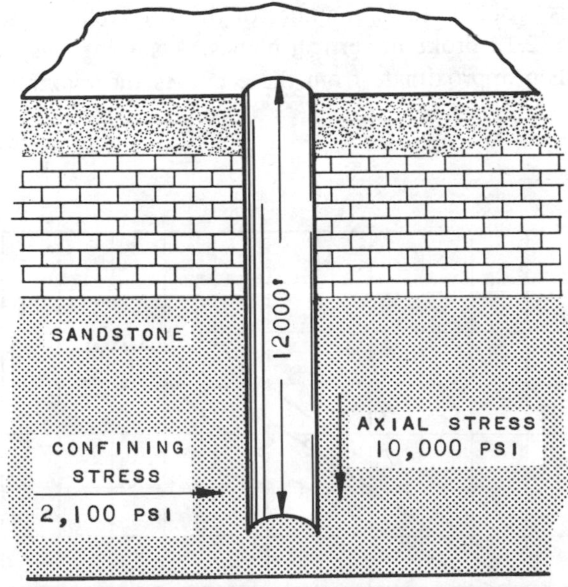

CASE II: CONFINING STRESS EQUAL TO EFFECTIVE OVERBURDEN TIMES POISON'S RATIO ÷ (1—POISSON'S RATIO)

Fig. 2.11 Two extremes of confining stress around wells.[2]

These theories may be clarified by illustration. In a well 12,000 ft deep, the maximum confining stress equals the effective overburden, which, in turn, is equal to 12,000 times the fracturing pressure gradient of 0.83, or approximately 10,000 psi (see Case I, Fig. 2.12).

The minimum value of confining stress for a well of the same depth is shown in Case II (Fig. 2.11). The effective overburden pressure of 10,000 $\times \dfrac{\nu}{1-\nu}$ or 0.21 when $\nu = 0.17$, gives a confining-stress value of 2,100 psi.

Evidence of the actual limits of the confining stress around wells may be found by analyzing the curves in Fig. 2.12. The curves show the maximum, minimum, and average values of the effective overburden, or fracture extension pressure, plotted against depth for 276 wells located in the Gulf Coast, Mid-Continent and West Texas areas. These data were obtained from cement squeeze jobs before the advent of hydraulic fracturing. The trend of the curves has been substantiated by subsequent data accumulated during the treatment of wells by hydraulic fracturing. These curves define the bottom-hole pressure required to inject a fluid into induced fractures. In other words, the curves show the confining stress that must be surpassed by the well-fluid pressure in order to hold the fracture planes apart to permit fluid entry.

In the case of horizontal fractures, it is generally agreed that the confining stress holding the fracture planes together is equal to the effective overburden at the depth of the fracture. In the case of vertical fractures, however, the confining stress holding the fracture planes together equals some function of the effective overburden.

It is probable that a number of the 276 wells plotted in Fig. 2.12 broke in vertical planes. Assuming this, we can then approximate from these curves the maximum and minimum ratios of vertical stress (effective overburden) to horizontal stress (confining stress). For example, at a depth of 4,000 ft the effective overburden on the minimum curve equals 2,200 psi. If it is assumed that this particular fracture is vertical, then the confining or horizontal stress is equal to 2,200 psi. At this same depth, the maximum vertical stress is equal to 4,000 psi (1 psi/ft $\times$ 4,000 ft). Then the minimum ratio of horizontal stress to vertical stress is 2,200/4,000, or 0.55.

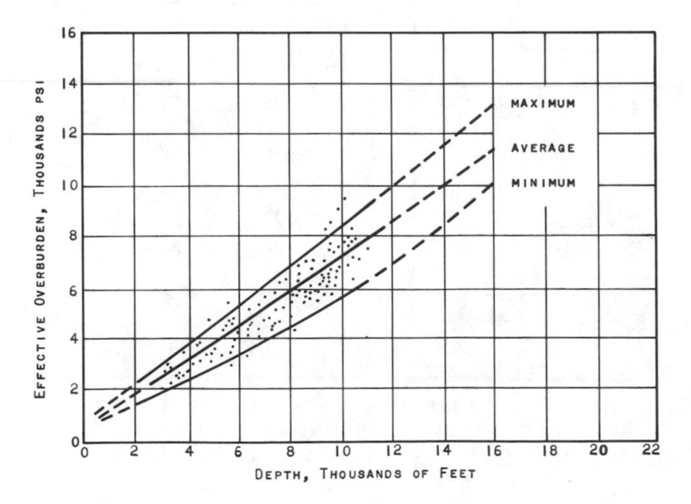

Fig. 2.12 Effective overburden vs depth.[7]

A more representative ratio, however, would be one obtained by comparing the maximum overburden (1 psi/ft) with the average effective overburden, which, in the case of the well 4,000 ft deep, would be 3,100/4,000, or 0.78; and at 10,000 ft it would be 7,100/10,000 or 0.71. At a depth of 10,000 ft the ratio of minimum to maximum is 5,800/10,000 or 0.58.

If a well fractured vertically at a pressure of 1.0 psi/ft or more, this would suggest that even in the earth's upper crust a column of rock may act as a column of fluid, with its overburden felt in the same magnitude in both the horizontal and vertical planes. Other investigators have suggested that the differences in the stresses in the earth's crust, except in the upper layers, gradually equalize throughout long geologic periods. In other words, the ratio of horizontal stress to vertical stress approaches a value of 1.0 with increased depth. Therefore, in deep wells the ratio of horizontal confining stress to effective overburden logically can be expected to become larger and to approach 1.0 in the deepest wells.

In brief, then, this study of the earth's overburden pressure shows that the minimum pressure required to inject fluid into a well fracture may be as low as 0.5 times the actual weight of the overburden at that depth, and that the maximum pressure is equal to the effective weight of the overburden of approximately 1 psi per foot of depth.

It has been observed that the pressure of the fluid saturating a rock formation has a marked effect on the pressure required to initiate and to extend a fracture in that rock. Formations with high pressures generally require higher pressures to initiate fractures than do formations with lower pressures since the borehole pressure required to achieve a differential pressure great enough to part the formation will be higher because of the higher formation pressure. It has been observed that as the pressure declines in a given reservoir, the fracturing pressure also declines. Conversely, in waterflood projects the fracture-opening pressure increases as more and more water is injected and as the reservoir pressure is increased.

2.5 Comparison of Calculated and Observed Bursting Pressure of Thick-Walled Cylinders of Rock

Scott *et al.*[7] performed tests under two sets of conditions. In one type of test, thick-walled cylinders of rock formations were ruptured by injecting drilling mud that had a zero filtrate rate; thus, relatively little fluid penetration was attained. In the other type of test, similar rock formations were ruptured by injecting, under steady-state flow conditions, a penetrating mineral oil. When the bursting tests were completed, the theoretical rupturing pressures were calculated. In these calculations, it was assumed that failure would occur soon after any stress exceeded the elastic limit of the rock. The value of the ultimate strength was used as the value of the stress at the elastic limit. Table 2.1 shows the observed and calculated rupturing pressures. No relationship between the calculated and observed bursting pressure is evident, indicating that the inhomogeneity, anisotropy, and permeability of rock preclude the use of thick-walled-cylinder equations for calculating rupturing pressure.

All of the cylinders that were ruptured by a nonpenetrating fluid fractured in vertical planes across the horizontal or nearly horizontal bedding planes of the rock formations. The cores that were ruptured with a penetrating fluid fractured parallel to the bedding planes, regardless of their position with respect to the longitudinal axes of the cylinders. From these tests, Scott concluded that, because of the dissimilarity of the rock, no valid comparison of the magnitude of rupturing pressure of cylinders fractured with penetrating fluid and those fractured with nonpenetrating fluid can be made from the data in Table 2.1. However, if a fluid does penetrate a rock, the rupturing pressure is low compared with the pressure when the fluid does not penetrate.

In their experimental work on Green River shale, McLamore and Gray[20] showed that the Walsh-Brace theory (the single-plane-of-weakness theory) and the variable-cohesive-strength theory fit the observed data over the entire range of α (where α is the orientation angle between the plane of anisotropy and the applied load).

All three theories were derived to describe the strength behavior of brittle materials that fail by fracturing. However, if maximum strength is considered to be analogous to fracture strength, they can also be used to describe the strength behavior of materials that behave in a nonbrittle manner.

The strength data of this study indicate that the Walsh-Brace theory and the single-plane-of-weakness theory describe very well the strength behavior of rocks that have a linear type of anisotropy, but that they should be used only with caution on rocks that possess a planar type of anisotropy. The modification of the variable T_o and tan Φ theory can be applied to both types of anisotropic rocks by considering the proper anisotropy "factor" and evaluating the anisotropic behavior of T_o (cohesive strength of the material) and tan Φ (internal fricton).

2.6 General Discussion

The preponderance of research on rock mechanics has been performed on materials that were homogeneous and isotropic. Fracturing in oil wells takes places in oil-bearing formations of basically sedimentary origin. Some formations were laid down as beds — for example, sandstones — while some have shown no bedding planes, as exemplified by the massive limestones. Where shale is involved in the bedding, a very sharp cleavage often results. If the beds of sediment were laid down and then left completely undisturbed over geologic time, there would be no other anisotropies. However, many sedimentary formations have been subjected to orogenic movement that caused cracks of all sizes as well as faults and joints.

These anomalies affect the mechanical behavior of rocks. A number of investigators[3,20] have shown that strength properties and failure in both tension and compression are functions of orientation with respect to bedding planes as well as of confining pressure.

Most investigators believe that the Griffith failure theory applies quite well to isotropic materials. However, for anisotropic materials, a modification of either the Griffith theory, the Walsh-Brace theory, or the Coulomb-Navier linear-envelope-failure theory seems to fit best the available experimental data. It appears, however, that much of the experimental data on the

TABLE 2.1 — COMPARISON OF OBSERVED AND CALCULATED INTERNAL RUPTURING PRESSURE OF CYLINDERS OF ROCK

Cylinder Radii (inches)					Internal Bursting Pressure (psi)					
					Calculated					Observed
Internal (r_i)	External (r_e)	Type of Rock Formation	Tensile Strength (psi parallel to bedding plane)·	Shear Strength* (psi)	Max. Principal Stress (Eq. 2.7)	Max. Shear (Eq. 2.8)	Max. Strain (Eq. 2.10)	Max. Strain Energy (Eq. 2.11)	Plastic $\sigma_v = 0$ (Eq. 2.12)	
Nonpenetrating Fluid										
0.19	2.0	Neat Cement	182	1400	179	1390	143	115	494	700
0.19	1.75	Sandstone	504	3000	495	2960	396	342	1290	4200
0.19	1.75	Sandstone	378	2000	368	1970	298	240	968	5550
0.19	2.13	Shale	20	500	20	20	16	13	56	886
3.0	Infinite	Sandstone	360	1800	360	1800	288	228	infinite	>3500
Penetrating Fluid										
0.19	1.0	Sandstone	200	850	186	820	151	122	386	258
0.19	1.0	Sandstone	120	400	112	385	91	73	232	166

*Estimated from average rock strength figures compiled by U. S. Bureau of Standards.

linear-envelope theory was obtained from tests on shale and slate, with less work done on sandstones and limestones, while Scott's experimental work[7] used sandstones only.

An interesting aspect of fracturing is whether the formation acts as an elastic, brittle, ductile, or plastic material. Some laboratory work has been performed at atmospheric pressure while other work has been performed under triaxial stress. The general state of stress underground, according to Hubbert and Willis,[3] is that in which the three principal stresses are unequal. In tectonically relaxed areas the least stress should be horizontal and should produce normal faulting. Then if a fracture forms perpendicular to the axis of the least stress, it will be vertical. On the other hand, if orogenic forces are active, the least stress could be vertical, in which case it probably would be equal to the effective overburden stress and would result in horizontal fractures.

The injection pressure during the formation of a vertical fracture is less than the theoretical overburden pressure (1 psi/ft). Theoretically, at least, to create a horizontal fracture the injection pressure must be equal to or greater than the effective overburden pressure. In some limited areas, Scott *et al.* theorized, the vertical stress can be less than the overburden pressure because strong, competent overlying beds accept the overburden load, making it possible for a horizontal fracture to form at pressures less than the theoretical overburden pressure.

2.7 Summary

Eleven theories to predict material failure are reviewed and methods of calculating the principal stress are presented. The results of a study of the relationship of the earth's effective overburden and the pressure acting to prevent rupture of bores by application of internal pressure, along with observations made while bursting, by internal pressure, thick-walled cylinders of rock formations and shallow wells in sandstone outcrops lead to the conclusion that the breakdown or rupture of rock surrounding a bore cannot be predicted by the maximum-principal-stress, maximum-shearing, maximum-strain, maximum-strain-energy, or the plasticity theories of relationship of pressure and stresses in thick-walled cylinders of homogeneous, isotropic, impermeable material.

Under the conditions of average relationship of effective overburden and confining stress around a bore, breakdown of rock formations by the penetration of fluids may be expected at a lower pressure than that required for breakdown with a nonpenetrating fluid.

Experimental work on shales indicates that the Walsh-Brace and the single-plane-of-weakness theories describe the strength behavior of rocks having a linear type of anisotropy, but should be used with caution on rocks having a planar type of anisotropy.

References

1. van Poollen, H. K.: "Theories of Hydraulic Fracturing", *Quarterly,* Colorado School of Mines, Golden (July, 1957) **52,** No. 3, 113-131.

2. Card, David C., Jr.: "Review of Fracturing Theories", UCRL 13040, Colorado School of Mines Research Foundation, Inc., Golden, Colo. (April 16, 1962) 14-20.

3. Hubbert, M. K. and Willis, D. G.: "Mechanics of Hydraulic Fracturing", *Trans.,* AIME (1957) **210,** 153-166.

4. Miles, A. J. and Topping, A. D.: "Stresses Around a Deep Well", *Trans.,* AIME (1949) **179,** 186-191.

5. Teplitz, A. J. and Hassebroek, W. E.: "An Investigation of Oil Well Cementing", *Drill. and Prod. Prac.,* API (1946) 76.

6. McGuire, W. J., Jr., Harrison, E. and Kieschnick, W. F.: "The Mechanics of Fracture Induction and Extension", *Trans.,* AIME (1954) **201,** 252-263.

7. Scott, P. P., Jr., Bearden, W. G. and Howard, G. C.: "Rock Rupture as Affected by Fluid Properties", *Trans.,* AIME (1953) **198**, 111-124.

8. Westergaard, H. M.: "Plastic State of Stress Around a Deep Well", *J. Boston Soc. Civil Eng.* (Jan., 1940).

9. Fenner, R.: "Untersuchung zur Erkenntnis des Gebirgsdrucks: Glueckauf" (Aug.-Sept., 1938).

10. Timoshenko, S.: *Strength of Materials,* 2nd ed., D. van Nostrand Co., Inc., New York (1941) Part I, 281 and 305; Part II, 474-476, 480.

11. Nadai, A.: *Theory of Flow and Fracture of Solids,* 2nd ed., McGraw-Hill Book Co., Inc., New York (1950) **I**, 208.

12. Seely, F. B.: *Resistance of Materials,* 3rd ed., John Wiley and Sons, Inc., New York (1947) 272.

13. Kirsch, G.: "Die Theorie der Elastizitat und die Bedurfnisse der Festigkeitslehre", *Zeitschr. des Vereines Deutscher Ingenieure* (1898) **XLII**, No. 29, 797.

14. Lamé and Clapeyron: "Memoire sur l'equilibre interieur des corps solides homogenes: *Memoirs presents par divers savans*" (1833) **4.**

15. Jaeger, J. C.: *Elasticity, Fracture and Flow,* 2nd ed., John Wiley and Sons, Inc., New York (1962) 208.

16. Griffith, A. A.: "The Phenomena of Rupture and Flow in Solids", *Philos. Trans.,* Royal Soc. (1920) **221A,** 163.

17. Cottrell, A. H.: *Theoretical Aspects of Fracture in Fracture:* Averbach, B. L. *et al.,* Eds., Technology Press and John Wiley and Sons, Inc., New York (1959).

18. Gilman, J. J.: "Fracture", *Proc.,* Intl. Conference, Swampscott, Mass., M.I.T. Press, Cambridge, Mass. (1959) **II**, 193.

19. Gilman, J. J.: "Fracture in Solids", *Scientific American* (Feb., 1960) **CCII,** 94.

20. McLamore, R. and Gray, K. E.: "The Mechanical Behavior of Anisotropic Sedimentary Rock", paper No. 66-Pet-2, presented at the Petroleum Mechanical Engineering Conference, ASME, New Orleans, La., Sept. 18-21, 1966.

21. Walsh, J. B. and Brace, W. F.: "A Fracture Criterion for Brittle Anisotropic Rocks", *J. Geophys. Research* (1964) **LXIX,** No. 16, 3449.

22. Jaeger, J. C.: "Shear Failure of Anisotropic Rocks", *Geologic Magazine* (1960) **XCVII,** 65-72.

23. Yuster, S. T. and Calhoun, J. C., Jr.: "Pressure Parting of Formations in Water Flood Operations", *Oil Weekly,* (March 12, 1945) Part I, 34; (March 19, 1945) Part II, 34.

24. Orowan, E.: "Fatigue and Fracturing of Metals", paper presented at M.I.T. Symposium, Boston, Mass., June, 1950.

25. Rixe, F. H.: "Review of Oil Well Fracturing Theories", Pan American Petroleum Corp. (March 16, 1967) unpublished.

26. Guest, J. J.: *Phil. Mag.* (1900) **50,** 69.

27. Perry, J. H.: *Chemical Engineers Handbook,* McGraw-Hill Book Co., Inc., New York (1950) 1238-1240.

Chapter 3

Determination of Wells Applicable for Hydraulic Fracturing

3.1 Introduction

In order to produce gas or liquids from a well at a higher rate following a hydraulic fracturing treatment, the reservoir must contain enough fluids in place and also must have adequate potential gradients available to move the fluids to the well following the creation of a high-permeability fracture. To determine the applicability of a well for a hydraulic fracturing treatment, the cause for the present low productivity must be analyzed, or, in the case of a newly completed well, available reservoir and well data should be studied to determine possible fluid content and reservoir pressure.

As discussed by Gladfelter, Tracy and Wilsey,[1] low productivity in a well may be the result of any one or a combination of the following causes.

1. A severe permeability reduction may exist near the well. Because all fluids must flow through the formations adjacent to the well, a reduction in permeability here can result in a noncommercial well even though a considerable amount of recoverable oil still remains. By removing this permeability block, a large increase in productivity will result. The spectacular results of acid wash treatments, relatively small-volume fracturing operations, and other similar measures that affect only the reservoir rock near the well, attest to this fact.

2. There may be a substantial amount of recoverable oil in place, but the formation permeability is so low that the oil cannot be recovered at economical rates using conventional completion methods. As suggested by Wilsey and Bearden[2], large increases in recovery from this kind of reservoir can be obtained by producing deep-penetrating fractures.

3. Reservoir pressure has been depleted even in the interwell area; i.e., there is insufficient reservoir energy remaining to expel more oil. In this situation, fracture stimulation generally will not increase productivity enough to be profitable unless gravity drainage rates can be greatly increased.

The key to determining whether or not a well should be fractured is to ascertain if any one of these conditions is the cause of the low productivity. One method we favor for this evaluation is the analysis of pressure buildup data. These data may be obtained by shutting the well in, or in the case of a new completion, by using a drill-stem or wireline tester.

3.2 Recognizing Near-Bore Permeability Reductions from Pressure Buildup Tests

The subject of SPE Monograph Vol. 1, *Pressure Buildup and Flow Tests in Wells,* is treated here briefly for the sake of continuity and as further background for the present study. Many authors have shown that the average permeability of the drainage area of a well can be determined by analyzing the pressure buildup behavior of a well when it is shut in.[3-9] The average permeability of the reservoir, as measured by this method, is not affected by the permeability near the well. The effective permeability, which is a combination of the permeability near the well and the average permeability in the drainage area, can be measured by a productivity-index test.[8,9]

Pressure Buildup in Infinite Homogeneous Reservoirs

The theory for the pressure buildup behavior of a well that is producing a single, slightly compressible fluid from an infinite homogeneous reservoir was presented by Horner.[3] According to him, the equation for buildup when oil is the only phase flowing is

$$p_{ws} = p_i - \frac{162.6 q\mu B}{kh} \log \frac{t + \Delta t}{\Delta t} , \quad \text{(3.1)}$$

where p_{ws} is the pressure in the well during buildup in psi.

Fig. 3.1 shows a graph of this equation, along with actual field data for a new well in an oil reservoir. For this case, the theory and practice agree. In analyzing this curve, note that the slope of the curve is equal to the coefficient of the logarithm term in Eq. 3.1. Therefore,

$$kh = \frac{162.6\, q\mu B}{m} \quad \text{(3.2)}$$

Extrapolation of the straight-line section to an infinite shut-in time, $[(t + \Delta t)/\Delta t] = 1$, gives the value for p_i as shown.

Bounded Homogeneous Reservoirs

The pressure buildup behavior of a well in a bounded homogenous reservoir is given by

$$p_{ws} = p_i - 162.6\frac{q\mu B}{kh} \log \frac{t + \Delta t}{\Delta t} + \frac{Y(t + \Delta t) - Y(\Delta t) - Y(t)}{2.303} \quad (3.3)$$

This equation differs from Eq. 3.1 only by the addition of the $Y(t)$ terms, which account for the boundary effect and are functions of the shape of the drainage boundary of a well and of the production time. For a square drainage boundary, the boundary effect causes the pressure buildup curve to bend over, as shown by the field example in Fig. 3.2.[9] The extrapolated value from the straight-line section is called p_i; in a bounded reservoir, p_i is best defined as this extrapolated pressure. The final static value of the pressure is called $\bar{p}$, the average pressure. A plot of $(p_i - \bar{p})/(70.6q\mu B/kh)$ is given in Fig. 3.3 for one well in the center of a square drainage area. Note that the difference between p_i and $\bar{p}$ increases as production time increases.

The function, $Y(t)$, is related to the function plotted in Fig. 3.3 by

$$Y(t) = \frac{0.00331kt}{\phi\mu cA} - \frac{p_i - \bar{p}}{70.6\ q\mu B/kh} \quad (3.4)$$

The quantity $(p_i - \bar{p})/(70.6q\mu B/kh)$ is plotted in Ref. 6 for a number of cases. In that paper the quantity is called $(p^* - \bar{p})/(q\mu/4\pi kh)$.

If there are other wells in a reservoir, production at these wells causes each well to be surrounded by a drainage boundary. On one side of this boundary, fluid flows toward that well, and on the other side toward another well. For some time after it is closed in, a well can be treated as if its drainage boundary still exists. Thus, a well surrounded by other wells will have a buildup curve qualitatively similar to that one well in a bounded reservoir. For very long closed-in times this is not true.

The method described by Matthews *et al.*[6] may be used to find the value for the average pressure $\bar{p}$ for reservoirs that have not been closed in long enough to exhibit the "bend over" shown in Fig. 3.2.

Wellbore Damage

Wellbore damage caused by the invasion of mud, cement filtrate, etc., will distort the pressure buildup curve at an early closed-in time, as shown in Fig. 3.4.[11] The wellbore damage will cause an extra drop in pressure distribution at the wellbore (see Fig. 3.5[12]).

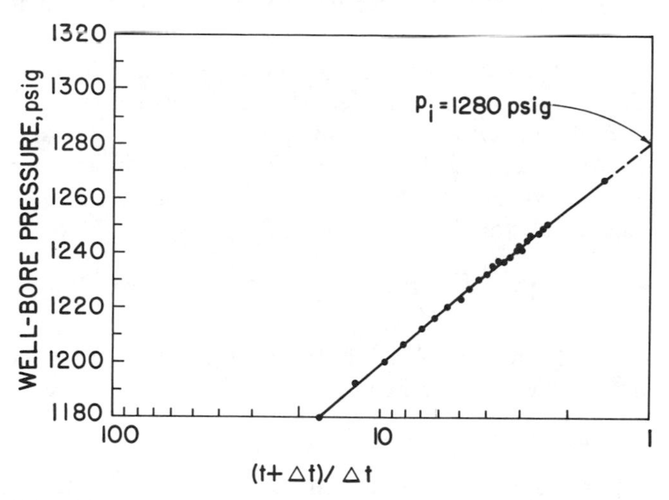

Fig. 3.1 Pressure buildup for a nearly ideal reservoir. (from Horner[3])

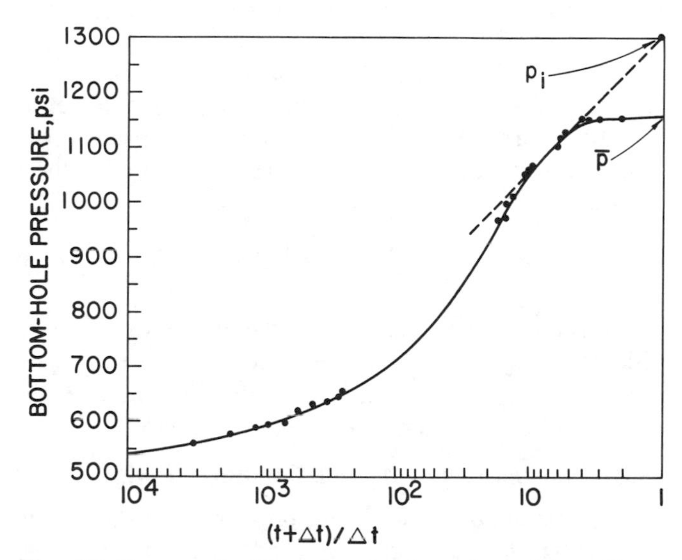

Fig. 3.2 Pressure buildup showing boundary effect. (from Mead[10])

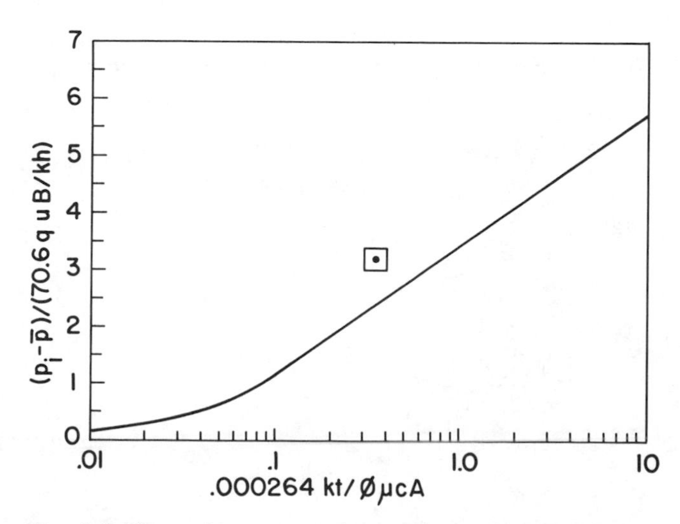

Fig. 3.3 Correction to p_i for well in square drainage area.[9]

This extra drop in pressure has been called a "skin effect". Shortly after shut-in, the well pressure should rise by the amount of Δp_{skin} shown on Fig. 3.5. Thus, the amount of the skin effect should be evident from the difference between the flowing pressure and the pressure shortly after shut-in. Quantitatively, the skin effect s, as defined by van Everdingen[11] and Hurst[12] can be calculated from

$$s = 1.151 \left[\frac{p_1 - p_{wf}}{m} - \log\left(\frac{k}{\phi \mu c r^2_w}\right) + 3.23\right]; \quad (3.5)$$

and the pressure drop in the skin or damaged zone near the well can be calculated by

$$\Delta p_{skin} = m \times 0.87 s \quad (3.6)$$

The pressure p_{wf} is measured before closing in, and the value for p_1 must be taken on the straight-line portion of the pressure buildup curve 1 hour after closing in. If the buildup curve is not straight at 1 hour, the straight-line portion must be extrapolated backwards, as shown in Fig. 3.4. Note that a scale for $(t + \Delta t)/\Delta t$, and also a scale for Δt, are used on Fig. 3.4 to facilitate extrapolation to p_1. The efficiency of the completion can be determined by comparing the actual productivity index, J, and the ideal (no skin). The ratio of these two quantities is

$$\text{Flow Efficiency} = \frac{J_{actual}}{J_{ideal}} = \frac{p_i - p_{wf} - \Delta p_s}{p_i - p_{wf}} \quad (3.7)$$

This ratio is quite similar to the condition ratio of Gladfelter *et al.*[1] Although the use of p_i in this equation is strictly correct only for an infinite reservoir, the error introduced is small.

3.3 Determination of Well Condition Ratio

Most methods of analyzing pressure buildup data to determine the cause of low productivity require that wells being tested be shut in until the flow of fluids into the wellbore is small compared with the producing rate prior to shut-in. In some cases, shut-in periods for testing far exceed 48 hours. Economics often prohibits shutting in wells this long. Gladfelter, Tracy and Wilsey presented a method of analysis that does not require long shut-in periods.[1]

If a reduction in permeability exists near the well, the permeability determined from a productivity-index test will be lower than that measured from a pressure buildup test. Thus, a ratio of the permeabilities as determined from a productivity-index test (k_{PI}) and from a pressure buildup test (k_{BU}), will reveal a reduction in permeability near the well. An expression indicating the flow conditions near the well can be written as condition ratio, CR $= k_{PI}/k_{BU}$. When a permeability reduction exists near the well, this ratio will be less than 1. If an increase in effective permeability exists near the wellbore, such as might result from an acid job or an earlier fracture treatment, the ratio will be larger than 1.

Determining Permeability from Buildup Tests

Unfortunately, to determine the average formation permeability from uncorrected buildup test data, it is necessary to wait until the flow of fluids into the well becomes small compared with the producing rate prior to shut-in. Otherwise, the pressure buildup data must be adjusted to take into account the flow of fluids into the well.

This adjustment is indicated by

$$\Delta p_n' = \Delta p_n \left(\frac{q}{q - \bar{q}_n}\right) = m \log \Delta t_{ws,n} \quad (3.8)$$

From inspection of Eq. 3.8, we can see that the slope of the corrected buildup pressure vs log shut-in time is directly related to the average formation permeability (k_{BU}) and that this permeability can be determined by

$$k_{BU} = \frac{162.6\, q \mu B}{mh} \quad (3.9)$$

Fig. 3.6 gives an example of calculating k_{BU} from buildup data measured during the first 5 hours after the well is shut in.

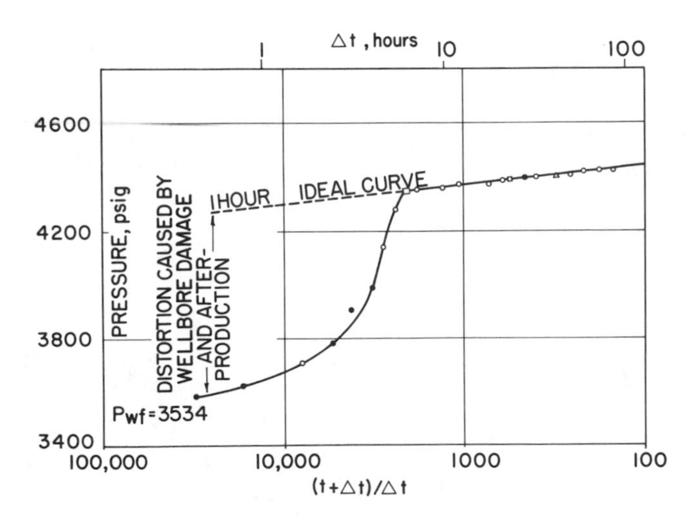

Fig. 3.4 Pressure buildup showing effect of wellbore damage and after-production.[9]

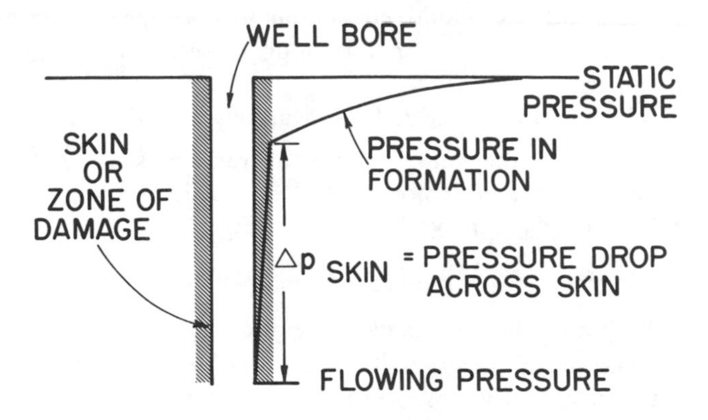

Fig. 3.5 Pressure distribution in a reservoir with a skin. (from van Everdingen and Hurst[5])

Determining Permeability from Productivity-Index Test

A permeability may be determined from a productivity-index test by producing a well at as nearly a constant rate as possible for as long as possible, or until the producing bottom-hole pressure has ceased to change significantly. Under these conditions, the combination permeability (interwell permeability plus wellbore effects) can be approximated by

$$k_{PI}=\frac{q\mu B\ \ln\left(\frac{r_e}{r_w}\right)}{0.007073h(p_{ws}-p_{wf})}, \quad (3.10)$$

where p_{ws} is the static reservoir pressure in psia.

The static pressure, p_{ws}, is calculated using pressure buildup data as already discussed. If we use pressure buildup data collected just after the well is shut in, these methods must be modified. To allow for the flow of fluids into the well, the pressures must be adjusted by plotting $\Delta p_n'$ from Eq. 3.8 vs log $(t+\Delta t)/\Delta t$ and then proceeding as previously outlined. To get both the average interwell permeability and the static buildup pressure, $\Delta p_n'$ is plotted vs log $(t+\Delta t/\Delta t$, where t is the time since last shut-in. Fig. 3.7 gives an example of the static pressure determined by this means. Although Fig. 3.7 shows p^* (the pressure extrapolated to infinite shut-in time) to be very nearly the same as the static pressure, this will not always be so.

The static reservoir pressure determined by this method may be in error because of a lack of specific information about the shape of the well's drainage area. The error, however, will not be large enough to be significant in measuring condition ratio.

In highly permeable formations where low productivity is caused by a very large reduction in permeability adjacent to the well, it may be practical to measure the approximate static reservoir pressure by shutting in a well. In many formations, however, this cannot be done because of the long shut-in period required. The length of time needed to obtain essentially static reservoir pressure is increased by the following factors: (1) thin, low permeability formations, (2) high viscosity oil, (3) high production rates prior to shut-in, (4) large radius of drainage, and (5) long producing time prior to shut-in.

The rate of pressure buildup in a well may be very slow, and yet the bottom-hole pressure may differ greatly from the static reservoir pressure. Gross errors in estimating the static reservoir pressure can be made if it is assumed prematurely that the bottom-hole pressure has reached the static pressure. A frequent cause of the errors is that the change in pressure over a period of a few hours is less than the accuracy of the instrument measuring the bottom-hole pressure. For this reason, and because it is impractical to shut in wells for a long period of time in order to measure static pressure, the foregoing method should be used to estimate static pressure.

Method of Determining Condition Ratio

To apply Eq. 3.8 to field data, Gladfelter *et al.*[1] suggested that the following steps be taken to facilitate the analysis of the data.

1. Plot on coordinate paper the actual field-measured bottom-hole pressure (psia) vs time in hours, and draw a *smooth* curve. This minimizes scatter resulting from errors of observation.

2. Plot the height of fluid column in the well vs time on coordinate paper and draw a *smooth* curve. If reliable sonic fluid-level data cannot be taken because of foaming in the annulus, it is necessary to calculate the height of the fluid column as it changes with time.

$$\text{Height of fluid} = \frac{(\text{Bottom-hole pressure} - \text{casing pressure})}{\text{Fluid gradient}} \quad (3.11)$$

3. Starting at zero shut-in times, read off values of bottom-hole pressure and fluid height from each of the curves in Steps 1 and 2 at 1-hour intervals and tabulate these values. Time intervals less than or greater than 1 hour can be used, depending upon how fast the well builds up (Fig. 3.8).

4. Evaluate the rate of change, or rise, of the fluid column in feet per hour at each time interval, n. This is now q_n.

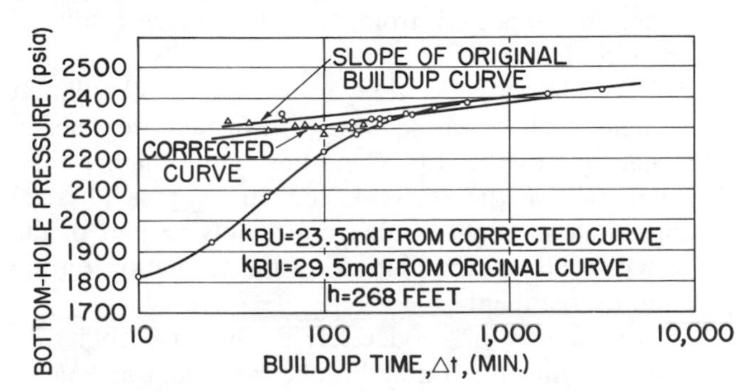

Fig. 3.6 Bottom-hole pressure vs buildup time, Well No. 2 (from Gladfelter *et al.*[1]).

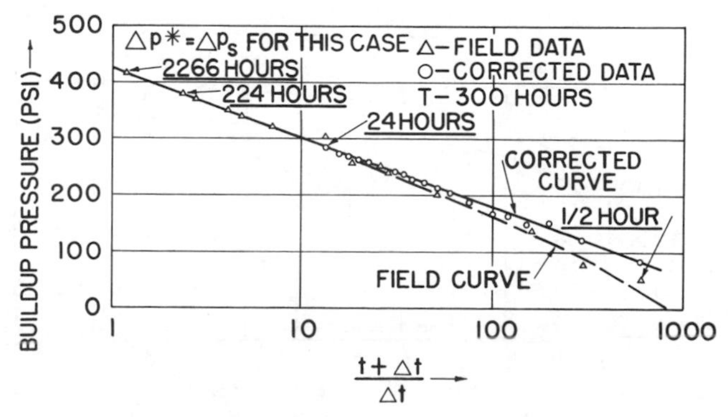

Fig. 3.7 Buildup pressure vs $\log(t+\Delta t)/\Delta t$, Well No. 1. (from Gladfelter *et al.*[1])

5. Convert the original producing rate, q, from barrels per day to feet of rise in the casing annulus per hour; i.e.,

$$q_{\mathrm{ft/hr}} = q_{\mathrm{bbl/day}} \left[\frac{1}{24\ (\text{casing annulus capacity, bbl/ft})}\right]. \quad (3.12)$$

If tubing and casing are equalized, use casing volume capacity minus tubing and rod-metal volume.

6. Calculate $q - q_n$ at each successive time interval—i.e., Step 5 minus Step 4.

7. Calculate $q/(q - q_n)$ at each successive time interval—i.e., Step 5 divided by Step 6.

8. Subtract the bottom-hole pressure at zero shut-in time from the bottom-hole pressure at each of the time intervals, i.e., the values as read in Step 3. This gives Δp_n, the buildup pressure, as a function of buildup time, t_n.

9. Multiply Step 7 by Step 8 (i.e., $[q/(q - q_n)]$ (Δp_n) for each successive time interval, which will give the corrected buildup pressure, $\Delta p_n'$, as a function of buildup time.

10. Plot, on semi-log paper, $\Delta p_n'$ (Step 9) vs the log of the shut-in time.

11. Measure the slope of the curve from Step 10 in psi per cycle. This will give the value of m to use in obtaining k_{BU}.

12. Calculate k_{BU} using Eq. 3.9 and the value of m just obtained in Step 11.

13. The condition ratio can be calculated by dividing the k_{PI} (millidarcies) from a productivity-index test, Eq. 3.10, by the k_{BU} from buildup data, (Step 12) —i.e., condition ratio $= k_{\mathrm{PI}}/k_{\mathrm{BU}}$.

Some mistakes might be made in calculating the condition ratio. To compute both k_{PI} and k_{BU}, we must assume values of q, μ, B, and h. Fortunately, any errors in estimating or measuring these quantities cancel out, provided the same value is used in the calculation of both k_{PI} and k_{BU}.

Since it is not possible to know r_e/r_w precisely, the value substituted for ln (r_e/r_w) in calculating k_{PI} may be significantly in error if only a short test is run. However, if the production rate has been settled for a period of several days or weeks before the well pressure is measured, the $\ln(r_e/r_w)$ can be approximated by $\ln(2 \times$ average distance to offset wells). Not producing the well long enough when measuring the productivity index causes the calculated condition ratio, CR, to be too high.

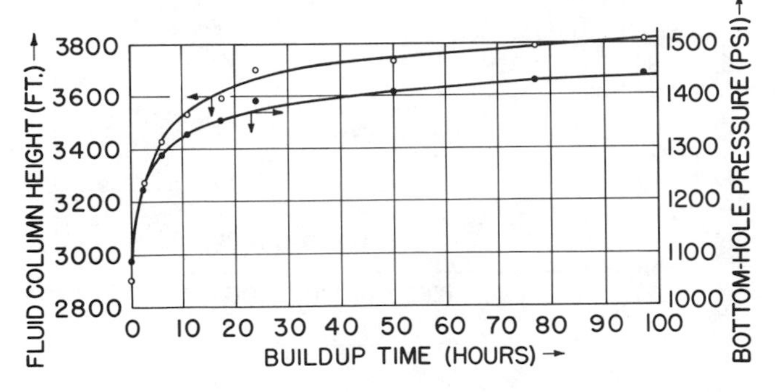

Fig. 3.8 Steps 1 and 2 initial plot of field data, Well No. 1. (from Gladfelter *et al.*[1])

3.4 Determining from Buildup and Condition Ratio Data the Applicability of Wells for Fracturing

Effect of Low Permeability on Fracturing Response

A well may not produce at attractive rates even though there is a high reservoir pressure and no reduction in permeability exists near the wellbore. The low production rate may be caused by the reservoir's being too tight to produce at economical rates without some special stimulation, such as reservoir fracturing. Based on the data of Wilsey and Bearden,[2] it was found that if the slope of the pressure buildup curve is greater than about 50 psi/cycle/B/D producing rate (with buildup curve plotted on semilog coordinates), and if substantial reservoir pressure exists, a well will give a desirable response to a deeply penetrating or a reservoir-fracturing treatment.

Determination of State of Pressure Depletion

Analyzing a well's pressure buildup data to get static pressure will show whether or not the reservoir has been depleted. If little reservoir pressure exists, most of the primary oil has already been recovered and there is little point in stimulating this kind of well. Fracture treatments on wells in a low-pressure reservoir may result in a temporary increase in production rate, but the rate will soon decline and the stimulation treatment may not be an economic success.

Estimation of Possible Increase in Productivity From Fracturing Using Condition Ratio

The increase in well productivity to be expected from a particular stimulation treatment can be found by comparing the condition ratio measured before the treatment with the condition ratio that normally results from similar treatments. For example, if a condition ratio of 2 can be expected after a stimulation treatment, and the condition ratio of a well is 0.2 before treatment, a tenfold increase in stabilized producing rate can be expected from this well as a result of stimulation treatment.

The condition ratio that results from stimulation treatments such as oil squeezes, surfactant treatments, etc., can at best be 1.0. Condition ratios higher than 1.0 require that the reservoir rock undergo a physical change such as might result from dissolving part of the rock with acid or from rupturing the formation during a fracture treatment.

In formations of moderately high permeability, a condition ratio of 2.0 may be about as high as can be expected from ordinary stimulation treatments. In formations of low permeability, condition ratios much higher than this can result from fracture treatments, depending upon reservoir conditions.

In low permeability formations, condition ratios of about 5.0 are common. Despite this variation in condition ratio that can result from stimulation treatments, experience gained from measuring this factor in a few stimulated wells will provide information to indicate the condition ratio that can be expected. The use of condition ratio to estimate increases in productivity to be expected from a stimulation treatment is subject to the errors that may arise in trying to determine what that condition ratio will be. However, using the condition ratio does remove the much more important error resulting from not knowing the well conditions before treatment. Thus, a minimum of experience in an area makes it possible to predict the increases in productivity that will result from a stimulation treatment.

Estimation of Static Reservoir Pressure

The estimation of static reservoir pressure from early pressure buildup data is illustrated by Fig. 3.7. Data collected during the first 24 hours after shutting in the well were used to indicate the same static pressure as measured after 2,266 hours of shut-in.

As previously discussed, the static reservoir pressure

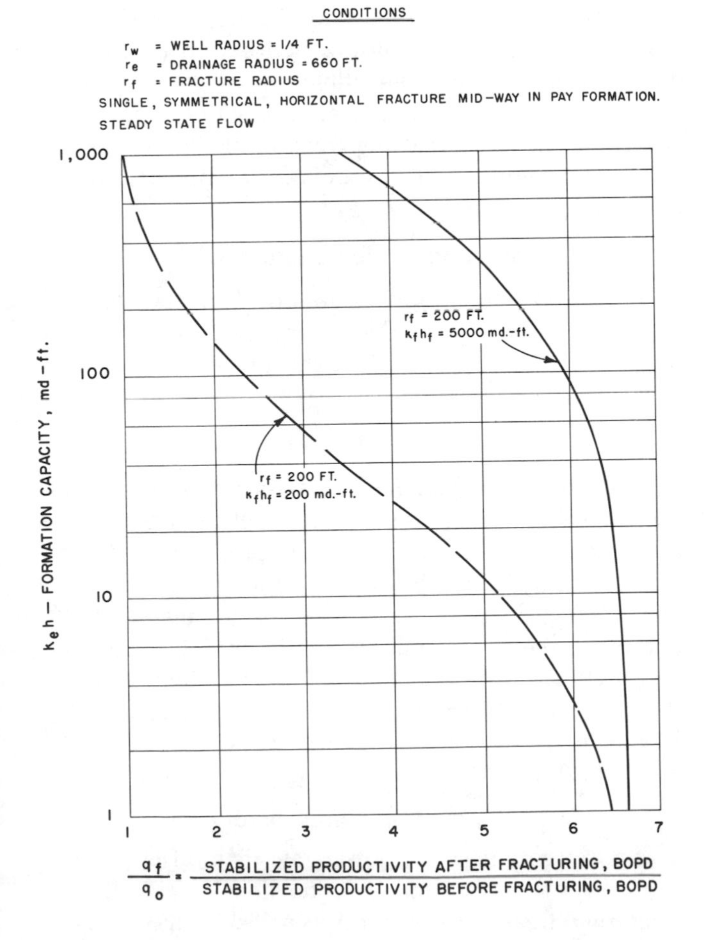

Fig. 3.9 Effect of fracture capacity on post-fracturing productivity.

can be used to determine the likelihood of a successful fracturing treatment.

3.5 Effect of Formation Capacity *(kh)* on Applicability of Wells for Treatment

The formation-fluid-carrying capacity (formation thickness × permeability) has a major effect on the response that might be expected from various types and sizes of hydraulic fracturing treatments. Fig. 3.9 is a plot of formation capacity vs stabilized productivity ratio after fracturing divided by productivity before fracturing. It shows a circular horizontal fracture with a 200-ft radius for a fracture with a fluid-carrying capacity of 200 md-ft and one of 5,000 md-ft. The data for these plots were calculated from Darcy's equation for radial flow of a homogeneous fluid flowing through a formation with discontinuous radial variations in permeability. For these calculations the following assumptions were made: (1) the formation is homogeneous and isotropic, (2) the fluids are incompressible and homogeneous, (3) the system is in a steady state of flow, (4) no wellbore impairment exists, and (5) there are no gravity effects.

These curves demonstrate that with high capacity formations, higher capacity fractures are required to produce satisfactory results from a hydraulic fracturing treatment.

Fig. 3.10 (prepared like Fig. 3.9 and using the same assumptions) is a plot of the estimated production increase from fracturing; the parameters are the fracture penetration and the fracture:formation-capacity contrast. The interrelationship of the fracture area (penetration) and the fracture capacity is clearly depicted on these plots. Note that as the ratio of fracture capacity to formation capacity increases, deeper penetrating fractures yield much greater increases in stabilized well productivity. Note also that if maximum benefit is to be obtained from the deeper penetrating fracture systems, the fracture:formation-capacity contrast must be high; that is, high capacity fractures must be made. The optimum combination of fracture area (penetration) and fracture capacity is basic to any treatment design.

3.6 General Criteria for Selecting Wells for Fracturing Treatment

Criteria for selecting wells for hydraulic fracturing treatment have been published by many authors.[13-17] The following criteria are believed to be applicable in most cases. However, no general rule should be used when data are available from buildup or drawdown tests, drill-stem tests, core analyses or the like.

State of Depletion of the Producing Formation

Fracture treatments are successful in increasing oil and gas production rates because they increase permeability in the vicinity of the wellbore. If a formation is depleted of reservoir energy, fracturing generally will not increase oil production enough to justify the expense of the treatment. Larger and more sustained production increases can be expected if fracturing treatments are used early in the life of the field. Results have shown, however, that successful treatments can be performed in old fields where pressures are comparatively low, and where gravity drainage rates are enhanced by fracturing.

Formation Composition and Consolidation

Limestone, dolomite, sandstone, and conglomerate pay formations can be fractured successfully; and although the treatments generally are not considered applicable in unconsolidated formations, there have been some successful ones.

Formation Permeability

As the permeability of the formation approaches the permeability of the created fracture, the possible production increase approaches zero. Therefore, a larger production increase, expressed in percent of natural production, can be expected from fracturing low permeability pay zones than from fracturing high permeability pay zones.

Formation Thickness

Calculations and early experience indicated that better results could be expected in thin producing zones than in thick ones. Vertical fracturing, multiple fracturing, the large size of jobs, and high injection rates have reduced the importance of formation thickness to the extent that it is no longer a major consideration in the selection of wells for fracture treatment.

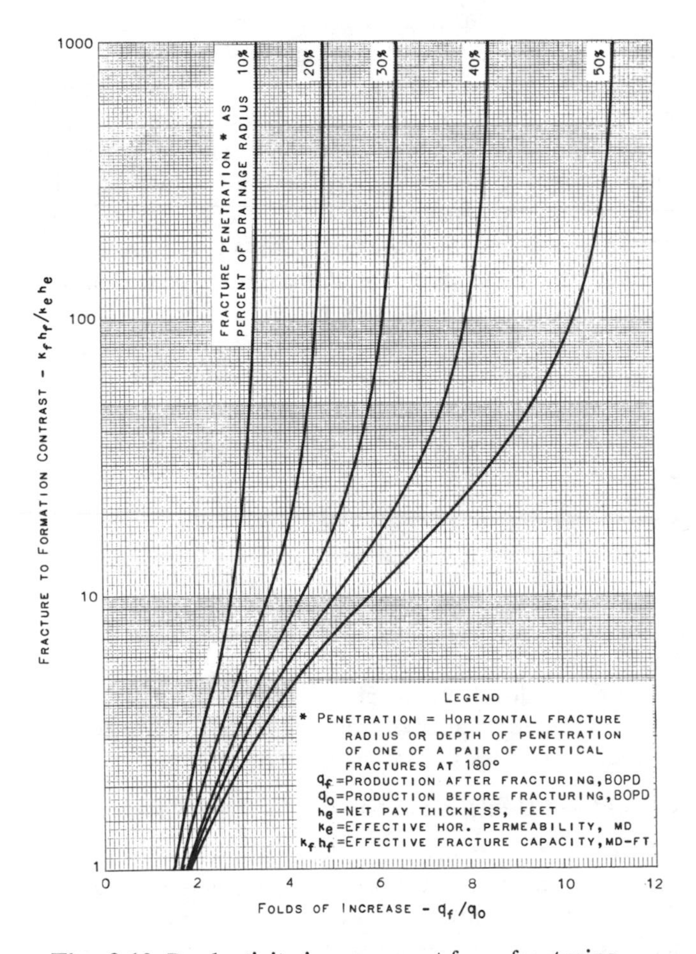

Fig. 3.10 Productivity improvement from fracturing.

Previous Workovers

From the study of early fracturing treatments it was reasoned that a well in which the permeability had already been improved by some type of treatment would not respond enough to a fracture treatment to make the operation profitable. Wells from which production increases were obtained by small-size fracture treatments are good candidates for retreatment, and wells that have been acidized or shot can also be fractured successfully. New techniques, large-size jobs, high injection rates, and improved and more economical carrying agents are responsible for highly successful fracture treatments of wells that have been treated previously by other stimulation methods.

Isolation of the Zone To Be Treated

Injected treating fluids will follow the path of least resistance. No oil production increase can be expected if a fracture is created in cement, shale, or coal, instead of in the producing zone. The practice of extending perforations a few feet beyond the upper and lower limits of a pay zone to insure perforation of all productive zones is not compatible with fracturing. Perforating only within the indicated limits of a pay zone is recommended so that the treating materials and the resulting fractures will be confined to the productive portions of the zone.

Condition of Well Equipment

Bottom-hole treating pressures of 1 psi per foot of depth should be expected, and working pressure ratings of well equipment should be adequate to withstand these pressures.

Production History of the Well

Wells with comparatively flat or steep production decline curves offer good opportunities for fracture treatment. A comparatively flat production decline curve indicates that the well is draining a large area and that the rate of drainage can be increased by improving the permeability near the wellbore. A comparatively steep production decline curve indicates that the well may be draining a limited area and that the production rate can be increased by extending the radius of drainage.

Offset Production History

If a well produces at a lower rate than offset wells, it can expect to have a larger production increase from a fracturing treatment than other wells in the field. A comparatively low rate of production indicates that the effective permeability near the well is less than the permeability in adjacent areas in the same pool. Fracture treatment will likely increase permeability near the well and thereby increase the production rate.

Location of Water-Oil and Gas-Oil Contacts

Creating or extending fractures into a water- or gas-bearing zone would increase water-oil or gas-oil ratios without increasing oil production. No change in GOR or WOR should be expected from a fracture treatment unless the fracture is created in a water or gas zone. Experience has indicated that vertical fractures can be created unintentionally and can extend many feet into water-bearing formations below the producing zone. Results of fracturing in water-drive reservoirs and waterflood projects indicate that the danger of adversely affecting the producing WOR is greatest in bottom-water-drive reservoirs or when bottom water is present. Efforts should be made to avoid vertical fractures that extend into the water zone. Oil and water are usually increased proportionately by fracturing in edge-water-drive reservoirs and in lateral-drive waterfloods.

3.7 Summary

Before a well is selected for a hydraulic fracturing treatment, it must be determined that the reservoir contains sufficient fluids in place and has adequate potential gradients or formation pressure available to produce at higher rates following the creation of a high permeability fracture. The cause for low productivity must also be determined so that the right type of fracturing job can be applied.

One of the best methods for evaluating the condition of the formation adjacent to the well and the condition of the reservoir in the interwell area is to analyze pressure buildup data. By analyzing the pressure buildup behavior of a well when it is shut in and by using a productivity-index test, the average permeability of the formation both near the well and in the reservoir can be determined. From this information any wellbore damage or skin effect can be calculated and the condition ratio or flow efficiency of the well can be determined.

With this information, the cause of the low productivity of a well can be determined. It usually is caused by one of three situations, and for each one a different course of action is required. (1) It may be the result of a permeability reduction near the well. The removal of this block by a small fracturing treatment will often result in a substantial production increase. (2) It may be a result of low permeability throughout the reservoir. If so, and if substantial recoverable oil is in place, production can be greatly increased by deep-penetrating fractures achieved with large-volume, high-injection-rate fracturing treatments. (3) Sometimes the reservoir pressure has been depleted even in the interwell area. In this case, fracturing generally will not increase productivity enough to make a treatment profitable.

The key to determining whether or not a well is a good candidate for hydraulic fracturing is diagnosing the well to find the cause for its low productivity. Buildup, drawdown, or drill-stem tests, core analyses, offset well data and other information can be used to accomplish this. After diagnosis, the optimum hydraulic fracturing treatment can be designed for the well.

References

1. Gladfelter, R. E., Tracy, G. W. and Wilsey, L. E.: "Selecting Wells Which Will Respond to Production-Stimulation Treatment", *Drill. and Prod. Prac.*, API (1955) 117.
2. Bearden, W. G. and Wilsey, L. E.: "Reservoir Fracturing — A Method Of Oil Recovery from Extremely Low Permeability Formations", *Trans.*, AIME (1954) **201,** 169-175.
3. Horner, D. R.: "Pressure Buildup in Wells", *Proc.*, Third World Pet. Cong., E. J. Brill, Leiden (1951) **II.**
4. Platt, George: "A Study of Transient Pressure Behavior in Porous Media", unpublished MS thesis, U. of Tulsa, Tulsa, Okla. (1952).
5. van Everdingen, A. F. and Hurst, W.: "The Application of the Laplace Transformation to Flow Problems in Reservoirs", *Trans.*, AIME (1949) **186,** 305-324.
6. Matthews, C. S., Brons, F. and Hazebroek, P.: "A Method for Determination of Average Pressure in a Bounded Reservoir", *Trans.*, AIME (1954) **201,** 182-191.
7. Jacob, C. E.: *Trans.*, AGU, Part II (1940).
8. Miller, C. C., Dyes, A. B. and Hutchinson, C. A., Jr.: "Estimation of Permeability and Reservoir Pressure from Bottom Hole Pressure Build-Up Characteristics", *Trans.*, AIME (1950) **189,** 91-104.
9. Matthews, C. S.: "Analysis of Pressure Build-up and Flow Test Data", *J. Pet. Tech.* (Sept., 1961) 862-870.
10. Mead, H.: "Another Concept for Final Buildup Pressure", paper 1111-G presented at SPE 33rd Annual Fall Meeting, Houston, Tex., Oct. 5-8, 1958.
11. van Everdingen, A. F.: "The Skin Effect and Its Influence on the Productive Capacity of a Well", *Trans.*, AIME, (1953) **198,** 171-176.
12. Hurst, W.: "Establishment of the Skin Effect and Its Impediment to Fluid-Flow Into a Well Bore", *Pet. Eng.* (Oct., 1953) B-6.
13. Kaufman, M. J.: "Well Stimulation by Fracturing", *Pet. Eng.* (Sept., 1956) B-53.
14. Clark, J. B.: "A Hydraulic Process for Increasing the Productivity of Wells", *Trans.*, AIME (1949) **186,** 1-8.
15. Clark, J. B., Fast, C. R. and Howard, G. C.: "A Multiple-Fracturing Process for Increasing Productivity of Wells", *Drill. and Prod. Prac.*, API (1952) 104.
16. Maly, Joe W. and Morton, Tom E.: "Selection and Evaluation of Wells for Hydrafrac Treatment", *Oil and Gas J.* (May 3, 1951) No. 52, 126.
17. Clark, R. C., Freedman, H. G., Bolstead, J. H. and Coffer, H. F.: "Application of Hydraulic Fracturing to the Stimulation of Oil and Gas Production", *Drill. and Prod. Prac.*, API (1953) 113.

Chapter 4

Fracture Area

Hydraulic fracturing technology has been characterized by a continuing trend toward larger-volume, higher-injection-rate fracturing jobs. The trend, however, has been limited somewhat by the pressure restrictions and friction losses encountered in tubular goods installed in wells. To select the best fracturing fluids available, those factors must be analyzed that control or determine the fracture extension that can be produced with various types and volumes of fluids and at various injection rates.

This chapter presents a discussion of the effect of fracture penetration on well productivity, develops formulas for calculating the extent of fracturing, and presents a method of determining the optimum fluid characteristics for maximum fracture extension for various types of fracturing fluids. Using the methods presented, we can determine the treatment that is best from the standpoint of productivity. The calculations involved are important since fracturing services are available from many service companies, each offering a wide selection of fluids. Variations in both cost and physical characteristics of these materials affect the economics of the operation and the effectiveness of the treating procedure.

4.1 Effect of Fracture Areal Extent on Well Productivity

Analytical and electrical model studies have shown the influence of fracture penetration on both the flush and stabilized production that may be obtained from a given reservoir with a given fracture system. Such investigations[1-3] have shown that a fracturing treatment is influenced not only by the flow capacity of a hydraulically created fracture but also, to a large degree, by the areal extent of the fracture. The benefits derived from fracturing may be classified as flush production and post-flush production.

Flush Production

The effect of horizontal fracture penetration on the flush production immediately following a fracturing treatment has been illustrated by analytical calculations.[1] Typical results of these calculations are presented in Fig. 4.1, which is a plot of producing rate vs time on a logarithmic scale for the conditions shown.

The curves on Fig. 4.1 indicate that higher rates of production are maintained for a longer time when deeper fractures are made, and that the fracture penetration has little effect on initial flush production rate. Calculations from this plot show that the higher rates of production can be maintained until essentially all of the recoverable oil is produced from the reservoir flanking the fractures.

Post-Flush Production

The result of fracture penetration on steady-state productivity (after decline of the flush production) has been demonstrated by electrical model studies.[2,3] Fig.

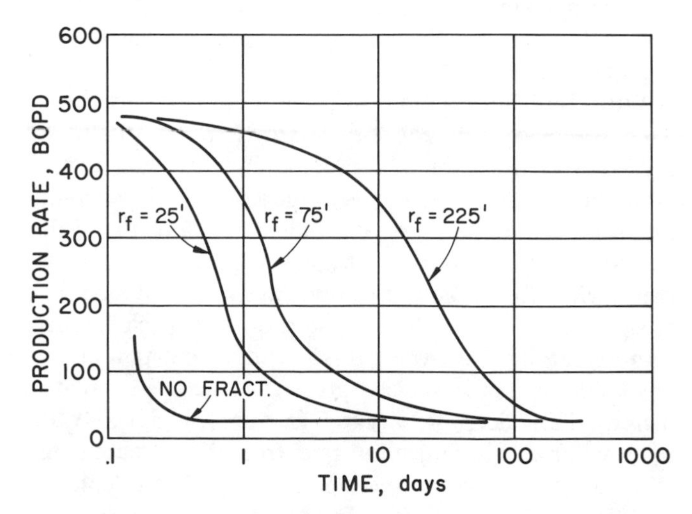

Fig. 4.1 Effect of horizontal fracture radius on production rate.

4.2 is a plot of productivity ratio (productivity after fracturing divided by productivity before fracturing) vs fracture extent. It shows the effect of increased fracture extent on productivity ratio for a single horizontal fracture in the center of the pay and for a vertical fracture, the plane of which bisects the borehole and extends the same distance into the reservoir as does the horizontal fracture. This plot reveals that the productivity ratio increases as the fracture extent increases.

It follows from this that both flush and stabilized well productivity are increased as the fracture length increases and that the total production can be obtained in less time.

Deeply penetrating fractures not only result in sustained increases in flush and post-flush production, but also may result in increased ultimate recovery.[1,4-6] This increase has been evidenced by extension of the economic producing life of a well or field, and the stimulation of initial production from fields in which the relationship between permeability and differential pressure did not permit economical production prior to fracturing.

4.2 Calculation of Fracture Area

The importance of fracture penetration on well productivity has been pointed out. We must also consider the effects of fracturing fluid characteristics and reservoir fluid and rock characteristics on the areal extent of a fracture. To establish these effects, R. D. Carter[7] derived equations for calculating the areal extent of hydraulically created horizontal or vertical fractures. The following assumptions were used in this derivation.

1. The fracture is of uniform width.
2. The flow of fracturing fluid from the fracture into the formation is linear and the direction of flow is perpendicular to the fracture face.
3. The velocity of flow into the formation at a point on the fracture face depends on the length of time this point has been exposed to flow.
4. The velocity function $v(t)$ is the same for every point in the formation. However, zero time for a given point is defined individually as the time at which the fracture and, consequently, the fracturing fluid reaches that point.
5. The pressure in the fracture is constant and is equal to the sand-face injection pressure.

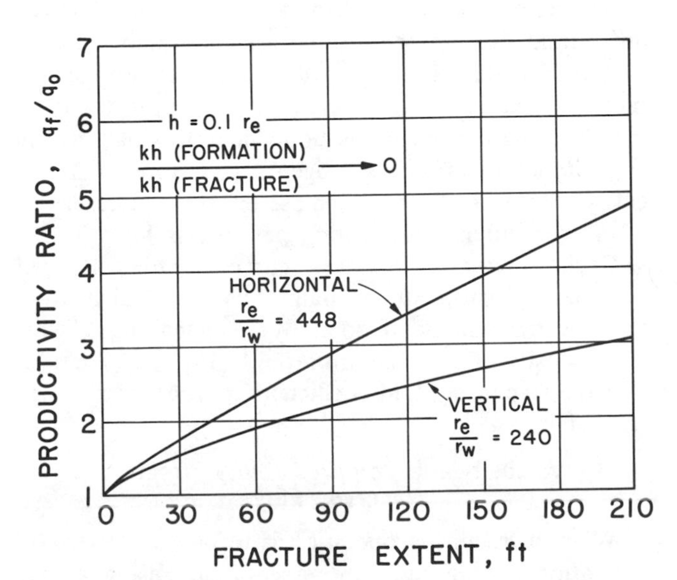

Fig. 4.2 Effect of fracture penetration on productivity.

The volume rate at which fluid leaks from the fracture into the formation may be expressed by

$$i_L(t) = 2\int_o^{A_{ff}(t)} v\,dA_{ff} \quad . \quad . \quad . \quad . \quad . \quad . \quad . \quad . \quad (4.1)$$

The quantity A is a function of time, and the value of v at the time t, corresponding to a given element dA_{ff} formed at the time λ, is $v(t-\lambda)$. Also, since A_{ff} is a function of time, $dA_{ff} = \frac{dA_{ff}}{d\lambda}\,d\lambda$. Eq. 4.1, therefore, can be written

$$i_L(t) = 2\int_o^{t} v(t-\lambda)\,\frac{dA_{ff}}{d\lambda}\,d\lambda \quad . \quad . \quad . \quad . \quad . \quad . \quad (4.2)$$

The rate at which the volume of the fracture is increased is given by

$$Q_f = W\,\frac{dA_{ff}}{dt} \quad , \quad . \quad . \quad . \quad . \quad . \quad . \quad . \quad . \quad . \quad . \quad . \quad (4.3)$$

where W is the fracture clearance.

The rate of fluid injection is equal to the sum of the rate at which fluid is lost to the formation and the rate of volume increase of the fracture itself. That is,

$$i = i_L + Q_f \quad . \quad . \quad . \quad . \quad . \quad . \quad . \quad . \quad . \quad . \quad . \quad . \quad . \quad (4.4)$$

Substituting Eqs. 4.2 and 4.3 into Eq. 4.4,

$$i = 2\int_o^{t} v(t-\lambda)\,\frac{dA_{ff}}{d\lambda}\,d\lambda + W\,\frac{dA_{ff}}{dt} \quad . \quad . \quad . \quad . \quad (4.5)$$

Eq. 4.5 can be solved for $A(t)$ by means of the Laplace transformation, once the forms of $v(t)$ and $Q(t)$ are given. The ease with which a solution is obtained, of course, depends on the forms taken by $v(t)$ and $q(t)$.

Let $v(t) = \frac{C}{\sqrt{t}}$, and $i(t) = i =$ constant. Then, substituting this into Eq. 4.5 and using the Laplace transform method of solution, results in

$$A_{ff}(t) = \frac{iW}{4C^2\pi}\left[e\left(\frac{2C\sqrt{\pi t}}{W}\right)^2 \cdot \operatorname{erfc}\left(\frac{2C\sqrt{\pi t}}{W}\right) + \frac{4C\sqrt{t}}{W} - 1\right] \quad . \quad . \quad . \quad . \quad (4.6)$$

$$A = \frac{iW}{4\pi C^2}\left(e^{x^2} \cdot \text{erfc}\,(x) + \frac{2}{\sqrt{\pi}}\,x - 1\right) \quad . . . (4.7)$$

where

$$x = \frac{2C\sqrt{\pi t}}{W} \quad . . . . (4.8)$$

If desired, the spurt loss V_{sp} may be included in the calculation of fracture area by using the spurt loss as an increased fracture clearance where

$$W^1 = W + 2\,\frac{V_{sp}/A_c}{30.48} = W + \frac{V_{sp}}{15.24A_c}\ ; \quad . . . (4.9)$$

the 30.48 is a conversion factor to change the V_{sp}/A_c from cc/sq cm to cu ft/sq ft.

The coefficient C* will be referred to as the *fracturing fluid coefficient*. In a given type of flow system this coefficient depends on the characteristics of the fracturing fluid used and on the characteristics of the reservoir fluids and rock. Whereas a high coefficient means poor fluid-loss properties, a low fracturing fluid coefficient means low fluid-loss properties and thus larger fracture area for a given volume and injection rate.

The calculated results using Eq. 4.7 show that pump rate and pumping time (volume) affect the fracture area or extent; i.e., as the pump rate or fracturing fluid volume (pumping time) is increased, the fracture area is increased.

4.3 Fracturing Fluid Coefficient

Since the fracturing fluid properties are reflected in Eq. 4.7 only through the fracturing fluid coefficient, C, a method must be established for determining this factor for various types of fracturing fluids.

The fracturing fluid coefficient, C, defines the three types of linear flow mechanisms that are encountered with fracturing fluids and for which Eq. 4.1 applies. The three types of linear flow mechanisms are: (C_{I}) — viscosity and relative permeability effect; (C_{II}) — reservoir fluid viscosity-compressibility effects; and (C_{III}) — wall-building effects. The first two mechanisms involve coefficients that can be calculated from reservoir data and fracturing fluid viscosity with the aid of presently available formulas. The third case involves fluid-loss coefficients for fluid-loss additives, which must be determined experimentally. Although each mechanism is considered to be acting alone by this equation, all may act simultaneously in a fracturing treatment so that the mechanisms may complement each other and increase the fluid's effectiveness.

High Viscosity Fracturing Fluids

A constant pressure injection of a high viscosity Newtonian fluid into an evacuated porous medium under conditions of linear flow will result in flow velocities given by

$$v = \frac{i}{A_{ff}} = \left(\frac{k\,\Delta p\,\phi}{2\mu t}\right)^{1/2} = \frac{C_{I}}{\sqrt{t}}, \quad . . . . . . (4.10)$$

where all factors except q and t are constant. Thus, for this case, the fracturing fluid coefficient, C_{I}, is equivalent to the square root of $k\,\Delta p\,\phi/2\mu$, where k is the formation permeability to the fracturing fluid, Δp is the difference in pressure between the fluid at the formation face and the fluid in the formation, μ is the viscosity of the fracturing fluid at bottom-hole conditions, ϕ the porosity of the formation, and v the flow velocity. All of the measurements are in metric units.

If the velocity is expressed in feet per minute, a conversion factor of 0.0469 must be used to define correctly the fluid-loss coefficient when k is expressed in darcies, t in minutes, Δp in psi, and μ in centipoises at bottom-hole fracturing temperatures; that is,

$$C_{I} = 0.0469\,\sqrt{\frac{k\,\Delta p\,\phi}{\mu}},\ \frac{\text{ft}}{(\text{min})^{1/2}} \quad . . . . . . (4.11)$$

The fracturing fluid coefficient can be readily determined by the use of Eq. 4.11 once the absolute viscosity of the fracturing fluid, the formation permeability, the treating differential pressure, and the formation porosity have been ascertained. For example, if the permeability of a formation to be treated is 0.01 darcy, the expected treating differential pressure is 1,000 psi, the formation porosity is 0.20 and the fracturing fluid viscosity is 500 cp; then

$$C_{I} = 0.0469\,\sqrt{\frac{0.01(1000)(0.2)}{500}}$$
$$= 2.96 \times 10^{-3}\ \text{ft}/(\text{min})^{1/2}.$$

The effect of fracturing fluid viscosity on a fracture area is demonstrated on Fig. 4.3, which presents a plot of fracture area vs fracturing fluid viscosity for an assumed set of reservoir and treating conditions. The two differential treating pressure curves show that increases in the absolute viscosity will cause increases in the fracture area. Also, a 50 percent reduction in the treating differential pressure is equivalent to doubling the viscosity of the fracturing agent. Therefore, high viscosity fracturing fluids are most effective in reservoirs where the differential treating pressure is low.

Each differential pressure curve of Fig. 4.3 will become asymptotic to a common area (equivalent to the injected volume divided by the clearance) when the viscosity of the fracturing fluid (Eq. 4.11) is such that the fracturing fluid coefficient approaches zero (no leak-off).

Viscosity and Compressibility Effects of Reservoir Fluids

When a fracturing operation is to be performed in a formation containing compressed fluids using a fracturing liquid that has physical properties identical with

*See Eqs. 4-10, 4-14, and 4-18 for the determination of the fracturing fluid coefficient for various flow mechanisms.

those of the reservoir fluid, the mechanism previously discussed would not be applicable. Under these conditions, the rate of fluid loss would be controlled by the viscosity, by the coefficient of compressibility of the fluid being injected, and by the reservoir fluid.

Eq. 4.12 relates rock permeability and porosity to pressure, time, and the compressibility of the fluids involved.

$$\left(\frac{k}{c\phi\mu}\right)\left(\frac{\partial^2 p}{\partial z}\right)=\frac{\partial p}{\partial t}\ , \quad . \ . \ . \ . \ . \ . \ . \ . \ . \ . \quad (4.12)$$

Rock compressibility, because of its negligible effect, is not considered.

If the assumption is made that the reservoir boundaries are infinite in the direction of the leakoff and that a constant pressure drop exists between the fluid in the fracture and the fluid at the outer extremities of the reservoir, the following equations[10] for flow velocity v result.

$$v\bigg|_{z=0}=-\frac{k}{\mu}\left(\frac{\partial p}{\partial z}\right)\bigg|_{z=0}=\frac{k}{\mu}\frac{\Delta p}{\sqrt{\pi\alpha t}}=\Delta p\sqrt{\frac{k\phi c}{\mu\pi t}}=\frac{C_{II}}{\sqrt{t}} \quad . \ . \ . \ . \ . \ . \ . \ . \ . \ . \ . \ . \ . \ . \ . \ . \quad (4.13)$$

Eq. 4.13 indicates that the fracturing fluid coefficient C_{II} for this case would be

$$C_{II}=0.0374\ \Delta p\sqrt{\frac{k\phi c}{\mu}},\ \frac{\text{ft}}{(\text{min})^{1/2}}\ , \quad . \ . \ . \ . \quad (4.14)$$

where Δp is expressed in psi, k in darcies, c in reciprocal psi, and μ in centipoises at bottom-hole fracturing temperature.

The fracturing fluid coefficient can be determined by use of Eq. 4.14 once the treating differential pressure, the formation permeability and porosity, and the reservoir fluid viscosity and compressibility coefficient are determined. For example, if the treating differential pressure is 1,000 psi, the formation permeability is 0.01 darcies, the porosity is 0.20, the compressibility coefficient is 1×10^{-5}, and the reservoir fluid viscosity is 2 cp, then $C_{II} = 0.0374\ (1000)\sqrt{\frac{0.01(0.2)(0.00001)}{2}} = 3.74 \times 10^{-3}$, ft/(min)$^{1/2}$.

The effect of reservoir fluid viscosity and compressibility on the area of a fracture is shown in Fig. 4.4. Since the reservoir fluid compressibility does not vary appreciably, this parameter is held constant at 1×10^{-5} psi^{-1} on this plot. The fracture area is plotted against the reservoir fluid viscosity with the compressibility held constant. Separate curves show the effect of differential treating pressures of 600, 1,000 and 1,400 psi on the areal extent of a fracture.

Wahl[11], through the concept of cumulative fluid loss, has demonstrated the effect of a formation thickness on fracture fluid leakoff in a horizontal fracture when the viscosity and compressibility effects of the reservoir are the dominating factors. Cumulative fluid loss per unit area V_c is obtained by integrating $v(t)$ over the desired time interval. Thus, for the bounded bed,

$$V_c(t)=\frac{8}{\pi^2}\,\frac{h_t}{2}\,\Delta p_t\phi c\sum_{n=1}^{\infty}\frac{1}{(2n-1)^2}\,e-\beta(2n-1)^2\,t \quad . \ . \ . \ . \ . \ . \ . \ . \quad (4.15)$$

To illustrate this, let us assume two beds 12-and 25-ft thick. If we further assume that the fracture divides the pay zone in half, the term $h_t/2$ is one-half the formation thickness.

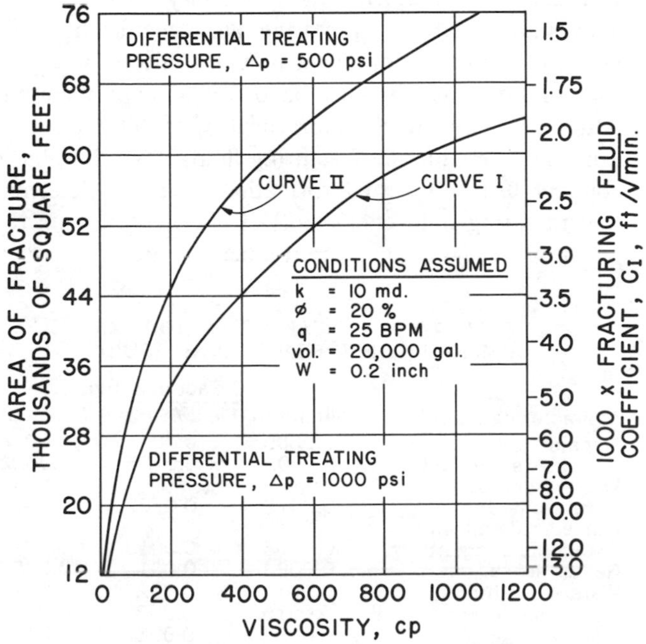

Fig. 4.3 Effect of fracturing fluid viscosity on area of fracture.

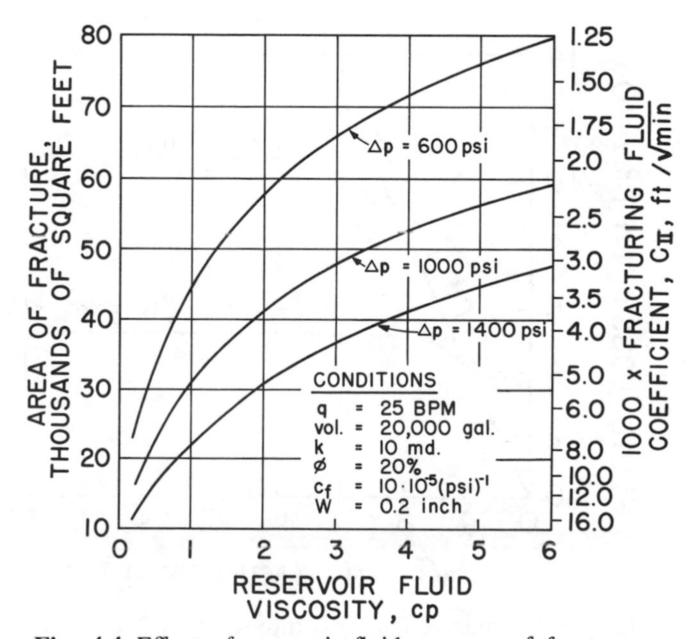

Fig. 4.4 Effect of reservoir fluid on area of fracture.

The results of these calculations are illustrated in Fig. 4.5 where cumulative fluid-loss volume V_c in thousandths of cubic feet per square foot is plotted against the square root of time in minutes. Plotting the equation of $V_c(t)$ for an infinitely thick sand yields a straight line with a slope of 1.06.

The curves for the bounded reservoirs follow the infinite case closely until a limiting value of V_c is approached. This limit equals the effective volume of oil times the product of the compressibility of the oil and the treating pressure differential. For $h_t/2 = 6$ ft, the maximum fluid-loss volume is 0.00518 cu ft/sq ft. For $h_t/2 = 12.5$ ft, the maximum figure is 0.0108 cu ft/sq ft. For an unbounded bed, there is no limit on the value of V_c.

The important feature of Fig. 4.5 is the slight deviation of the finite-reservoir curves from the C_{II} line until the maximum cumulative fluid-loss value $(V_c)_{max}$ is approached. In planning a fracturing operation it is sufficient to use the fracturing fluid coefficient C_{II} until $(V_c)_{max}$ is reached, although this approximation will result in a conservative figure for the fracture area formed during this time interval.

The characteristics of the fluid in the reservoir are of primary importance in calculating the areal extent of a fracture system. High-fluid-loss Newtonian oil of relatively low viscosity may be used successfully to extend a deeply penetrating horizontal fracture in a thin reservoir where the treating differential pressure is low, the viscosity of the reservoir fluid is high, or its coefficient of compressibility is low.

Wall-Building Fracturing Fluids

Work with water or oils containing fluid-loss additives has shown that the fluid-loss characteristics of these materials, when pressured against porous media,

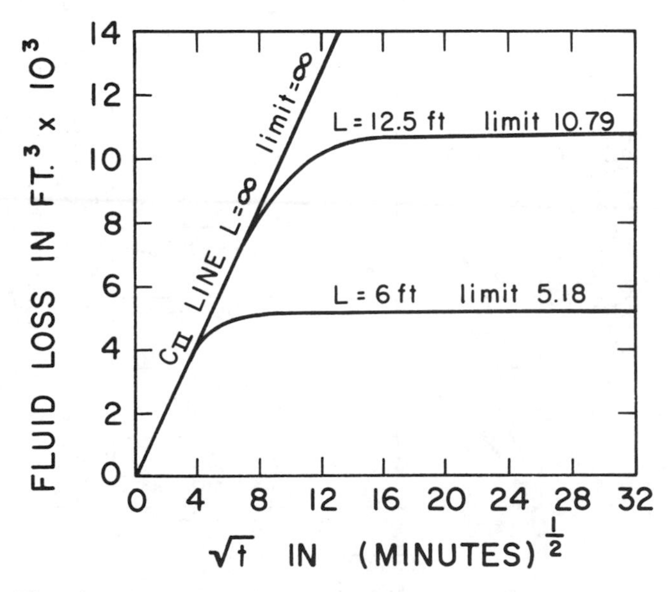

Fig. 4.5 Cumulative fluid loss through 1 sq ft of fracture surface area.[11]

can be determined graphically by plotting the experimentally determined cumulative filtrate volume vs the square root of flow time. This plot generally gives straight-line relationships for liquids containing effective fluid-loss additives when a constant pressure drop is imposed across the filter medium. Stated mathematically, the relationship is

$$V = m\sqrt{t} + V_{sp}, \quad . \quad . \quad . \quad . \quad . \quad . \quad . \quad . \quad . \quad (4.16)$$

where V_{sp} is the "spurt loss" or the volume represented by the intercept of the straight line on the y-axis at zero time. The relationship between the slope m and the fracturing fluid coefficient for this case is obtained by differentiating Eq. 4.16 with respect to time and dividing the result by the cross-sectional area A_c of the medium through which flow q takes place; that is, the velocity of flow v is equal to

$$v = \frac{q}{A_c} = \frac{1}{A_c}\frac{dV}{dt} = \frac{1}{2A_{ff}} \cdot \frac{m}{\sqrt{t}} = \frac{C_{III}}{\sqrt{t}} \quad . \quad . \quad . \quad (4.17)$$

Thus, we see that the fracturing fluid coefficient C_{III} is equivalent to the slope of the straight lines of Eq. 4.16 divided by twice the area of the test medium. During the differentiation process, V_{sp} dropped out and therefore had no effect on the fracturing fluid coefficient. The filtrate volume is normally measured in cubic centimeters, the time in minutes, and the area in square centimeters. To obtain the value of C that will yield the velocity in ft/(min)$^{1/2}$, a conversion factor of 0.0328 must be used; that is,

$$C_{III} = 0.0328\frac{m}{2A_c} = 0.0164\frac{m}{A_c}, \frac{\text{ft}}{\sqrt{\text{min}}} \quad . \quad . \quad (4.18)$$

The fracturing fluid coefficient for the two previous flow mechanisms can be calculated from known reservoir data and fracturing fluid viscosity. However, Eq. 4.18 shows that the fracturing fluid coefficient is dependent upon the slope of a plot of a filtrate vs square root of time, which must be obtained experimentally. Although many investigators have used filter paper as a means of evaluating fracturing fluid, it is advisable to conduct tests on actual formation cores to determine the fracturing fluid coefficients. Table 4.1 presents representative C_{III} data for frequently used fracturing

TABLE 4.1 — TYPICAL C* FACTORS OF COMMONLY USED FRACTURING LIQUIDS

Fracturing Liquid	C Factor Range		
	Minimum	Field Average	Maximum
Napalm gel	0.0005	0.001	0.0018
Viscous refined oil	0.0009	0.0125	0.0022
Lease crude oil (untreated)	0.0012	0.0762	—
Lease crude oil with fluid-loss agent	0.0008	0.0013	0.00146
Gelled lease oil	0.0004	0.001	0.020
Water-base fluids	0.001	0.002	0.0090
Acid-base fluids	0.0015	0.002	0.04
Acid emulsions	0.0005	0.001	0.018

*Wall-building fracturing fluids.

fluids. Note that while there are wide variations in the observed C factor, it is practical to improve this fluid characteristic by adding appropriate fluid-loss-control agents.

Experimental Procedure: Filtrate-vs-Time Tests (Static). The core to be tested is mounted in a lucite sleeve and sealed under pressure. Core wafers 0.5 cm thick are then sliced from the mounted core. The wafers are constructed so they can be placed in a high temperature-high pressure filter press and sealed with an O-ring. A photograph of a core wafer and the filter press in which the wafer is placed as shown on Fig. 4.6. To eliminate relative permeability effects, the core wafers are saturated under a vacuum, with a mineral oil or water, depending on the fracturing fluid, prior to the fluid-loss tests. The fluid-loss tests are conducted at the estimated bottom-hole treating temperature and differential pressure. The cumulative filtrate volumes ejected at 1, 4, 16, 25, and 30 minutes are plotted vs the square root of the corresponding flow times. The slope of the curves is then used to define the fracturing fluid coefficient.

Typical plots of the type of data obtained are presented on Fig. 4.7. This figure shows that many of the curves, regardless of core permeability, pass through the origin. Two of the curves plotted, however, have a positive y-axis intercept at zero time. Many of the experimental curves obtained to date have exhibited this characteristic. The engineer must recognize that a finite period of time exists before a filter cake builds up, during which time the loss of fracturing fluid in reservoirs is resisted by viscosity and compressibility effects alone.

The point where the fluid-loss curve intersects the ordinate on a fluid-loss plot is known as the spurt loss. Spurt loss is not accounted for in Eq. 4.18 and must be considered separately. Eq. 4.9 presents a method for including spurt loss in the fracture area calculations. Crittendon[12] suggested a different method for including spurt loss in the fracturing fluid coefficient:

1. Construct the fluid-loss plot in the conventional manner (Fig. 4.7).

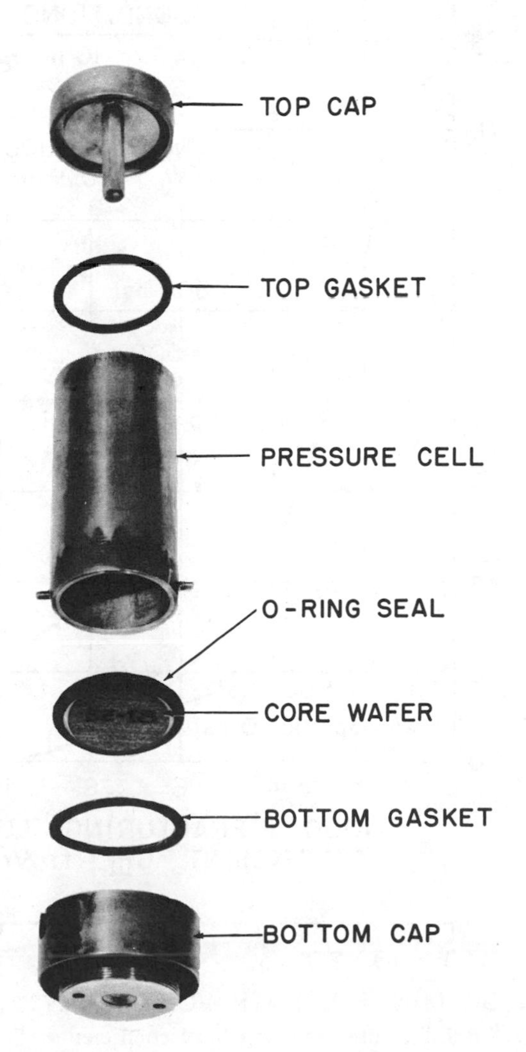

Fig. 4.6 Fracturing fluid filter loss apparatus.

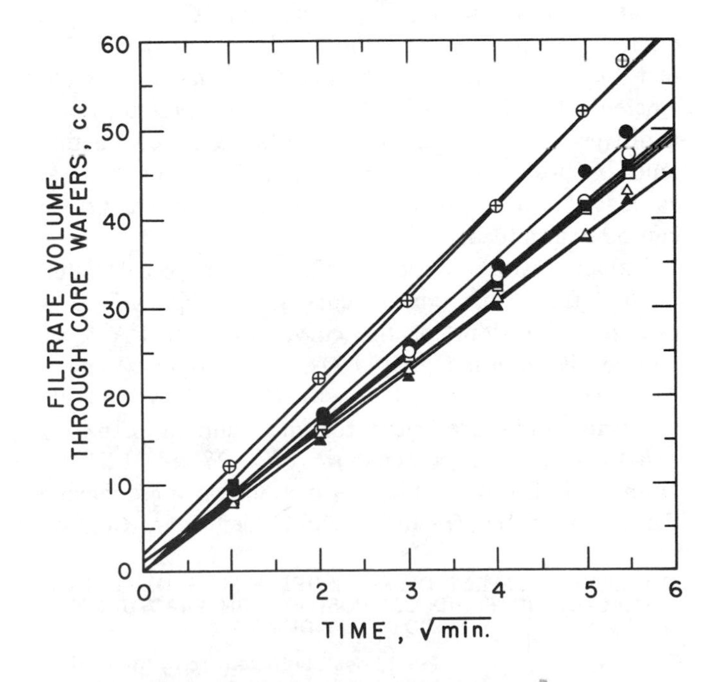

Fig. 4.7 Fluid-loss characteristics of paraffin base crude with Fluid Loss Additive A.

TABLE 4.2 — EFFECT OF PERMEABILITY ON FRACTURING FLUID COEFFICIENTS

Additive-Crude Oil System	Pressure Drop (psi)	Type of Sandstone	Permeability (md)	Fluid Loss Coefficient [cm/(min)½]
AA	900	Bandera	5.2	0.240
AA	900	Berea	99.0	0.281
AB	900	Bandera	12.0	0.154
AB	900	Nellie Bly	149.0	0.081
BC	400	Bandera	10.1	0.077
BC	400	Torpedo	244.0	0.060

2. Draw a line through the intercept of 0 and the plot of volume of fluid loss at a time equal to one-half of the total time that the fracturing fluid is being injected into the formation (Fig. 4.7, Point A).

3. Employ the slope of this constructed line in Eq. 4.18.

Since we must consider formation permeability and differential pressures in calculating the fracturing fluid coefficient for the other fluid flow mechanisms (Eqs. 4.11 and 4.14), we must also determine the effect of these reservoir properties on the coefficient determined for fracturing fluids with wall-building properties. Data selected to demonstrate the effect of permeability (pore area distribution) are presented in Table 4.2.

This table shows that rather wide variations in the value of fluid-loss coefficients occur between data for different fluids, but that for a particular fluid-loss-additive crude oil system, large variations in permeability have little effect on the fracturing fluid coefficient.

Data to show the effect of pressure differential on loss of fluid through core wafers are given in Table 4.3. Table 4.3 reveals that variations in the additive (fracturing fluid system — oil- or water-base) had a greater effect on the fracturing fluid coefficient (C_{III}) than did differential pressure variations of 100 to 900 psi. The fact that there were substantial variations in the coefficient (0.0015 to 0.193) for different pressures and additives demonstrates the importance of conducting fluid-loss tests with the prospective fracturing fluid at as near the anticipated treating pressure differential as can be determined.

Values for the fluid-loss coefficient can be calculated readily from the typical curves on Fig. 4.7. For example, the slope of the curve for core BA-2, as plotted, is approximately 49/6 or 8.16 cc/(min)$^{1/2}$. The core cross-sectional area is 18.4 sq cm, and the fracturing fluid coefficient for this fluid is 8.16/(2)(18.4) or 0.222 cm/(min)$^{1/2}$ or 7.28 $\times$ 10^{-3} ft/(min)$^{1/2}$ by Eq. 4.18. Fig. 4.8 demonstrates graphically the effect of the fracturing fluid coefficient, C_{III}, on fracture area. Also shown are the effects of variations in the fracturing fluid additive systems and of the treating pressure differential on the fracturing fluid coefficient and on the fracture area. For example, an increase in the differential treating pressure in system AB from 500 psi to 1,500 psi decreased the fracture area from 93,000 sq ft to approximately 51,000 sq ft. Differences may be noted also in systems AA and BA. Both the system and the pressure differential may cause

TABLE 4.3—EFFECT OF PRESSURE DIFFERENTIAL AND FRACTURING FLUID COMPOSITION ON FRACTURING FLUID COEFFICIENT

Pressure Drop (psi)	Fluid Loss Coefficient [cm/(min)½]* Additive—Fracturing Fluid Systems: Crude Oil AB	Crude Oil BB	Water CD
900	0.154	0.052	0.0030
400	0.193	0.027	0.0026
100	0.144	0.019	0.0015

*All cores Bandera sandstone

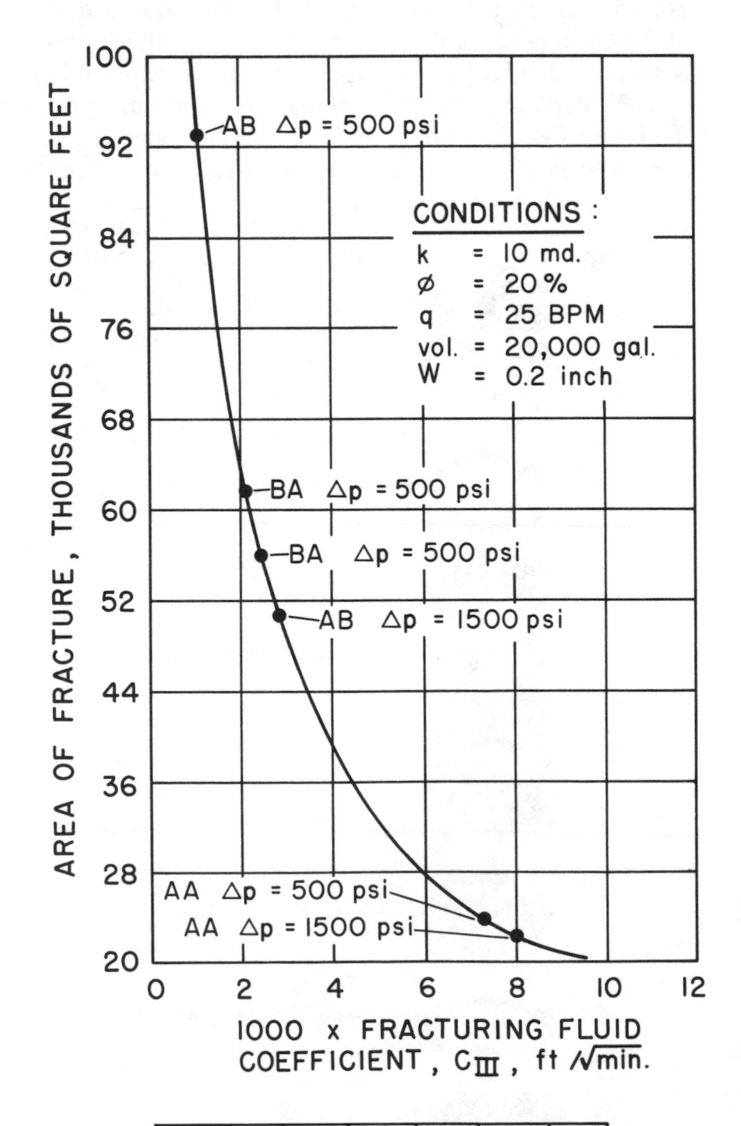

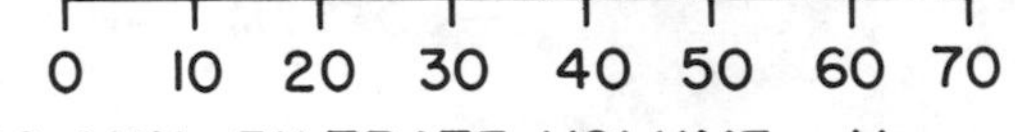

Fig. 4.8 Effect of fracturing fluid coefficients (fluid loss) on area of fracture.

a marked change in the fracturing fluid coefficient and fracture area. The effect of differential pressure is dependent upon the additive and upon the base fluid used.

Dynamic Fluid-Loss Testing. Hall and Dollarhide developed a dynamic fluid-loss testing method that is applicable to kerosene or water-base wall-building fracturing fluid.[13] The method involves the use of a specially designed test cell (Fig. 4.9) that permits the fluid-loss test to be conducted under flowing conditions. The dynamic tests show that fluid-loss volume is proportional to time rather than to the square root of time, as is shown by the static tests previously described. Spurt loss ordinarily is not affected by the flow velocity in the fracture, and pressure effects are minor. This work leads to a different equation for fracture area.

Fracture area calculations based on dynamic fluid-loss measurements yield fracture areas that are less than those obtained by using Eq. 4.7 and static fluid-loss coefficient C_{III}. The dynamic filter-loss tests and the area computation method that employs this factor are applicable only when the areal extent is controlled by the wall-building characteristics of the fracturing fluid. Hall and Dollarhide's fracture area calculations show that the fluid-loss behavior may be characterized by a spurt distance L_s and a leakoff velocity V_L; the leakoff velocity is constant with time but dependent upon the fracture-flow velocity and fracture width.[13]

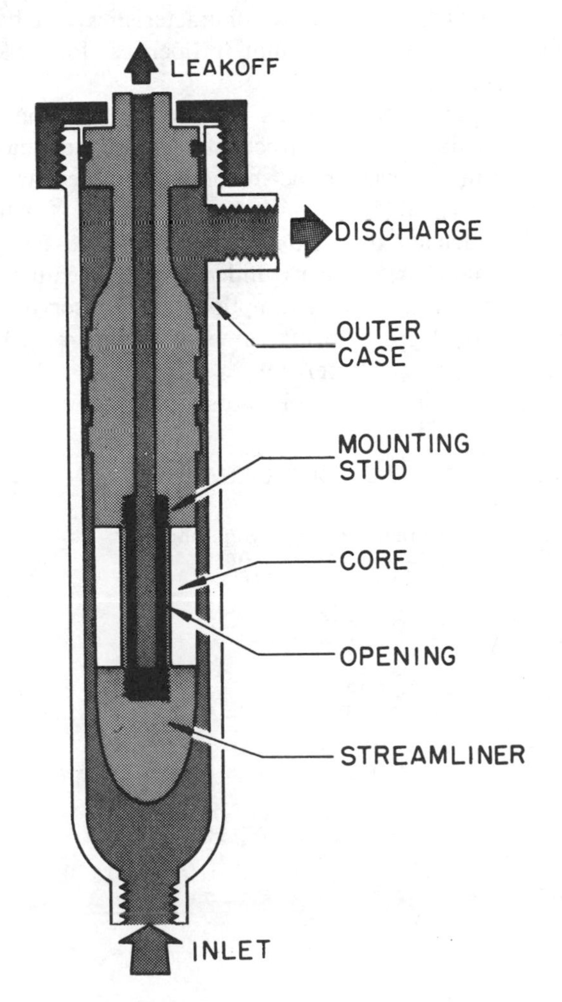

Fig. 4.9 Core holder and cell.[13]

This means that an exact summation of total fluid loss in a fracture would require knowledge of the pattern of fracture propagation (circular, rectangular, etc.), since this shape would determine the velocities in various areas. However, the approach of V_L to a limiting value indicates that a satisfactory average value might be chosen and used for the entire fracture area. If this approximation is used, an equation for fracture area may be derived by following the method set forth in Eq. 4.7, but treating V_L as a constant. The resulting equation for fracture area is

$$A_{ff} = \frac{i}{2V_L}\left(\frac{e^{yt}-1}{e^{yt}}\right), \quad \ldots\ldots\ldots (4.19)$$

where

$$y = \frac{2V_L}{w + 2L_{sp}}, \text{ and}$$

$$V_L = \frac{\text{leak off rate (cu ft/min)}}{\text{filter area (sq ft)}},$$

The quantity $i/2V_L$ represents the fracture area on whose two surfaces the total leakoff would equal the injection rate. Theoretically, this is the maximum fracture area attainable and consequently will be called $(A_{ff})_{max}$. In a conventional treatment with the fluid-loss agent injected continuously, the equation could not hold true when the area approaches $(A_{ff})_{max}$; the agent would deposit in the areas of very low fracture-flow velocity, thereby decreasing V_L and permitting further fracture extension. However, practical treatment design would not call for approaching $(A_{ff})_{max}$ so closely. The quantity in parentheses in Eq. 4.19 represents not only the fraction of $(A_{ff})_{max}$ attained at a time t, but also the fraction of the injection rate that is leaking off at a time t (not including the amount lost by spurt as new area is opened).

A comparison of the fracture area prediction obtained from Eqs. 4.7 and 4.19 is shown in Fig. 4.10. To calculate fracture area, it is presumed that the user of the Carter equation (Eq. 4.6) would insert the most

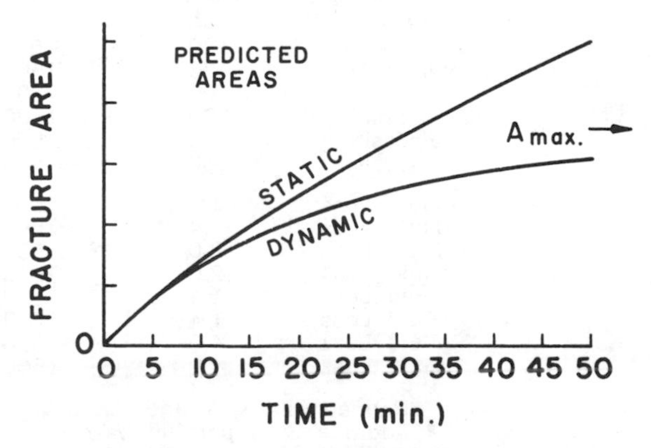

Fig. 4.10 Comparison of area predictions from static and dynamic fluid-loss tests.[13]

accurate fluid-loss data available. Generally speaking, dynamic data are more representative of fluid losses than are static data when mud flow in a wellbore is considered, but there is little apparent difference in the fluid loss in a fracture between the static and the dynamic tests. An additional point to consider is that there is very little information on dynamic fluid loss available for general use.

Composite Fracturing Fluid Coefficient

The fracturing fluid coefficient that controls the leak-off of fluid to the formation is actually a combination of all the mechanisms acting to prevent loss. Smith devised a method of combining the three coefficients (C_{I}, C_{II}, and C_{III}) that were presented as Eqs. 4.11, 4.14 and 4.18.[14]

Each of these coefficients acts simultaneously to varying extents and complements the other. The suggestion has been made that these coefficients are analogous to a series of electrical conductors and can be combined as such:

$$\frac{1}{C_t}=\frac{1}{C_{I}}+\frac{1}{C_{II}}+\frac{1}{C_{III}}, \qquad (4.20)$$

$$\frac{1}{C_t}=\frac{1}{0.0469\sqrt{\dfrac{k\Delta p\phi}{\mu}}}+\frac{1}{0.0374\Delta p\sqrt{\dfrac{k\phi c}{\mu}}}+\frac{1}{0.0164\dfrac{m}{A_c}} \qquad (4.21)$$

All of the variables used in determining the composite fracturing coefficient should be of the pay section that may actually be fractured. For vertical fractures this is usually the gross pay section and for horizontal fractures it is the net pay section. In addition, the permeability employed should be either an average vertical or an average horizontal permeability, depending on whether the well can be expected to fracture predominantly in a vertical or in a horizontal plane.

Most investigators[15] use a combined coefficient consisting of C_{I} and C_{II} for those conditions where leak-off is controlled by fracturing fluid viscosity and reservoir fluid viscosity and compressibility. This coefficient is calculated as follows:

$$C_{I}+C_{II}=\frac{C_{I}C_{II}}{C_{I}+C_{II}} \qquad (4.22)$$

For those conditions where leakoff is controlled by the wall-building characteristics of the fracturing fluid, the coefficient C_{III} (Eq. 4.18) is used alone.

Effect of Fracturing Fluid Coefficient on Fracture Area

There are three distinct flow mechanisms that control the loss of fluid to a formation during a fracturing treatment. Of the three, two are characteristic of the fracturing fluid and can be controlled. The third is characteristic of the reservoir fluids and is not subject to control with the fracturing fluid. The fracturing fluid coefficient also exerts a major influence on the pump rate and fracturing fluid volume required to effect a given fracture area.

The two controllable mechanisms are those involving the displacement of highly viscous fracturing fluids into a porous medium and those employing a low-penetrating fracturing fluid that has wall-building properties. Fracturing fluids with these characteristics can be designed to provide optimum properties for fracture extension.

The types of calculations and the comparisons that can be made of various fracturing fluids and fluid-loss mechanisms by use of the methods described are presented in Table 4.4. This table shows the fracturing fluid coefficients that are obtained with the two controllable flow mechanisms under various treating conditions, and where the assumptions are a reservoir fluid viscosity of 2 cp, a formation permeability of 10 md, and a porosity of 20 percent.

Fig. 4.11 is a plot of radius of fracture in feet

TABLE 4.4—EFFECT OF FLUID TYPE ON FRACTURING FLUID COEFFICIENTS

Controlling Variable	Viscosity (cp)	Treating Differential Pressure (psi)	Type of Fracturing Fluid	Fracturing Fluid Coefficient (ft/$\sqrt{\text{min}}$)	Reference Abscissa (Fig. 4.11)
Viscosity	30	500	RX*	C_{I}0.00855	13
Viscosity	30	1500	RX	0.01482	—
Viscosity	100	500	RY*	0.00469	8
Viscosity	100	1500	RY	0.00811	12
Viscosity	500	500	RZ*	0.00210	2
Viscosity	500	1500	RZ	0.00362	6
Fluid Loss	1.4	500	BA**	C_{III}0.00238	4
Fluid Loss	1.4	1500	BA	0.00213	3
Fluid Loss	2.0	500	AB***	0.00096	1
Fluid Loss	2.0	1500	AB	0.00282	5
Fluid Loss	1.4	500	AA†	0.00728	10
Fluid Loss	1.4	1500	AA	0.00805	11
Fluid Loss	4.0	500	BB††	0.00590	9
Fluid Loss	4.0	1500	BB	0.00443	7

*Refined oils X, Y and Z.
**Additive B in paraffin base crude oil.
***Additive A in mixed base crude oil.
†Additive A in paraffin base crude oil.
††Additive B in mixed base crude oil.

(assuming a single horizontal circular fracture) and the percent of fracturing fluid lost to the formation vs the fracturing fluid coefficient in ft/$\sqrt{\text{min}}$ for a 20,000-gal fracturing treatment with an injection rate of 2.5 and 25 bbl/min.

Each example in Table 4.4 is shown as a point on the abscissa of Fig. 4.11. The points are numbered both in the table and on the figure to indicate the relative position on the abscissa of each of the curves. This figure shows that as the fracturing fluid coefficient decreases, the fluid loss decreases and the fracture radius increases.

The data presented in Table 4.4 and plotted on Fig. 4.11 are typical for fracturing fluids used in 1969. The plotted fracturing fluid coefficient values, which vary from 1×10^{-3} to 9×10^{-3}, show the need for careful selection of the fracturing fluid to obtain the most effective treatment. This ninefold variation in fracturing fluid coefficient resulted in a calculated variation in fracture radius, at a pump rate of 25 bbl/min, ranging from 85 to 180 ft for the conditions specified in Table 4.4 and Fig. 4.11. Fracturing fluids of the same type and price, with the same pumpability characteristics, may vary widely in effectiveness.

Other factors such as pressure loss due to friction, pump truck requirements, and fluid cost may offset the benefits to be gained by using one fracturing fluid as opposed to another — for example, high viscosity fluids vs low viscosity, low-fluid-loss fluids. These factors must be considered carefully from an economical as well as from an operational standpoint when an optimum fluid is being selected.

Optimum Fluid Characteristics

Fig. 4.11 indicates that the loss of fluid to the formation — that is, the fluid that is not effective in extending the fracture — is high when the fracturing fluid coefficient is high. Even with the best commercially available fracturing fluid tested, the loss of fluid may be as high as 40 percent, depending on the treating conditions.

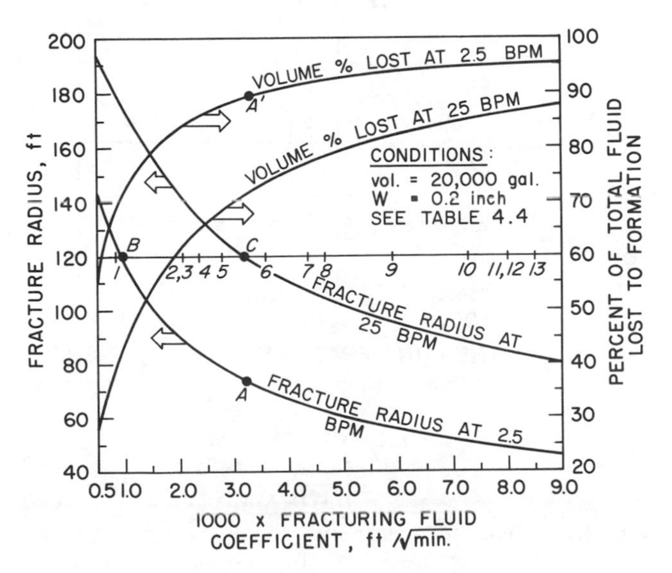

Fig. 4.11 Effect of fracturing fluid coefficient and pump rate on fracture extent.

The ideal fluid has a low fracturing fluid coefficient, and sufficient viscosity to permit successful placement of the propping agent. It also should be capable of being pumped to the formation face with a minimum of hydraulic horsepower. These characteristics may be obtained with very high viscosity fluids pumped at low rates or with very low fluid-loss, moderate-viscosity fluids pumped at higher rates.

4.4 Relationship of Pump Rate to Fracturing Fluid Coefficient

Two major factors contributing to the cost of a fracturing job are the pumping equipment and the fracturing fluid. A comparison of these two variables reveals that decreasing the fracturing fluid coefficient has the same effect on the fracture area as increasing the pump rate. For example, decreasing the coefficient from 3.25×10^{-3} to 1.0×10^{-3} ft/$\sqrt{\text{min}}$ (points A to B, Fig. 4.11) at 2.5 bbl/min results in a 62 percent increase in fracture radius and an 18 percent decrease in fluid loss. A tenfold increase in pump rate (points A to C, Fig. 4.11) is required to accomplish the same results without changing the fracturing fluid coefficient.

Fig. 4.12 is a plot of fracture areal extent vs fracturing fluid coefficient with pump rate as a parameter. A

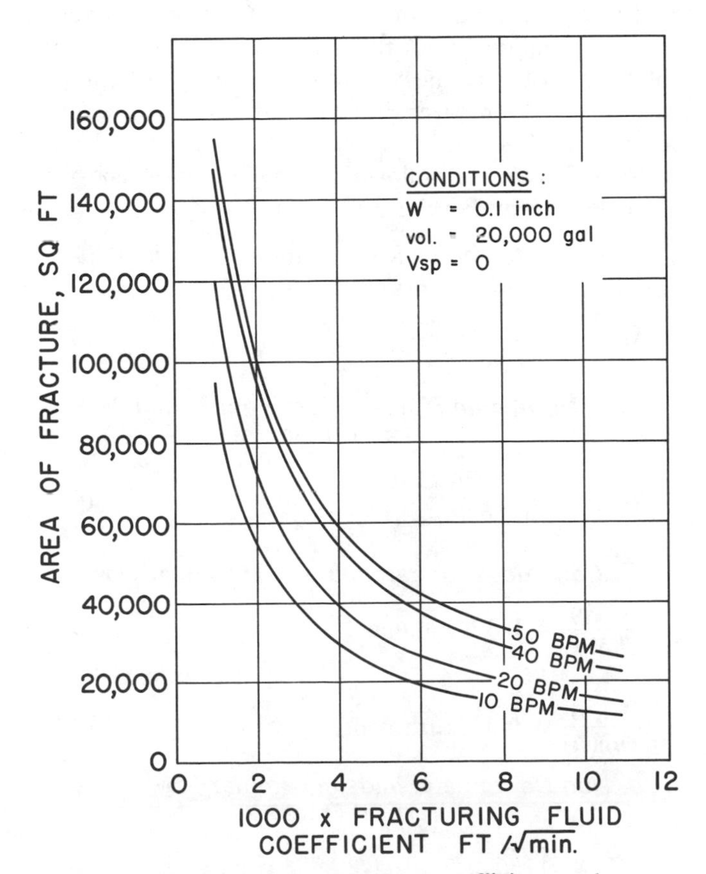

Fig. 4.12 Effect of fracturing fluid coefficient and pump rate on fracture areal extent.

treatment volume of 20,000 gal, a fracture width of 0.1 in., and a zero spurt loss were assumed in making the area calculations using Eq. 4.7. These data again show the importance of both pump rate and fluid coefficient on fracture area. Increase in pump rate and decrease in coefficient (decrease in loss of fluid to the formation) both result in an increase in fracture area generated. In general, an increase in pump rate will probably not be economical when a poor quality fracturing fluid (high coefficient) is used. A much greater fracture area can be obtained by improving the fluid (decreasing the coefficient). However, increasing the pump rate of a good quality fluid (low C factor) generally will improve greatly the efficiency of the fracturing. The method of obtaining the optimum fracturing treatment (that is, the greatest productivity increase per dollar expended) is discussed in a later chapter.

The foregoing examples point out that the effectiveness of any treatment depends upon the choice of the proper fracturing fluid. The examples also show that reducing the fracturing fluid coefficient (by using a fluid-loss reducing additive or by increasing the viscosity) is economically sound when the fracturing fluid coefficient is reduced enough either to yield a significant increase in fracture radius at the same pump rate or to permit a major reduction in pump rate, yet yield the same fracture extent.

4.5 Fracture Width

Smith,[14] in his treatment of this facet of the fracture area calculation, states that fracture width is a function of rock elasticity, injection rate, fracturing fluid properties, and fracture size. Fracture widths, obtained during treatment, for restricted vertical fractures can be determined from the following formulations, which were first derived by Perkins and Kern.[16].

Newtonian Fluids in Laminar Flow in a Vertical Fracture

$$\overline{W}_F = 0.25\left[\frac{i_B(\mu)(L)}{E}\right]^{1/4} \quad \ldots \ldots \ldots \ldots (4.23)$$

Newtonian Fluids in Turbulent Flow in a Vertical Fracture

$$W_F = 0.4\left[\frac{(i)^2(\gamma)(L)}{(E)(h_f)}\right]^{1/4} \quad \ldots \ldots \ldots \ldots (4.24)$$

Conditions for Laminar and Turbulent Flow

$$\frac{(i)(\gamma)}{(h_f)(\mu)} < 0.32, \text{ laminar,} \quad \ldots \ldots \ldots \ldots (4.25)$$

$$\frac{(i)(\gamma)}{(h_f)(\mu)} > 0.32, \text{ turbulent} \quad \ldots \ldots \ldots \ldots (4.26)$$

Non-Newtonian Fluids in Laminar Flow in the Vertical Fracture

$$\overline{W}_F = 8\left[\left(\frac{2'}{3\pi}\right)(n'+1)\left(\frac{2n'+1}{n'}\right)^{n'}\right.$$

$$\left.\left(\frac{0.9775}{144}\right)\left(\frac{5.61}{60}\right)^{n'}\right]^{1/(2n'+2)}\left(\frac{i^{n'}k'Lh_f^{1-n'}}{E}\right)^{1/(2n'+2)}, \quad (4.27)$$

where $\overline{W}_F$ is the average crack width at the wellbore in inches.

The widths of horizontal fractures, during treatment, from Newtonian fluids in laminar flow in the fracture can be obtained from the following formulation:

$$\overline{W}_F = 0.15\left[\frac{(i)(\mu)(r_f)}{E}\right]^{1/4} \quad \ldots \ldots \ldots \ldots (4.28)$$

Eqs. 4.27 and 4.28 are the same as those presented by Perkins and Kern[16] except that these expressions give the *average* width of an elliptically shaped fracture, whereas Perkins and Kern's formuations give the *maximum* width.

Figs. 4.13 through 4.16 are graphic solutions of Eqs. 4.23 through 4.26, respectively. To solve these equations and to use Fig. 4.15, values of Young's moduli for various types of rocks must be known. Since a 1.5-fold to threefold variation in Young's moduli may exist in rocks of the same type, the use of average value of Young's moduli appears justified. Table 4.5 presents

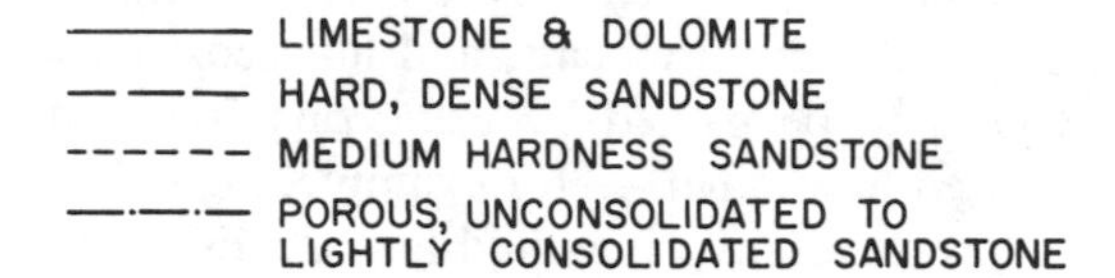

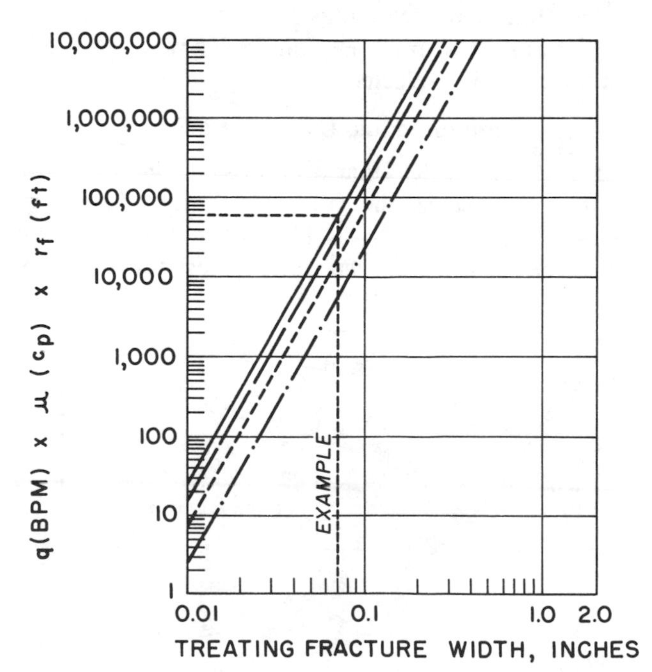

Fig. 4.13 Chart for determining pumping rates and treating fracture widths for restricted vertical fractures resulting from Newtonian fluids in laminar flow in the fracture.[14]

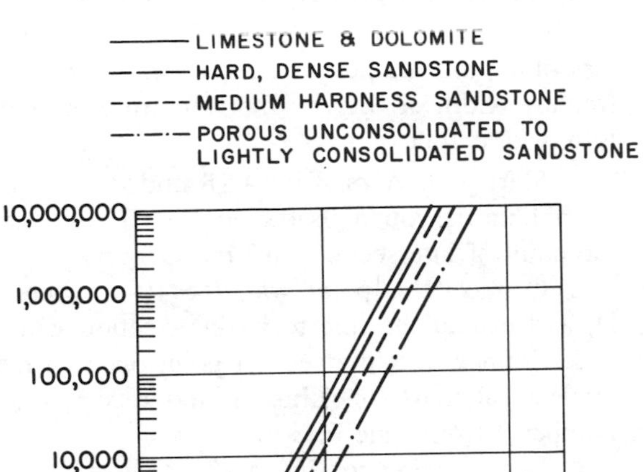

Fig. 4.14 Chart for determining pumping rates and treating fracture widths for restricted vertical fractures resulting from Newtonian fluids in turbulent flow in the fracture.[14]

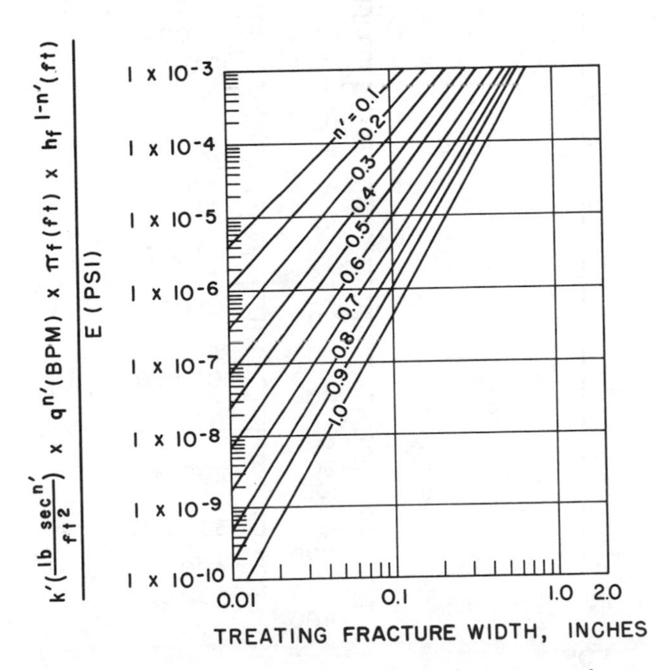

Fig. 4.15 Chart for determining pumping rates and treating fracture widths for restricted vertical fractures resulting from non-Newtonian fluids in laminar flow in the fracture.[14]

TABLE 4.5—VALUES OF YOUNG'S MODULI OF FORMATION ROCKS

Type of Rock	Range of Young's Moduli (psi $\times 10^6$)	Average Value of Young's Moduli (psi $\times 10^6$)
Limestone and dolomite	8.0 to 13.0	10.50
Hard, dense sandstone	5.0 to 7.5	6.25
Medium-hardness sandstone	2.0 to 4.0	3.00
Porous, unconsolidated to lightly consolidated sandstone	0.5 to 1.5	1.00

values of Young's moduli that can be used.

Laboratory testing has established that flow in horizontal fractures during treatment is generally laminar except for the area immediately around the well. Thus, for calculating fracture widths for horizontal fractures, laminar flow generally will be assumed, and Eq. 4.28 will be used. Flow in vertically oriented fracture systems may be either laminar or turbulent, depending on the vertical extent of the fracture. To determine the flow regime, use Eqs. 4.25 and 4.26.

Fracture width calculations generally aid in establishing the required injection rate for placement of a given size proppant. Injection tests have demonstrated that to assure placement of the propping agent in the fracture the fracture widths must be two to three times the maximum diameter of the proppant particle. As a result, most investigators assume that fracture width during placement is 0.10 in.

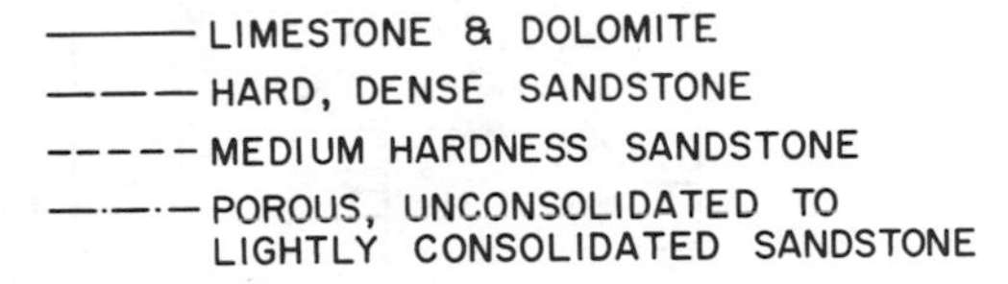

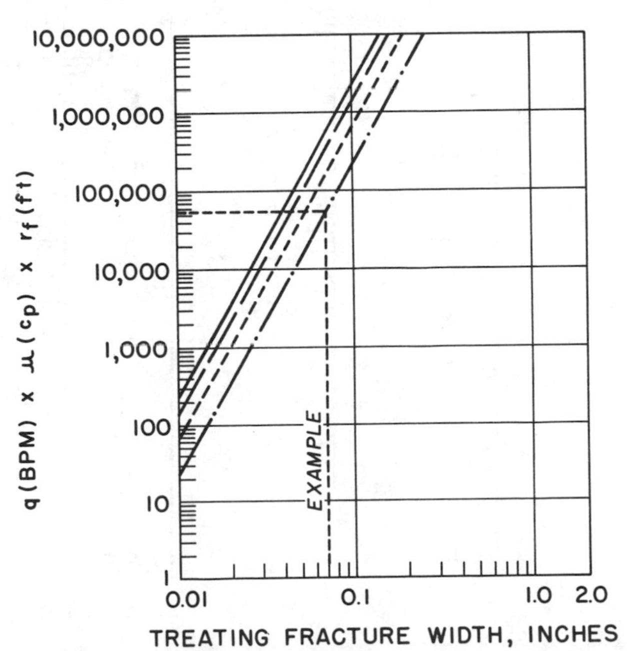

Fig. 4.16 Chart for determining pumping rates and treating fracture widths for horizontal fractures resulting from Newtonian fluids in laminar flow in the fracture.[14]

4.6 Nomograph Solution of the Fracture Area Formula

Several investigators have prepared nomograph solutions that greatly facilitate the use of Eq. 4.7. The nomographs shown as Figs. 4.17 and 4.18 were prepared by Dowell[15] and permit a determination of the fracturing fluid volume and pump rate required to produce a given fracture area with various fracturing fluid characteristics. They also permit calculation of the fracture area that will be produced with any given fracturing fluid volume and pump rate. After the areal extent of the fracture has been determined, Figs. 4.19 and 4.20 may be used to calculate fracture penetration for radial and linear fractures, respectively.

For example, if a well with a 100-ft-thick pay formation is to be fractured with 50,000 gal of a fracturing fluid having a coefficient of 0.002 ft/$\sqrt{\text{min}}$, a spurt loss of 10 ml pumped at 50 bbl/min, and an assumed fracture width of 0.1 in., the fracture area may be determined as follows.

1. Start with A on Fig. 4.18 and draw a line from 50 bbl/min through 0.002 on Line B and then mark the point of intersection on Line C.

2. Connect this point with the 10-ml point on Line D_I and extend the line to its intersection with Line E.

3. Connect the intersection point on Line E with the 50,000 gal mark on Line H and record the transfer number 9 from Line F.

4. Now turning to Fig. 4.17, draw a line from the 50 bbl/min point on Line A through the 0.002 point

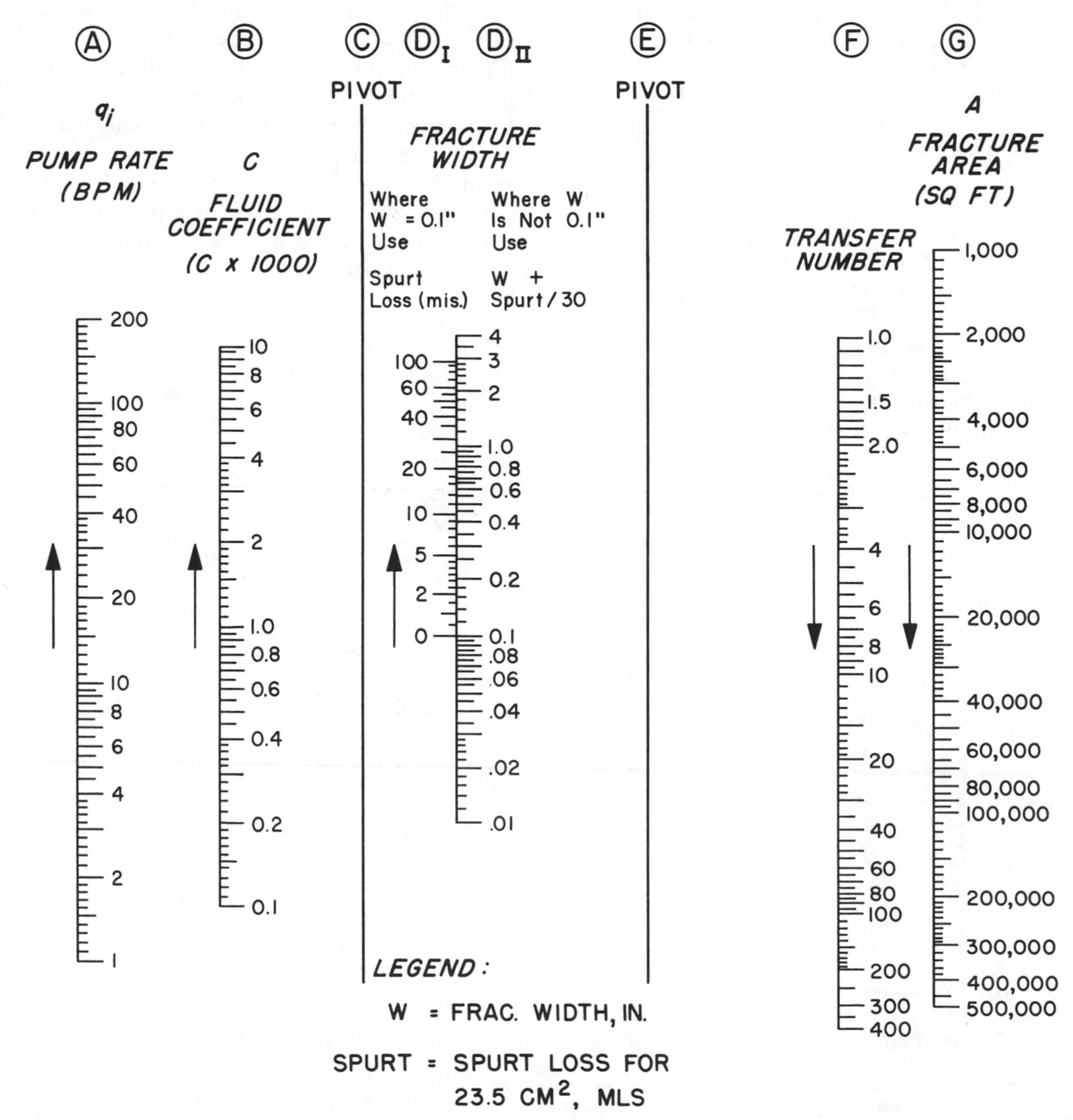

Fig. 4.17 Fracture-area: volume relationship.[15] (To determine fluid volume from known fracture area, start with Fig. 4.17, transfer to Fig. 4.18.) Eq. 4.7.

on Line B and extend this line until it intersects Line C.

5. Draw a line from the intersection point on Line C through the 10-ml spurt-loss point on Line D_I. Extend this line and mark its intersection on Line E.

6. Mark the transfer number 9 from Fig. 4.18 on Line F of Fig. 4.17.

7. Connect these two points and extend the line to its intersection with G.

This shows that a fracture area of 100,000 sq ft would be created.

To determine the fracture penetration for a vertical fracture, refer to Fig. 4.19, which shows a 100,000-sq. ft area to be equivalent to a radial fracture with a radius of 175 ft. Also, refer to Fig. 4.20 to see that the 100,000 sq ft of fracture area will be equivalent to a vertical fracture penetration of 500 ft into the 100-ft formation.

4.7 Proppant Transport:Fracture Area Relationship

A computer program[17] for correlating the principal parameters of hydraulic fracturing is available for use

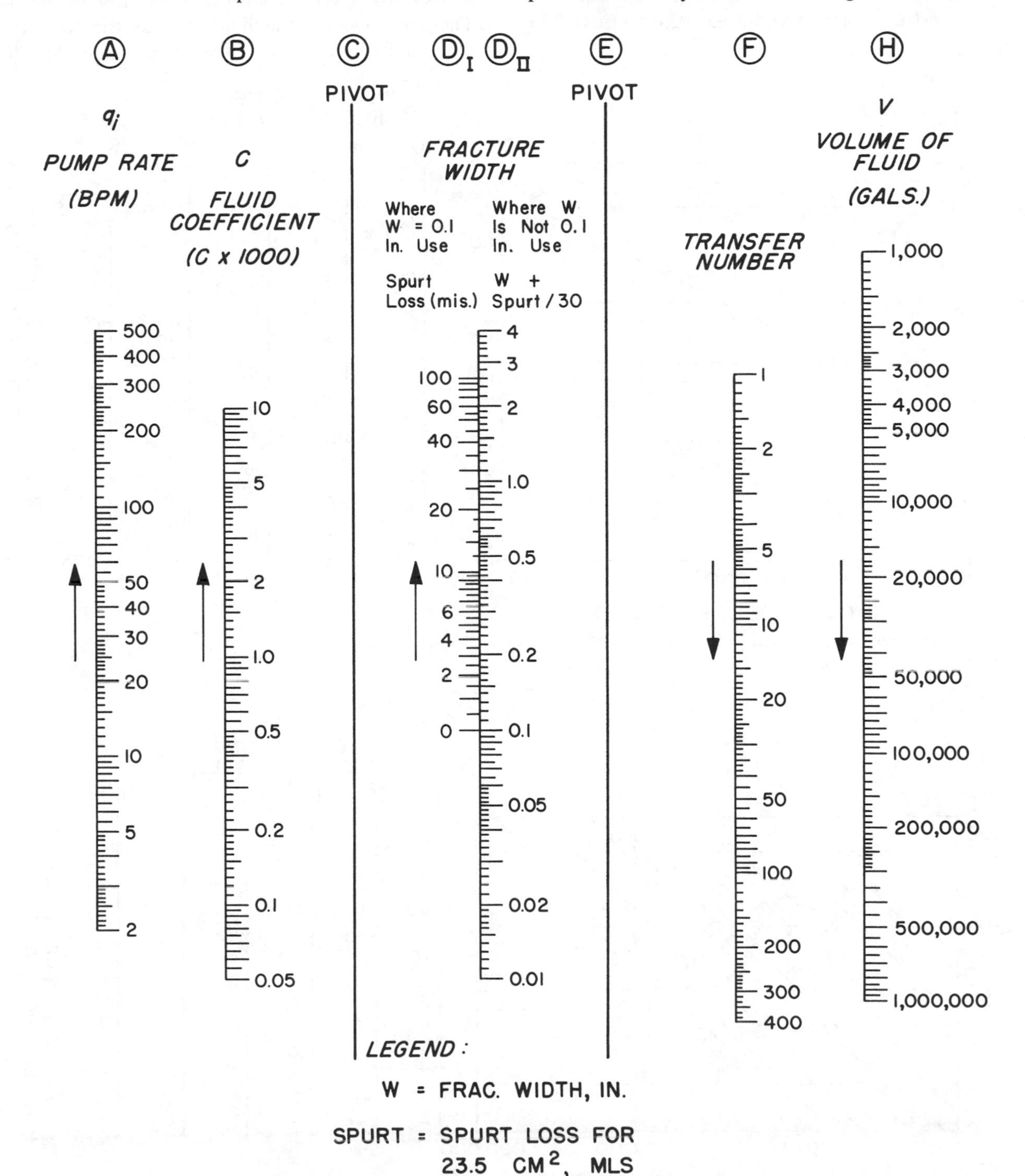

Fig. 4.18 Fracture-area: volume relationship.[15] (To determine fluid volume from known fracture area, start with Fig. 4.17, transfer to Fig. 4.18.) Eq. 4.7.

in designing horizontal hydraulic fracturing treatments. This program combines the area equation presented by Howard and Fast,[7] the width equations presented by Perkins and Kern,[16] and the Wahl[18] propping agent transport equations.

The computer logic used by Lowe *et al.*[17] (Fig. 4.21) consists of several sets of calculations designed to match fracture width, injection rate and fracture area. First, the fracture radius and average fracture width at the completion of fluid injection are calculated simultaneously by Carter's fracture radius equation (Eq. 4.6) and the fracture width equations (Eqs. 4.23 through 4.28).

To determine proppant movement in the fracture, it is necessary to establish the fluid velocity in the fracture as a function of time and penetration. A table of velocity values generated by the computer program is stored for subsequent use. The time span from the beginning of fluid injection until the end of pumping is divided into 100 equal increments. The fracture radius at each of the 100 incremented time steps is calculated. The total radius is also divided, by interpolation, into 100 equal increments. Some of the fracturing fluid that arrives at a given radial location is lost to the matrix. The remainder of the fluid extends the fracture. The local fluid velocity as a function of radius and time is

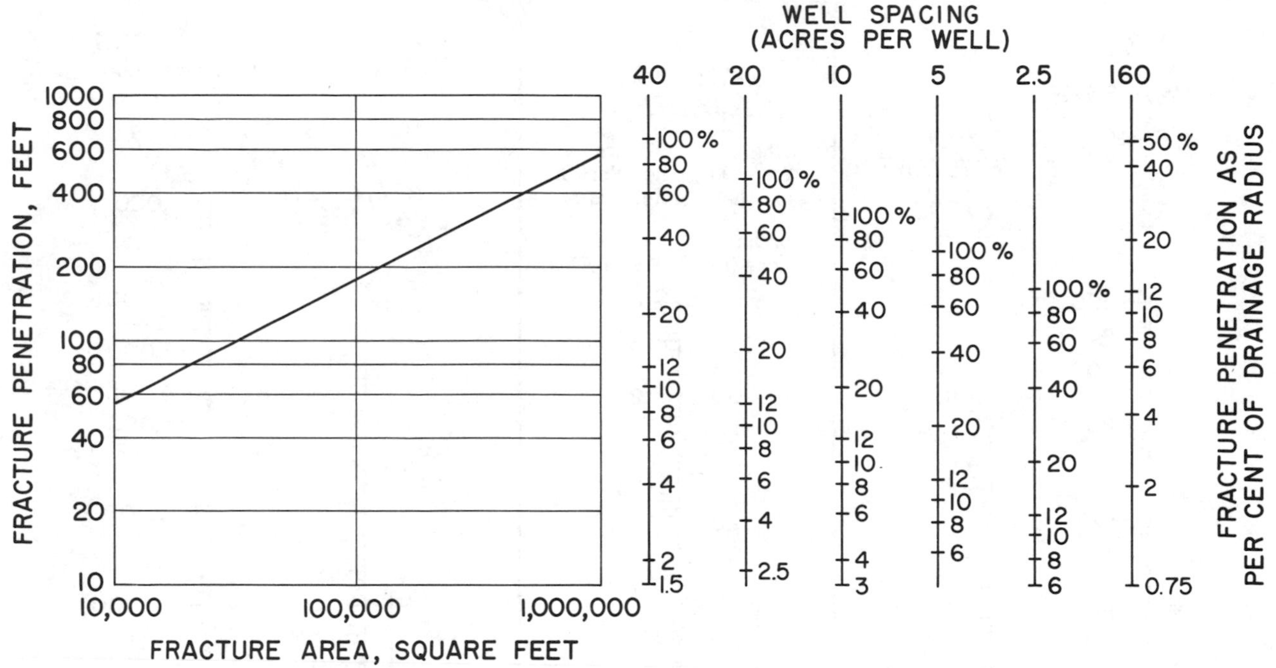

Fig. 4.19 Area:penetration relationship for radial fractures.[15]

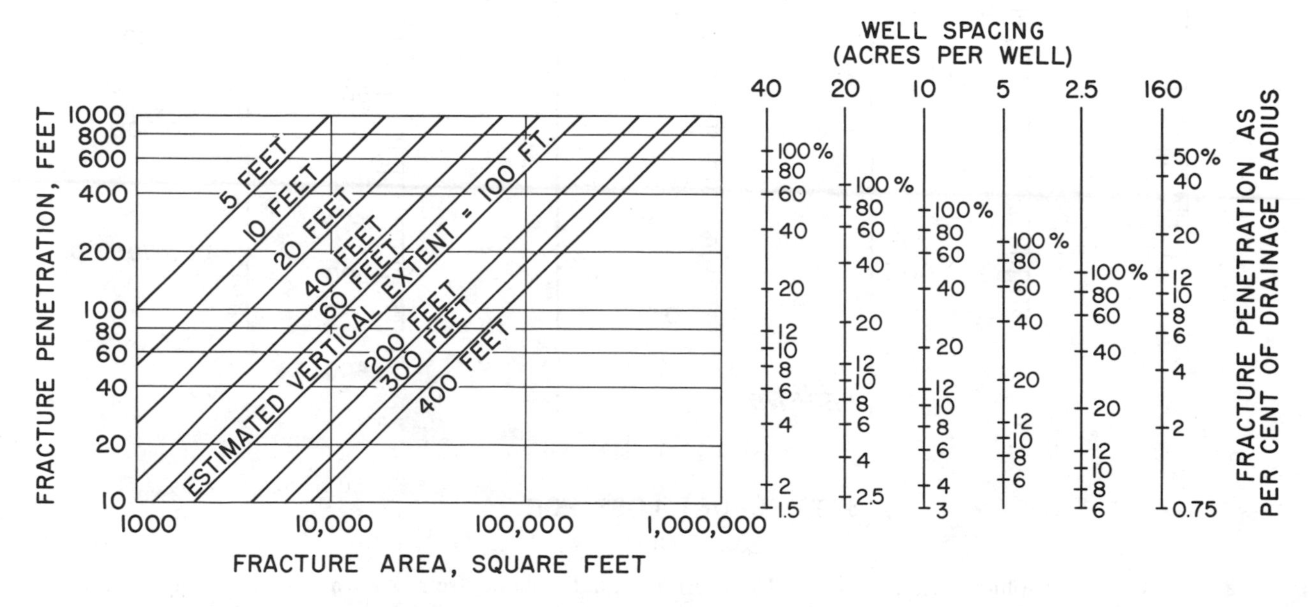

Fig. 4.20 Area:penetration relationship for vertical fractures.[15]

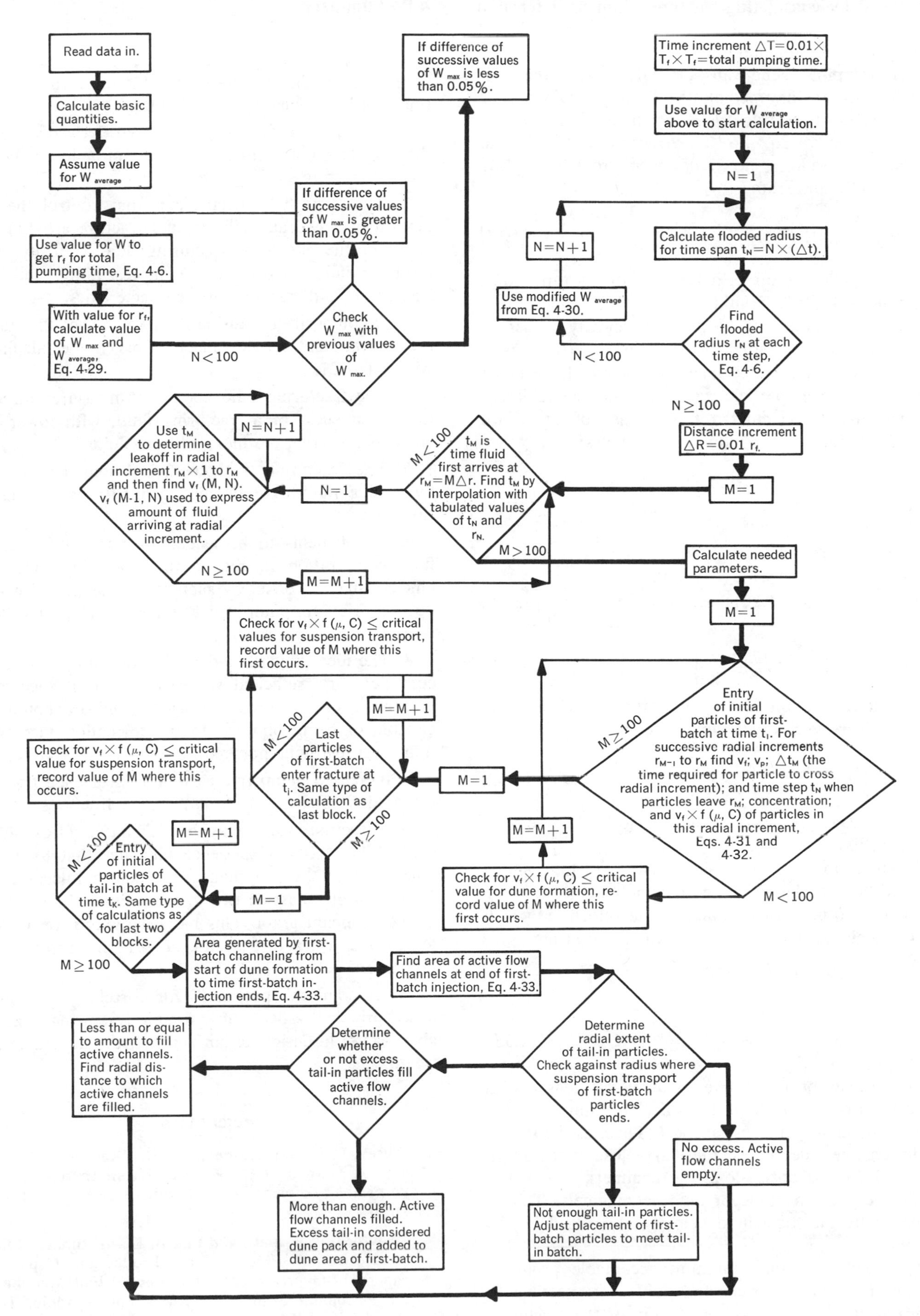

Fig. 4.21 Logic flow diagram.[17]

calculated by considering fracture width as a function of time and by accounting for the fluid that leaks off through the incremental fracture-face areas.

For propping agent transport calculations, the fracture widths at intermediate radii are required. The radial extent of the fracture at intermediate time t_x was found by use of Eq. 4.6. The fracture widths W_{fx} at intermediate radii r_x were obtained from Eqs. 4.23 through 4.28 and from

$$W_{fx} = \frac{3}{2} W_f \left(\frac{r_x}{r_f}\right)^{1/4} \quad (4.29)$$

When the fluid leakoff as a function of fracture radius and time is known, a fluid velocity table may be calculated. This table would provide velocity values at 100 equally spaced radial distances and at 100 equally spaced time increments. Once the velocity table is set up, the average particle velocities in the different batches may be calculated by the use of the Lowe *et al.*[17] relationship for individual particles; i.e.,

$$\frac{v_{pa}}{v_f} = 0.525 N_{\mathrm{Re}}^{[0.072(1.75 - d_{pa}/W_f)]} \left(\frac{\rho_{pa}}{\rho_f} - 1\right)^{-0.085}$$

$$\exp[9.5(10^{-5})(0.19 + d_{pa}/W_f)$$
$$(110 - \tau)\tau - 1.15(0.7 - d_{pa}/W_f)]$$
$$\exp[0.096(d_D - 1.26)] - 0.761 v_f^{1.85} \quad (4.30)$$

$$v_f = \frac{\sin\theta}{v_f}\left[\frac{3.58 d_D(\rho_{pa}/\rho_f)}{C_D}\right]^{1/2} \quad (4.31)$$

The concentration of the propping particles in the moving slurry varies with fluid velocity, as

$$C_{pa} = C_i \left(\frac{v_{pa}}{v_f}\right)_w \Big/ \left(\frac{v_{pa}}{v_f}\right), \quad (4.32)$$

where $(v_{pa}/v_f)_w$ is the ratio of particle to fluid velocity at the wellbore.

Where proppant is deposited in the fracture, slurry will still flow in channels. The channels move, carrying proppant into new fracture areas. The velocity at which the channels advance is given by the relationship

$$\frac{v_b}{v_f} = \min \begin{cases} (v_f)_{\max}/v_f(\text{in channel}) \\ \text{or} \\ 2.32\left(\frac{v_{pa}}{v_f}\right)\left(\frac{d_{pa}}{W}\right)^{-0.197}\left(\frac{\rho_{pa}}{\rho_f}\right)^{1.12} C_{pa}^{0.137} \end{cases} \quad (4.33)$$

The spread of the active channels decreases with distance because of fluid leakoff. The slurry in the channels is in suspension transport. Thus, as the spread of the channels decreases, the slurry loses particles to the fixed dune deposit along the channels, thus maintaining suspension transport in the channels. This is governed by the local fluid velocity and the fluid viscosity.

By establishing the minimum acceptable proppant concentration in the fracturing fluid, it is possible to determine the effective propped area of the fracture.

4.8 Summary

1. The effectiveness of a fracturing treatment is dependent upon the extent of the fracture produced; that is, the greater the fracture extent, the more the production that is obtained from a fracture.

2. An equation (Eq. 4.7) has been derived to relate those factors that control extension of a hydraulically created fracture.

3. The three flow mechanisms that control the extension of a hydraulically created fracture are (a) the result of the effects of fracturing fluid viscosity, (b) reservoir fluid viscosity and compressibility, and (c) fracturing fluid wall-building characteristics.

4. A composite fracturing fluid coefficient that takes into account all three flow mechanisms can be calculated with Eq. 4.21.

5. The fracturing fluid coefficient measures directly the effectiveness of a fracturing fluid, with lower coeffcients reflecting the more efficient fluids.

6. The fracturing fluid coefficient may be lowered by adding fluid-loss-reducing agents or by increasing the viscosity.

7. The benefits to be gained by varying fracturing fluid composition to reduce the fracturing fluid coefficient may be offset by such factors as pressure loss due to friction, pump truck requirements, and fluid cost.

8. The ideal fluid should have a low fracturing fluid coefficient and sufficient viscosity to permit successful placement of the propping agent, and it should be capable of being pumped to the formation face with minimum hydraulic horsepower.

9. Increased pump rate as well as decreased fracturing fluid coefficient effects larger-area fractures.

10. Fracture width may be calculated (Eqs. 4.23, 4.24, 4.27, 4.28). Experience indicates that to insure placement of the propping agent during injection the fracture width must be two to three times the diameter of the proppant grain. This has led many investigators to assume a fracture width of 0.1 in. during fluid injection.

11. A computer program[17] for correlating the published principal aspects of hydraulic fracturing is available for use in designing horizontal hydraulic fracturing treatments.

References

1. Wilsey, L. E. and Bearden, W. G.: "Reservoir Fracturing—A Method of Oil Recovery from Extremely Low Permeability Formations", *Trans.*, AIME (1954) **201,** 169-175.

2. Crawford, P. B. and Landrum, B. L.: "Estimated Effect of Horizontal Fractures on Production Capacity", paper 414-G presented at the Annual Fall Meeting of the Petroleum Branch of AIME, San Antonio, Tex., Oct. 17-20, 1954.

3. Crawford, P. B. and Landrum, B. L.: "Effect of Unsymmetrical Vertical Fractures on Production Capacity", *Trans.*, AIME (1955) **204,** 251-254.

4. "Fracturing: Big and Getting Bigger", *Petroleum Week* (May 13, 1955) 18.

5. Campbell, J. B.: "The Effect of Fracturing on Ultimate Recovery", paper 851-31-L presented at the Mid-Continent District Meeting, API Div. of Production, Tulsa, Okla. (1957).

6. Garland, T. M., Elliott, W. C., Jr., Dolyn, Pat and Dobyns, R. P.: "Effects of Hydraulic Fracturing Upon Oil Recovery from the Strawn and Cisco Formations in North Texas", RI 5371, USBM (1957).

7. Howard, G. C. and Fast, C. R.: "Optimum Fluid Characteristics for Fracture Extension", *Drill. and Prod. Prac.,* API (1957) 261.

8. Churchill, R. V.: *Modern Operational Mathematics in Engineering,* McGraw-Hill Book Co., Inc., New York (1944) 107.

9. Muskat, M.: *Flow of Homogeneous Fluids Through Porous Media,* McGraw-Hill Book Co., Inc., New York (1937) 145.

10. Churchill, R. V., *op cit.,* 111.

11. Wahl, H. A.: "Fracture Design in Liquid Saturated Reservoirs", *J. Pet. Tech.* (April, 1963) 437-442.

12. Crittendon, B. C.: "The Mechanics of Design and Interpretation of Hydraulic Fracture Treatments", *J. Pet. Tech.* (Oct., 1959) 21-29.

13. Hall, C. D., Jr., and Dollarhide, F. E.: "Effects of Fracturing Fluid Velocity on Fluid-Loss Agent Performance", *J. Pet. Tech.* (May, 1964) 555-560.

14. Smith, J. E.: "Design of Hydraulic Fracture Treatments", paper SPE 1286 presented at SPE 40th Annual Fall Meeting, Denver, Colo., Oct. 3-6, 1965.

15. *FracGuide Data Book,* Dowell Div. of Dow Chemical Co., Tulsa, Okla. (1965).

16. Perkins, T. K. and Kern, L. R.: "Widths of Hydraulic Fractures", *J. Pet. Tech.* (Sept., 1961) 937-949.

17. Lowe, D. K., McGlothlin, B. B., Jr., and Huitt, J. L.: "A Computer Study of Horizontal Fracture Treatment Design", *J. Pet. Tech.* (April, 1967) 559-569.

18. Wahl, H. A.: "Horizontal Fracture Design Based on Propped Fracture Area", *J. Pet. Tech.* (June, 1965) 723-731.

19. Lowe, D. K. and Huitt, J. L.: "Propping Agent Transport in Horizontal Fractures", *J. Pet. Tech.* (June, 1966) 753-764.

Chapter 5

Fracturing Fluids and Additives

5.1 Introduction

There are many types of fluids available for use in hydraulic fracturing. To select the proper fluid for a specific well, it is necessary to understand the properties of the fluids and how those properties may be modified to accomplish various desired effects.

5.2 Fracturing Fluid Properties

The properties that a fracturing fluid should possess are low leakoff rate, the ability to carry a propping agent, and low pumping friction loss. The fluid also should be easy to remove from the formation; it should be compatible with natural formation fluids; and it should cause a minimum of damage to the formation permeability. The degree to which these characteristics exist must be considered in the design and selection of a fracturing fluid.

Low leakoff rate is the property that permits the fluid to physically open the fracture and one that controls its areal extent. The rate of leakoff to the formation is dependent upon the viscosity and the wall-building properties of the fluid.[1,2] These two properties were detailed in the previous chapter as controlling fracture area. Viscosity and wall-building properties can be controlled with appropriate additives.

The ability of the fluid to carry the propping agent is important and also can be controlled by additives. Essentially, this property of a fluid is dependent upon the viscosity and density of the fluid and upon its velocity in the pipe or fracture. Density and velocity are not hard to describe; however, viscosity is difficult to measure and describe properly since many fracturing fluids are non-Newtonian. Two different fluids, such as an emulsion and gelled water, can appear to have the same viscosity by one measurement, but may have widely varying abilities to carry propping agents in suspension. This aspect of propping agent suspension is frequently overlooked. The rate of movement is a major factor in the ability of fluid to carry the propping agent. Plain water with its low viscosity will carry proppants satisfactorily if it is pumped at a high rate.

Friction loss has become more important in recent years because it is controllable and because higher pumping rates have proved effective in fracturing treatments. The ability to reduce friction loss in pumping has been one of the governing factors in the present trend toward the use of water-base fracturing fluids.

To achieve the maximum benefits from fracturing, the fracturing fluid must be removed from the formation. This is particularly true with the very viscous fracturing fluids such as viscous oils, gels, or emulsions. Most of the gelled-oil and water-base fracturing fluids have built-in breaker systems that reduce the gels to low viscosity solutions upon exposure to the temperatures and pressures existing in the formations. When the viscosity is lowered, the fracturing fluid may be readily produced from the pay formation and no flow restrictions remain. Generally, viscous oils such as the residual fuel oils are sensitive to heat and to crude oil dilution so that they flow easily from the well. Emulsions usually are diluted by formation fluids, have their emulsifying agents adsorbed on the rock surfaces, or are destroyed by contact with acid or heat. This allows the remaining liquids to flow from the well.

5.3 Oil-Base Fluids

Napalm Gels

The first fluids used in fracturing operations were oil-base fluids prepared by thickening gasoline with Napalm, an aluminum fatty acid salt. Kerosene, diesel oil, or crude oil gelled with this material was the basis of the "Hydrafrac" process. This gel imparted viscosity to the fracturing fluid and also reduced the fluid loss. It improved the ability of the oil to carry the propping agent into the formation and it reduced in viscosity when it reverted to a sol under bottom-hole conditions.

This oil gel was subsequently followed by viscous refined oil (such as API No. 5 and No. 6 residual fuel

Editor's Note: This chapter was prepared with the assistance of A. W. Coulter and H. E. David, Dowell Div. of the Dow Chemical Co.[25]

oil), lease crude oil, fatty acid-soap gels, and, finally, by emulsions and acid-base fluids.

Viscous Refined Oil

A viscous refined oil offered many advantages in fracturing and for a number of years was the most common fracturing medium. Typical specifications for a refined fracturing oil are given in Table 5.1.

These oils won acceptance because they were readily available from refineries and because they could be resold when produced back from the well. Thus, when the value of the produced fluid is taken into account, the over-all cost of refined oil might be as low as 3¢/gal. Refined oils also had the advantage of becoming less viscous as the temperature increased or when they were diluted with crude oil. Therefore, they could be readily produced from the formation.

Lease Crude Oils

In some areas where the lease oil was viscous it could be used for fracturing. It was soon learned that many lease oils either already had a low fluid loss, or could develop a low fluid loss if a suitable material was added to them. The most common of these materials now used is presently marketed by Continental Oil Co. under the trademark Adomite® Mark II and is offered by all service companies. This material, which is essentially an alkaline earth salt of a sulfonated alkylbenzene is effective in crude oil, refined oils, and emulsions.[3,4] Its predecessor was Adomite® which was essentially the sodium salt of the same material. The forerunners of these fluid-loss agents were asphalts, blown asphalt, gilsonite and still bottoms.[5]

Friction-Reducing Additives

Other effective additives for oil-base fracturing fluids are the friction-reducing agents. These chemicals are used principally in water-base fracturing fluids. It is important to note, however, that two types of agents are now used in oil-base fluids for the same purpose. One type is a fatty acid soap-oil gel that, although it increases apparent viscosity, reduces friction of fluids being pumped at a high velocity or in the turbulent region. The second type is a linear high-molecular-weight hydrocarbon polymer that performs approximately the same function, and is effective at low concentrations.

Gelled Lease Oils

The gelled lease oils became popular because they were readily available, inexpensive, and showed reduced friction properties. In addition, these materials are compatible with the formation, since they are simply formation oil plus a very small amount of fatty acid soap. A fatty acid soap gel is formed in situ in the crude oil by adding a fatty acid and caustic to the base oil. This increases viscosity, reduces fluid loss of the fracturing fluid, and reduces friction during pumping. Such gels can be prepared continuously or in batches by mixing a strong alkali, such as 30 to 50 percent caustic, with a fatty acid dissolved in oil. The reaction forms soap micelles that impart the desired properties.[6] The micelles are droplets of water surrounded by the fatty acid molecules in the oil system. The micelles form a head-to-head alignment with the polar portion of the molecule in the water and the hydrocarbon portion of the molecule in the oil. Alternating layers of the molecules around the droplets interact to form a gelled structure. The degree of gel dispersion or agitation in the base oil influences the apparent viscosity of the mixture. Passing the fracturing fluid through a homogenizer after it is gelled increases its apparent viscosity.

The gelled oil or soap-oil systems should not be used at high temperatures. At about 200°F these gels tend to become a grease-like semisolid. The sensitivity of the systems to gelation depends upon the total water content and upon the nature of the oil. A high-gravity oil or distillate should generally be limited to a maximum temperature of 175°F. The nature and extent of this problem is discussed more fully in a paper by Hendrickson *et al.*[6]

One of the advantages of the soap-oil gel system is its sensitivity to water. Contact with brine or water in the formation dissolves the soap and, unless some emulsion problem develops, reduces the system's viscosity. Normally, the gelled oil will be produced as plain oil.

Fluid-Loss-Control Additives

The choice of the proper fluid-loss-control additive is important in successful treatment design. Ideally, a fluid-loss additive should be: (1) effective at low concentrations, (2) easy to produce from the formation, (3) relatively inert and compatible with reservoir fluids, and (4) acceptable in the pipeline oil. The concentration of the fluid-loss agent should be determined, when possible, by laboratory tests on the formation rock in question, under bottom-hole conditions of temperature and differential pressure. Experience has shown that when testing is not possible a nominal concentration still can be used.[7]

One aspect of fluid-loss control resulting from the use of a gas phase in fracturing was reported by Foshee and Hurst.[8] They have shown that gas included with

TABLE 5.1 — TYPICAL SPECIFICATIONS FOR REFINED FRACTURING OIL

API gravity	6 to 25°F
Viscosity	300 to 900 SSU at 100°F or 50 to 300 cp at 100°F
API fluid loss	25 to 100 ml in 30 minutes
Sand falling rate	less than 7 ft/min at 150°F
Asphaltenes	less than 0.75 percent
Emulsion breakout time	less than 30 minutes at reservoir temperature

oil and water is comparable with solid fluid-loss agents in obtaining a low-fluid-loss fracturing fluid. Gas as a fluid-loss agent has the advantages of ease of removal, inertness or compatibility, and acceptance of subsequent oil production.

Dynamic fluid-loss measurements reported by Hall and Dollarhide[9] seem to follow rather closely the pattern of earlier static tests, with the possible exception that the rate of leakoff is not in all cases directly proportional to the square root of time but may vary between that and a straight-time relationship. Their work showed that a spearhead method of treatment may be effective, and they reported that once the filter cake has been established from a low-fluid-loss fluid, the filter cake is resistant to damage by a flow of a plain fracturing fluid. Under certain conditions, a treatment employing a spearhead of fluid with a high concentration of fluid-loss additive followed by plain fluid can be more economical.

5.4 Water-Base Fluids

One of the most pronounced recent trends in fracturing has been the increased use of water-base fluids. At one time, it was considered foolhardy to pump water into producing formations. The formations were considered to be water-sensitive, and it was thought that any invasion of fresh water into these formations might shut off the flow of oil. However, in spite of conflicting recommendations from the laboratories, field engineers with an eye for economy used water for fracturing; their success led to increased use of water. Now over two-thirds of the wells are being fractured with water-base fluids,[10] which are suitable for all but a few low pressure, oil-wet formations. There are several advantages to pumping water-base fluids and using water-base fluids for fracturing. Perhaps one of the most important is the fact that it is not a fire hazard. Other advantages are that water is available in most areas, the cost is low, and these fluids may be effectively treated with friction reducers. Water has a low viscosity, and therefore is easier to pump. This becomes more important at high injection rates, particularly when the fracturing fluid is in turbulent flow. Another advantage of water is its higher specific gravity, which creates a higher hydraulic head and provides greater suspending power for the propping agent. Finally, because it is easy to modify, water is very versatile.

Friction-Reducing Additives

As stated earlier, one of the advantages of using water is that the friction from pumping is much less. Fracturing treatments have been performed at rates in excess of 500 bbl/min, and such treatments would not have been possible if water had not been used and if it had not been treated for suppression of turbulence. The friction-loss-reducing agents now being used in water are essentially polyacrylamides and partially hydrolized polyacrylamides.[11] Other materials, such as guar gum and cellulose derivatives, act in a similar manner but are not so effective. These materials, however, are usually less expensive and therefore are used more widely. Depending on the demands that are to be made upon the fracturing fluids, the proper water-gelling agent or friction-reducing agent can be chosen.

The action of the friction-reducing agents in water has been described by Ousterhout and Hall as a suppression of turbulence.[12]. Data plots tend to show that the presence of a small amount of a high-molecular-weight linear polymer, such as polyacrylamide, reduces the swirling and eddying while the fluid is in motion and thereby suppresses the turbulence. Plots of friction factor vs Reynolds number show a tendency for the laminar flow curve to continue almost in a straight line through the transition zone instead of moving up to the turbulent flow curve. The success of friction reducers has been thoroughly demonstrated and they are used throughout the industry. The proper quantity of the best polymer-type additives can reduce pumping friction associated with water by as much as 75 percent.

Specific Gravity Advantage

Frequently overlooked is the advantage of the increase in specific gravity provided by a water-base fluid as opposed to an oil-base fluid. In a 10,000-ft well, the difference between a specific gravity of 0.7 for oil and 1.1 for brine is 0.4. The brine is actually one and one-half times as heavy as the oil. The hydraulic head will increase from 3,040 to 4,770 psi simply because of the density difference. This will be a bottom-hole as well as a wellhead pressure difference of approximately 1,730 psi, which reduces by a corresponding amount the hydraulic horsepower required in a treatment. At high treating pressures, such a reduction in pressure becomes important not only because it makes it possible to stay within the limits of the equipment available, but also because it reduces the actual cost of the operation. Service companies figure the cost per horsepower on the basis of the pressure required. A reduction of 20 to 30 percent in the cost of the treatment would not be unusual in switching from the oil-base to the water-base fluid.

Gelling, Fluid-Loss, Bactericide, and Scale-Removal Additives

The versatility of the water-base fluids when used with the additives that are now available makes them exceptionally attractive. The gelling agents, such as guar gum, hydroxyethyl cellulose, and polyacrylamide, are effective agents for increasing fracturing fluid viscosity in wells with bottom-hole temperatures of 200°F or greater.

The viscosity thus obtained not only influences leakoff but also is a prime factor in assuring good suspension and displacement of the propping agent. Many times, when low viscosity fluids are used, there is a danger of propping agent accumulation in the wellbore and subsequent screening-out at the fracture face. Not only is viscosity imparted by such gelling agents, but also lower fluid loss is obtained; and, by the addition of small amounts of inert solids such as silica flour, the fluid loss of the treated fluid can be reduced even further. There is also the alternative of using one of the ready-made fluid-loss or wall-building materials now on the market such as Adomite Aqua®, which is offered by Continental Oil Co. This is a mixture of starch, bentonite, silica flour, guar gum, surfactant, etc.

A bactericide is often necessary as an additive for water-base fluids. It has been shown by Hawsey *et al.*[13] that pumping untreated water into the reservoir may introduce bacteria. Several water-soluble bactericides are available and are easily used. Among these are the quaternary amines and amide-type materials, as well as the chlorinated phenols.

One of the components sometimes introduced into a fracturing treatment is a scale stabilizer such as the polyphosphate materials — sodium, magnesium, and calcium phosphates, and complexes thereof — that can be produced in various forms for use in wells to prevent precipitation of calcium carbonate and calcium sulfate. These materials perform in a manner described as a "threshold effect", wherein the growth of calcium carbonate or calcium sulfate crystals is inhibited by the presence of the phosphate.[14] The polyphosphate actually delays the precipitation of the undesirable scale, rather than preventing it completely.

Surfactants and Alcohol

One of the most commonly used additives in a water-base fluid is the surface active agent or "surfactant." The surfactant may be added for a number of different reasons. Generally, however, it is added to reduce the interfacial tension and the resistance to return flow. Some surfactants, however, are added to provide a foam-stabilizing action. These can be effective in gas wells where gas bubbling through the water in the wellbore fills the hole with foam and allows the gas to unload the hole by flowing the foam from the wellbore. There is some controversy surrounding the use of foaming agents. Bernard *et al.*[15] report that there is a tendency for foam to build up in the matrix of the reservoir itself when a foaming agent has been incorporated into the fracturing fluid. The foam builds up in the small pores and flow channels in the reservoir, and the increase in capillary forces reduces the flow capacity, which reduces the effectiveness of treatment. However, surfactants are readily adsorbed on the rock surfaces and probably penetrate only a small distance into the rock. The controversy regarding the use of foaming agents has not been resolved and the question is contingent upon the nature of the rock, the pressure of the reservoir and whether or not gas is coming out of solution at any distance from the wellbore. These factors may determine if the gas could cause such foaming within the formation matrix.

McLeod and Coulter[16] have presented strong arguments for using alcohol to lower interfacial tension. Alcohol in concentrations of 10 to 20 percent can penetrate large volumes of the reservoir. However, its use in fracturing has been limited because it is expensive and because it causes difficulties in getting the desired properties of viscosity and fluid-loss control in the fracturing fluid. This is an area that no doubt will be investigated more fully in the future.

Guar Gum

All these factors are important. However, more emphasis should be given to guar-gum-thickened water-base fracturing fluid. This is the most widely used field fracturing agent. Guar gum is suitable for thickening both fresh and salt water, including saturated sodium chloride brines. At least two basic types of guar gum formulations are used to obtain a desirable gelled water-base fluid. These may be classified as materials suitable for batch mix operations and materials suitable for continuous mix operations.

The most widely used form is the continuous mix grade of gum, referred to as such because it hydrates rapidly and reaches a usable level of viscosity fast enough that it can be used continuously. When properly mixed, this material will provide viscosity, friction reduction, and fluid-loss properties rapidly enough to become effective on the way down the hole, even though its residence time in the surface mixing equipment may be less than a minute. Such a formulation gains its continuous mix properties from its particle size. The relative hydration rates for various cuts of guar gum are shown in Fig. 5.1.

In addition to a fine particle size, continuous mix guar gum may contain one or more breakers for the system and one or more inert materials to improve the flow properties. The breakers are usually oxidizing

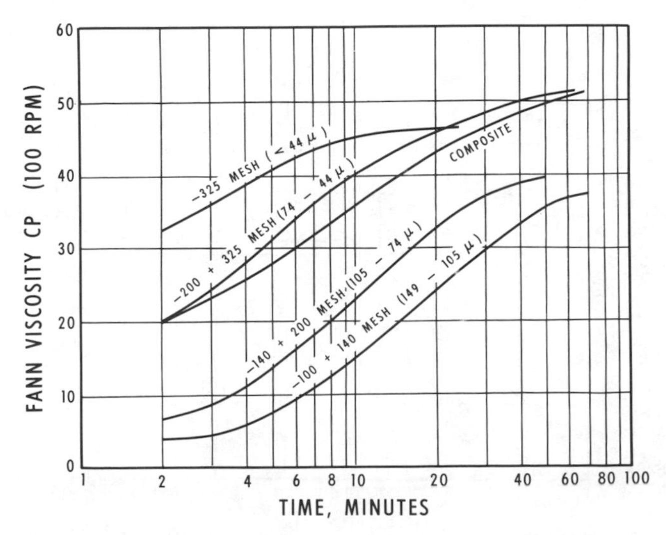

Fig. 5.1 Hydration rates of various cuts of guar gum, showing effect of particle size on hydration rate.[25]

agents — enzymes or acids — while the additives to improve free flow are hydrated amorphous silicas or similar products.

The easy mixing or batch mix materials are designed to take advantage of a unique property of the guar gums: they follow an unusual complexing reaction in the presence of a boron compound such as borax. In the presence of borax, guar gum can be dissolved in a slightly alkaline solution without increasing the viscosity of the solution. As the pH of the solution is lowered, the guar gum forms the usual gel and thickens the solution. Specific guar gum formulations are designed to take advantage of this property. They are alkaline mixtures of guar gum and borax with a delayed-action acid. An alkaline material such as lime is mixed with borax and guar gum, and is then mixed with a slow-dissolving, dry organic acid such as sulfamic acid. The lime and borax dissolve immediately, providing the alkaline borax medium. The guar gum becomes thoroughly wet and mixes in readily, and then the acid portion of the formulation slowly dissolves, lowering the pH and causing the guar gum to thicken the water in the normal manner.

This discussion has dealt only with the general operations conducted with guar gums. There are numerous variations that can be achieved by changing the breaker system and by changing the combinations and concentrations of breakers that are used in formulations. A typical viscosity performance and breakdown curve for one particular formulation with and without enzyme breaker is shown on Fig. 5.2. This curve is only an example, and would vary depending upon the exact formulation used.

5.5 Acid-Base Fluids

The acid-base fracturing fluids, in general, follow the pattern of the water-base fluids; that is, the important factors to consider are friction loss and fluid loss. Other matters of concern with acid-base fluids involve the concentration of acid and how the acid reaction may be influenced by additives or by the way the fluids are prepared.

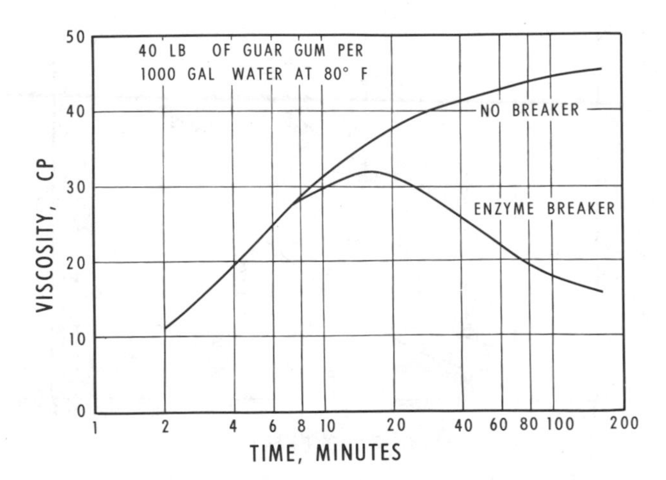

Fig. 5.2 Effect of enzyme breaker on a typical guar gum fracturing fluid.[25]

Friction-Loss Control

In regard to friction loss, the same considerations arise with acid as with water, and the same agents or similar agents are used. Guar gum will reduce the friction of acid-base fluid and is effective as long as the acid is at a low concentration. The guar gum is not stable in 15 percent hydrochloric acid and, as a result, can be used to reduce friction loss only at low temperatures and only on a continuous mix basis. Polyacrylamides, on the other hand, will effectively control the friction properties of acid, are more stable, and can be used at higher temperatures.

Fluid-Loss Control

Controlling fluid loss of acid-base fluids is difficult because of the action of the acid, not only on the fluid-loss-control agent, but also on the formation. Because of this, the fluid-loss-control agent must be effective and efficient. Two types of agents have been successful in controlling the fluid loss of acid: (1) blends of acid-resistant gums, silica flour, and oil-soluble resins, and (2) synthetic polymers that swell in acid but retain their particulate nature and are stable in the presence of hydrochloric acid. The agents now available are capable of restricting the flow of acid into both permeable matrix and small fractures. The photographs in Fig. 5.3 illustrate the effectiveness of fluid-loss-control agents in preventing acid flow into hairline fractures.

Gelled Acids

The gel or thickened acid, exclusive of the emulsions, employs for the most part a natural gum such as gum karaya or guar gum. The synthetic polymers and cellulose derivatives have not been competitive and are not generally used to thicken acid. Hydrochloric acid thickened with a material such as gum karaya will be held in the main fractures, resulting in a deep penetrating acid job. The use of this type of acid has declined, primarily because it is expensive and unstable at temperatures above 100°F. Two properties that these systems share, to some degree, with other systems are fluid-loss control and friction reduction. The fluid-loss control can be improved by the addition of silica flour.

Acid Emulsions

Acid emulsions, although perhaps not so popular as they were, are still useful at high temperatures because of the stability that can be built into them. Examples are acid-in-oil emulsions with a 60- to 90-percent internal (aqueous) acid phase. In making up the acid emulsions, various emulsifiers have been used for ease in preparation and to achieve the desired degree of stability. The emulsifier itself is typically a mixture of nitrogen-based organic compounds and nonionic surfactant-type materials.

The emulsifier is selected primarily on the basis of its ability to be stable at surface conditions and unstable at down-hole conditions.

The emulsifier either must be broken down by the acid on prolonged contact at high temperature, or must be adsorbed on the formation. The breakdown may result either from a combination of these two actions or from a reaction with the spent acid that causes it to lose its emulsifying properties.[17] The emulsifier must remain stable and in the system until the fracturing treatment is completed, and then it must lose its ability to be an emulsifier or to stabilize emulsions so that the fluid can readily be retrieved from the well.

Primary disadvantages of acid emulsions are the high viscosity and attendant high friction loss. A low-viscosity emulsion may not have the desired stability and flow properties, whereas a high-viscosity emulsion may be difficult to pump, especially down tubing. The acid emulsions are used primarily in high-temperature wells, since emulsification retards the reaction rate of the acid enough to enable the live acid to penetrate farther out into the fracture.

Fig. 5.3 Effectiveness of fluid-loss-control agents in hairline fractures.[25] (Upper photograph illustrates acid flow and etch pattern with 15 percent HCl. Lower photograph illustrates acid flow and etch pattern in core of same formation when acid contains fluid-loss additive.)

Chemically Retarded Acids

Chemically retarded hydrochloric acid has certain advantages over the other retarded acids in that it is inexpensive, slow reacting, and has a low viscosity; yet it retains the full reaction power of the hydrochloric acid. The retarding action of the chemical retarders depends on a wetting action. The carbonate rock is made oil wet, which provides an oil film to resist the attack of the acid. According to Knox *et al.*[18] the most effective surfactants for retarding reaction on limestone are the anionics such as alkyl phosphate, alkyl taurate and alkyl sulfonate. Due to economics, alkyl sulfonate is the material most widely used for this purpose. In carbonate formation, chemically retarded acids make it possible to fracture the formation and increase its fracture-fluid-carrying capacity a greater distance from the well than is possible with plain acid.

5.6 Selection of Fracturing Fluids

With a knowledge of the different fracturing fluids that are available, some guidelines can be established as to when and where they should be used. The selection of a fracturing fluid depends primarily on the nature of the formation encountered and the fluids in place in the formation. Not only the chemical nature (carbonate, sandstone or other), but also the physical nature of the rock itself must be considered. The chemical nature of the fluids in the reservoir — whether gas, oil, brine, etc. — and their physical nature, from the standpoint of what will influence their movement, also play a vital role in the selection of the fracturing fluid. Finally, the physical properties of the reservoir, such as temperature, pressure, wettability, and fluid saturation, can influence the choice of liquid.

Formation Properties

The first considerations are the chemical and physical properties of the rock being fractured and how these influence selection of the fracturing fluid. Of primary importance are the permeability and porosity of the formation. If the formation permeability is high and undamaged, there is little need for a fracturing treatment. However, during completion of the well, there generally is permeability damage, so a fracturing fluid must be selected that will not further reduce the permeability in the matrix rock. In many cases, the primary purpose of the fracture is to overcome damage caused by earlier exposure of the rock to drilling fluids, cement filtrate, etc.

Another important factor is the clay content of the rock. Oil-base fracturing fluids are frequently recommended to avoid permeability damage from clay effects. Several authors have published articles on the effects of different fracturing fluids on the permeability of rock containing clay.[19-23] Jones' study of aqueous fluids and their effect on permeability of rock containing clay has been one of the most exhaustive on this subject.[22] He showed that divalent cations can render the clay insensitive. Also, his analysis of the effect of varying ionic strengths on the permeability of the clay-bearing rock explains what may be expected from "water sensitive" formations. The addition of 0.5 percent calcium or potassium chloride to fresh-water fracturing fluid generally will prevent permeability reductions caused by clay effects.

The effect on productivity of different degrees of permeability reduction in a damaged zone is shown in Fig. 5.4. Although this specifically refers to damage to permeability by clay blocking, the principle applies for any kind of damage. The practice of selecting fluids for fracturing is not standardized, since everything from saturated salt water to fresh water is used. Perhaps the safest practice is to use formation brines, which should have very little detrimental effect. It has been advocated[21] that brine solutions be used whose salt content is at least 5 percent calcium or magnesium chloride. In some cases, potassium chloride brine is used to protect the water-sensitive formation.[10,24]

In contrast to the use of brines in water-sensitive formations is the use of fresh water where there is appreciable water-soluble salt in the rock. A refinement of this practice has been the use of very dilute hydrochloric acid (1 to 5 percent) to dissolve the sodium chloride and increase the permeability through the reaction of acid with the carbonates present. At one time, the special effectiveness of many large-volume, fresh-water treatments was attributed to the removal of salt from the formation. Perhaps part of the success of these treatments was due actually to the large volumes of water used, which resulted in deep penetrating fractures.

If the formation to be fractured is principally carbonate, the use of an acid-base fracturing fluid is usually considered. Many variations of acid are potentially useful. Generally, a long-reacting or retarded acid is desirable for reaching deep into the formation to increase fracture permeability. Emulsified acids are used at temperatures above 250°F, but below this temperature it is common to use chemically retarded acids and highly concentrated acid (28 percent). Present trends indicate that for fracturing limestone or dolomite formations high strength hydrochloric acid is especially useful.

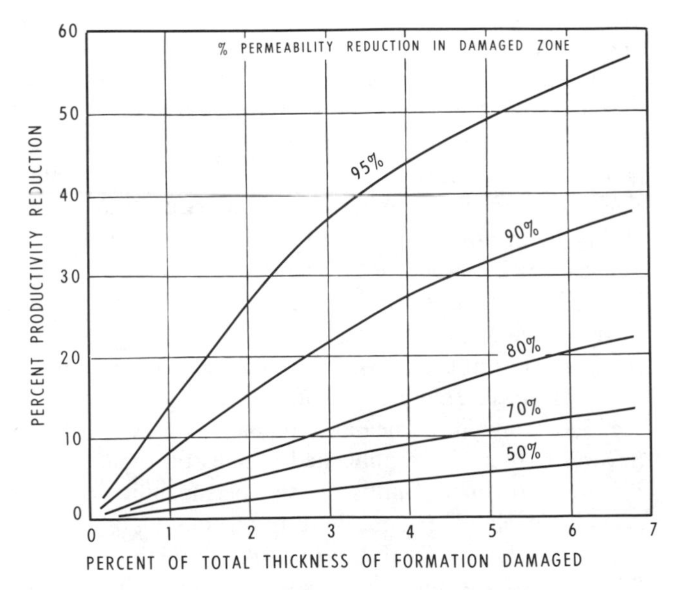

Fig. 5.4 Water-base fracturing liquid (after Jones and Fast[21]).

Bottom-Hole Temperature and Pressure

Bottom-hole temperature must be considered carefully in the selection of the fracturing fluid to be used, and in the selection of the type and concentration of additives that will be effective in the formation at the existing temperatures. The efficiency of oil, water, and acid-base fracturing fluids is affected by the temperature to which they are exposed. For example, the use of gelled lease oils is limited to temperatures of less than 200°F, depending upon the type of oil used, because of a reaction that results in the formation of a grease-like solid. With an increase in temperature there generally is an increase in the quantity of fluid and friction-loss-control additives required for oils; and with a decrease in temperature there is a decrease in the viscosity of refined residual oils.

Acid-base fracturing fluids are affected by the temperature to which they are exposed. For example, at high temperature the gelled acids break down much more rapidly; therefore, more thickening additive generally must be used in the deeper wells. At higher bottom-hole temperatures the friction-reducing additives, also, must be used in increased concentrations.

The bottom-hole treating temperature influences also the selection of water-base fracturing fluids, although probably less than other types of fluids. The gum and polymer thickening additives, and the fluid-loss and friction-reducing additives are generally increased for the treatment of higher temperature wells. In the highest temperature wells it is usually necessary to use synthetic polymers rather than guar or other natural gums.

Bottom-hole pressure is another property that should be given consideration. If the bottom-hole pressure is high, then a viscous fracturing fluid will be produced into the well following treatment. Another consideration, however, is that the density of the fracturing fluid should be high enough to help overcome the bottom-hole pressures and reduce the horsepower required for the fracturing treatment. As was pointed out earlier, a heavy fluid can contribute hydraulic head and thereby reduce the surface pressure and hydraulic horsepower necessary to obtain and extend a fracture when high bottom-hole pressures are encountered. In wells with low bottom-hole pressures, the first consideration is to select a fracturing fluid that can be removed easily from the formation. A low viscosity fluid should be considered such as a high gravity oil or alcohol or alcoholic water solution. These materials clean up more readily than do some of the other common fracturing fluids.

Formation Fluids

Another factor to consider in the selection of a fracturing fluid is the type of fluid in the formation. If a formation contains heavy oil and asphaltenes or some paraffinic materials, it is inadvisable to treat it with a high API gravity oil that might cause a precipitation of these heavy materials. In such cases, a crude oil containing aromatics would be the proper selection if an oil-base fluid were preferred for fracturing. However, if it can be adapted to the formation, a water-base fluid can assure control of asphaltene precipitation. In many ways water-base fluids have been superior to oil-base fluids. However, in preferentially oil-wet formations oil-base fluids may be required to prevent a reduction in the relative permeability to the oil and water blocking the formation.

One factor that should not be overlooked in the selection of fracturing fluids is the compatibility of the fluid with the fluids in the formation. This can be checked by a simple test in which the fracturing fluid is mixed with samples of oil and formation water to determine if any undesirable emulsions or precipitates form. In many cases, special consideration must be given to the fluids (cement or mud filtrate or a previous stimulation treatment fluid) that have been introduced into the formation by previous well operations. Frequently the most reasonable approach is to conduct laboratory tests with the fluids in question.

Another important factor to consider is the effect of formation fluids in gas-producing wells. The presence of liquid such as water can greatly hamper or impair gas flow. The use of alcohol in such wells or the use of acid or fracturing fluid containing alcohol will greatly enhance the return flow of gas to the well. McLeod and Coulter[16] pointed out that this can be particularly important when the formation in question contains cross-bedding and similar heterogeneity.

In selecting the base fracturing fluid for a particular formation, experience in the area and knowledge of individual well conditions can be very helpful. There is no substitute for a full knowledge of the well history and of exactly what fluids may have been used in the well. The deeper the well, with its higher bottom-hole temperature and pressure, the more expensive the fracturing treatment; this dictates that the operator exercise maximum care in designing the fracturing treatment that is best for the specific conditions. In many cases, this will mean laboratory testing and evaluation of potential fluids or samples from the formation to be treated. API RP 39, *Recommended Practice—Standard Procedure for Evaluating Fracturing Fluids,* presents procedures for evaluating fracturing fluids in the laboratory.

5.7 Summary

The properties of the rock, reservoir, and reservoir fluids affect the selection of a fracturing fluid. The chemical nature of the rock will basically determine the type of fluid, such as acid-base or non-acid-base. In limestone, dolomite, or variations with high solubilities, acid-base fluids are usually more effective. In sandstone or low-solubility rocks, water- or oil-base fluids generally are more economical.

The choice between oil- or water-base fluids will depend on the clay content or water sensitivity of the rock. Water-base fluids are preferred because they are less expensive and safer, and because they usually can be made compatible with water-sensitive rock. Only in a minority of cases will it actually be necessary to use oil-base fluids. This type of fluid usually will be required in low-pressure, preferentially oil-wet reservoirs where the use of water could result in water blocking of the formation.

The presence of unwanted minerals — iron compounds, anhydrite, etc. — can make it necessary to use certain additives in the fracturing fluid. Such minerals, due to their reaction with the fracturing fluid, can reprecipitate and cause plugging in the formation, in which case, special agents may be necessary.

The physical nature of the rock (the permeability, porosity, structure, and whether or not it is fractured, vugular, etc.) normally does not affect directly the selection of the type of fluid, but does affect the choice of the properties of the fluid. For example, in high permeability rock, viscous fluids or fluids with fluid-loss additives may be necessary. Acid-base, water-base, and oil-base fluids may be altered to suit the physical properties of the rock.

The fracturing fluid selected on the basis of the chemical nature of the producing formation generally can be altered with proper additives so that it will be compatible with the reservoir fluids. Emulsion problems, sludge-forming characteristics, and the fact that water blocks tend to form in gas wells must all be considered when matching a fracturing fluid to the reservoir fluid.

In those cases where the fracturing fluid selected on the basis of the chemical nature of the reservoir rock is not compatible with the reservoir fluids, it should be abandoned and a fluid selected that is compatible with produced liquids. The main reservoir properties to be considered are temperature and pressure. These properties also determine the additives required in a fracturing fluid rather than the type of fluid. For example, high reservoir pressure might dictate heavy or weighted fracturing fluids so that some benefit can be gained from the hydrostatic head. The high pressure might also dictate low injection rates that could require viscous fluids or retarded reaction rate acids. Temperature could affect the choice not only of the additives involved, but also of a secondary type of fluid. Extremely high temperature, for instance, might prohibit the use of a gelled-oil fluid or might seriously hamper the use of acid-base fluids.

Whenever possible, laboratory tests of rock and reservoir fluid properties in conjunction with a knowledge of the reservoir and field experience in the area should be used in a methodical approach to the selection of a fracturing fluid. Only when all properties and conditions existing in each individual well are taken into account can the most effective and efficient fluid be selected.

References

1. Howard, G. C. and Fast, C. R.: "Optimum Fluid Characteristics for Fracture Extension", *Drill. and Prod. Prac.*, API (1957) 261.
2. Wilsey, L. E. and Bearden, W. G.: "Reservoir Fracturing — A Method of Oil Recovery from Extremely Low Permeability Formations", *Trans.*, AIME (1954) **201,** 169-175.
3. Brown, J. L. and Landers, Mary M.: U. S. Patent No. 2,779,735 (Jan. 29, 1957).
4. Phansalkar, A. K., Roebuck, A. H. and Scott, J. B.: U. S. Patent No. 3,046,222 (July 24, 1962).
5. Stewart, J. B. and Coulter, A. W.: "Increased Fracturing Efficiency by Fluid Loss Control", *Pet. Eng.* (June, 1959) B-43.
6. Hendrickson, A. R., Nesbitt, E. E. and Oaks, B. D.: "Soap-Oil Systems for Formation Fracturing", *Pet. Eng.* (May, 1957) B-58.
7. Essary, Roy L.: "Fracture Treatments in S.E. New Mexico", *World Oil* (March, 1962) 99.
8. Foshee, W. C. and Hurst, R. E.: "Improvement of Well Stimulation Fluids by Including a Gas Phase", *J. Pet. Tech.* (July, 1965) 768-772.
9. Hall, C. D., Jr., and Dollarhide, F. E.: "Effects of Fracturing Fluid Velocity on Fluid-Loss Agent Performance", *J. Pet Tech.* (May, 1964) 555-560.
10. Black, Harold N. and Hower, Wayne E.: "Advantageous Use of Potassium Chloride Water for Fracturing Water Sensitive Formations", paper 851-39-F presented at Mid-Continent District Meeting, API Div. of Production Meeting, Wichita, Kans., March 31-April 2, 1965.
11. Root, R. L.: U. S. Patent No. 3,254,719 (June 7, 1966).
12. Ousterhout, R. S. and Hall, C. D., Jr.: "Reduction of Friction Loss in Fracturing Operations", *J. Pet. Tech.* (March, 1961) 217-222.
13. Hawsey, Jerry D., Whitesell, L. B. and Kepley, N. A.: "Injection of a Bactericide-Surfactant During Hydraulic Fracturing — A New Method of Corrosion Control", paper SPE 978 presented at SPE 39th Annual Fall Meeting, Houston, Tex., Oct. 11-14, 1964.
14. Alderman, E. N. and Woodard, G. W.: "Prevention of Secondary Deposition from Waterflood Brines", *Drill. and Prod. Prac.*, API (1957) 98.
15. Bernard, G. G., Holm, L. W. and Jacobs, W. L.: "Effect of Foam on Trapped Gas Saturation and on Permeability of Porous Media to Water", *Soc. Pet. Eng. J.* (Dec., 1965) 295-300.
16. McLeod, H. O. and Coulter, A. W.: "The Use of Alcohol in Gas Well Stimulation", paper SPE 1633 presented at SPE Third Annual Eastern Regional Meeting, Columbus, Ohio, Nov. 10-11, 1966.
17. *Petroleum Production Handbook,* Thomas C. Frick and R. William Taylor, Eds., McGraw-Hill Book Co., Inc., New York (1962) **2**, 47.
18. Knox, John A., Lasater, R. M. and Dill, W. R.: "A New Concept in Acidizing Utilizing Chemical Retardation", paper SPE 975 presented at SPE 39th Annual Fall Meeting, Houston, Tex., Oct. 11-14, 1964.
19. Gray, D. H. and Rex, R. W.: "Formation Damage in Sandstones Caused by Clay Dispersion and Migration", *Proc.,* Fourteenth National Conference on Clays and Clay Minerals (1965) **14**.
20. Hewitt, Charles H.: "Analytical Techniques for Recognizing Water-Sensitive Reservoir Rocks", *J. Pet. Tech.* (Aug., 1963) 813-818.
21. Jones, Frank O. and Fast, C. R.: U. S. Patent No. 3,179,173 (April 20, 1965).
22. Jones, Frank O.: "Influence of Chemical Composition of Water on Clay Blocking", *J. Pet. Tech.* (April, 1964) 441-446.
23. Monaghan, P. H., Salathiel, R. A., Morgan, B. E. and Kaiser, A. P., Jr.: "Laboratory Studies of Formation Damage in Sands Containing Clays", *Trans.,* AIME (1959) **216**, 209-215.
24. Smith, C. F., Pavlich, J. P. and Slovinsky, R. L.: "Potassium, Calcium Treatments Inhibit Clay Swelling", *Oil and Gas J.* (Nov. 30, 1964) 80.
25. Coulter, A. W. and David, H. E.: "Fracturing Fluids and Additives", unpublished report, Dowell Div., Dow Chemical Co. (Jan., 1967).

Chapter 6

Propping Agents for Hydraulic Fracturing

6.1 Introduction

Early experimental work in shallow wells demonstrated that a hydraulically formed fracture tends to heal — that is, to lose its fluid carrying capacity after the parting pressure is released — unless the fracture is propped.

For example, in one test a shallow well was drilled, cased, and cemented, then deepened to expose a section of sandstone. Water injectivity tests were conducted to determine the injection-rate:pressure relationship. The well was fractured with a Napalm-gasoline gel containing no propping agent. Post-fracturing tests were conducted to determine the effect of the Hydrafrac® (a service mark of Pan American Petroleum Corp.) treatment on well injectivity. Two months after treatment, without any intervening work on the well, a third water injection test was made to determine the permanence of the fracture treatment. Following this test, the well was refractured and a sand propping agent was injected into the fracture; then a series of injectivity tests was conducted to measure again the permanence of the treatment.

Fig. 6.1, a plot of injection pressure vs injection rate, shows the results of these tests. The well had a very low injectivity before the first fracturing treatment, and injectivity was increased greatly by fracturing the formation. After treatment, the fracture with no proppant partially closed, resulting in a reduction of effective well permeability. Following the injection of sand into the fracture, not only was the increase in fracture injectivity substantially greater than after the initial hydraulic fracture treatment, but also the increased injectivity was permanent.

The tendency of a fracture to heal has been observed in numerous field tests in which wells were hydraulically fractured both with and without a propping agent. Table 6.1 lists 10 wells in three Mid-Continent fields from which data of this type were recorded. The proppant used in these fields was flint shot Ottawa sand, 90-percent −20+40 mesh U. S. Standard Sieve. Wells in Fields A and B produced from a sandstone formation, and the wells in Field C produced from a dolomite reservoir. No specific information is available on the hardness of these reservoir rocks. In all instances, except for Well 3 in Field A, the difference in the production from the wells in which proppant was used and from those in which propping agent was not used is marked. The rate of production decline from the unpropped wells is much higher.

As a result of field studies, methods were devised for classifying formations so that the effectiveness of the propping agent could be evaluated on the basis of the characteristics of the rock and on the quantity and type of proppant in the fracture.

6.2 Theoretical Considerations of Factors Affecting Fracture Conductivity

Darin and Huitt,[1] in their study of fracture flow capacity, presented methods of calculating the permeability of fractures containing various amounts of proppant. They pointed out that fracture permeability varies between the one extreme of an open fissure and the other extreme of a "packed"* fracture.

*A "packed" fracture is one in which the space between the two fracture faces is completely filled with single or multiple layers of the propping agent.

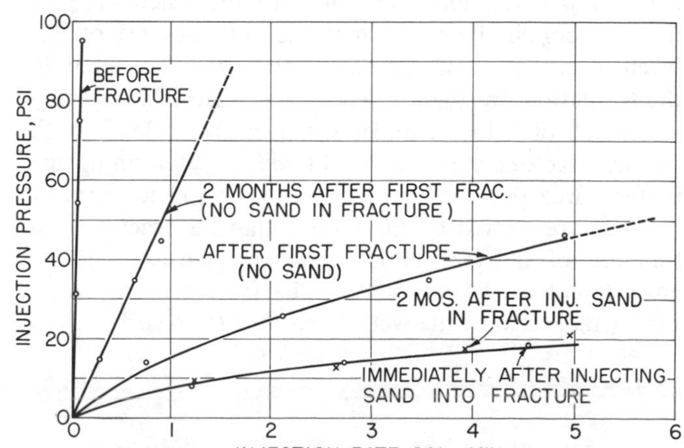

Fig. 6.1 Effect of propping agent on permeability of fractures in shallow sandstone wells.

Unpropped Fracture

The premeability of an open fracture is described by[2-4]

$$k=\frac{10^8 W_f^2}{12} \quad (6.1)$$

$$(kW_f)=4.47\times 10^9 W_f^3 \quad (6.2)$$

Eq. 6.1 presumes that the flow in the fracture is laminar.

Packed Fracture

When a fracture is filled with multilayers of proppant, the permeability may be expressed as[5]

$$k=\frac{\phi^3}{C_K S^2(1-\phi)^2} \quad (6.3)$$

Using earlier work by Carman,[5] Darin and Huitt[1] showed that

$$C_K=C_o\left(\frac{L_e}{L_B}\right)^2 \quad (6.4)$$

These investigators and Wyllie and Gregory[6] showed that $(L_e/L_B)^2$ was dependent upon the shape of particles in the bed, and should be about 2 for unconsolidated porous media, and C_K should have a value of 5. Thus, if C_K is assigned a value of 5 and W_f and S are expressed in inches, the fracture flow capacity* (md-ft) as obtained with Eq. 6.3 may be expressed as

$$(kW_f)=1.1\times 10^{10}\frac{W_f}{S^2}\frac{\phi^3}{(1-\phi)^2} \quad (6.5)$$

Partial Monolayer

A fracture propped with a partial monolayer of the propping agent is not properly described by Eqs. 6.2 or 6.5. Eq. 6.2 describes a fracture with 100 percent porosity. The surface area per unit volume of fracture and porosity, as described for a bed of particles in Eq. 6.5, must be changed to reflect a partial monolayer of proppants in the fracture.

Darin and Huitt,[1] in their discussion of the permeability of a propped fracture, showed that the reasons Eq. 6.5 *per se,* does not apply to the calculation of fracture conductivity are that the surface area of the fracture and the embedment of the propping agent in the formation are manifested by the fracture's tendency to "heal", or "close", under the overburden load. Both of these factors must be considered in determining the surface area described in Eq. 6.5. The surface area of the fracture wetted by the fluid within the fracture is a function of the size of the proppant particle and its embedment in the surfaces of the fracture. Therefore, the surface area of the wetted fracture (two surfaces) is

$$A_S=2-\frac{n_{pa}\,\pi d_i^2}{2} \quad (6.6)$$

*The term "fracture flow capacity" is commonly used in petroleum technology to indicate the ability of a fracture to transmit fluids. A more proper term is "conductivity." The two terms are used interchangeably in this monograph.

TABLE 6.1 — EFFECT OF FRACTURE CLOSURE ON WELL PRODUCTION

Field and Depth	Size of Fracture Job (gal)	Amount of Proppant Used* (lb)	Production (BOPD) After 1 Month	Production (BOPD) After 1 Year
Field A (2,500 ft) sandstone				
Well 1	1,000	1,000	40	30
Well 2	1,500	1,200	45	35
Well 3	1,200	0	42	27
Well 4	1,500	0	37	22
Field B (6,000 ft) sandstone				
Well 1	5,000	5,000	75	50
Well 2	5,500	0	80	30
Field C (7,000 ft) dolomite				
Well 1	5,500	5,500	85	70
Well 2	5,000	4,000	83	65
Well 3	7,500	7,000	90	60
Well 4	5,000	0	60	25

*Flint Shot Ottawa Sand, 90 percent −20+40 mesh U. S. Standard Sieve

The wetted area of the propping agent particles is given by

$$A_{pa}=n_{pa}(\pi d_{pa}^2-2\pi D d_{pa}) \quad (6.7)$$

The geometry of embedment is shown in Fig. 6.2,

$$D=\frac{d_{pa}-W_f}{2} \quad (6.8)$$

Substituting Eq. 6.8 for D in Eq. 6.3 results in

$$A_{pa}=n_{pa}\pi W_f d_{pa} \quad (6.9)$$

The total wetted area per square inch of fracture surface is

$$A_{wf}=2-\frac{n_{pa}\pi d_i^2}{2}+n_{pa}\pi W_f d_{pa} \quad (6.10)$$

Darin and Huitt[1] report that experimental data show a satisfactory correlation between the volume of the propping agent particles and the wetted area. In Fig. 6.2, the volume V_p of the propping agent particle (sphere) contained between the two planes formed by the fracture faces is

$$V_p=\frac{\pi d_{pa}^3}{6}-\pi D^2\left(d_{pa}-\frac{2D}{3}\right) \quad (6.11)$$

The volume of the propping agent V_p per unit area of fracture surface (one face) is therefore equal to

$$V_p=\frac{n_{pa}\pi}{6}(d_{pa}^3-6d_{pa}D^2+4D^3) \quad (6.12)$$

The wetted surface area S_p per unit volume of propping agent contained between the fracture faces is obtained by

$$S_p=\frac{A_{wf}}{V_p} \quad (6.13)$$

By knowing the volumes on the basis of a unit fracture area, the porosity of a partial monolayer ϕ_p is

$$\phi_p = \frac{W_f - V_p}{W_f} \quad . \quad . \quad . \quad (6.14)$$

By substituting Eqs. 6.13 and 6.14 into Eq. 6.5, the final form of the flow capacity equation (modified Kozeny-Carman relation) is obtained.

$$(kW_f) = 1.1 \times 10^{10} \frac{W_f}{S_p^2} \frac{\phi_p^3}{(1 - \phi_p)^2} \quad . \quad . \quad . \quad (6.15)$$

Experimental laboratory work is required to obtain embedment data for determining the various terms in this equation. Eq. 6.15[1] provides a means of estimating the fracture flow capacity obtainable with various sizes and density patterns of partial monolayers of the propping agent.

Huitt and McGlothin[12] showed that for formations in which the propping agent embeds rather than crushes under the overburden load, the embedment could be expressed by the relation

$$\frac{d_i}{d_{pa}} = B^{1/2} \left[\frac{W_p}{(d_{pa})^2}\right]^{m/2} \quad . \quad . \quad . \quad (6.16)$$

In this study,[12] a method of determining the extent of embedment by a simple ballpoint penetrometer test was described. From the results of tests on the depth of embedment D for a sphere of diameter d_{pa} under a load W_p, and by the geometry shown in Fig. 6.2,

$$d_i = 2(Dd_{pa} - D^2)^{1/2} \quad ; \quad . \quad . \quad . \quad (6.17)$$

and the constants m and B can be obtained from a plot of d_i/d_{pa} vs W_p/d_{pa}^2.

Huitt and McGlothin[12] showed that

$$W_f = d_{pa}\left[1 - B\left(\frac{W_p}{d_{pa}^2}\right)^m\right]^{1/2} \quad . \quad . \quad . \quad (6.18)$$

If the load W_p placed on each propping agent particle is replaced by the effective overburden pressure (psi) divided by the density pattern n_{pa} (number of propping agent particles per square inch), there results an equation for calculating the fracture width for the case of a partial monolayer.

$$W_f = d_{pa}\left\{1 - B\left[\frac{p_{OBe}}{(n_{pa}\, d_{pa})^2}\right]^m\right\}^{1/2} \quad . \quad . \quad . \quad (6.19)$$

Thus, by using Eqs. 6.15 and 6.19, the fracture flow capacity in a particular formation can be predicted for a given set of conditions, such as overburden pressure and size and density pattern of a propping agent.

6.3 Embedment Pressure

The mechanics of fracture propping is essentially one of either supporting the upper face of a horizontal fracture or preventing the closure of a vertical fracture by means of small spherically shaped particles distributed between the fracture faces. The degree of embedment depends upon the load applied to the proppant (effective overburden), and upon the diameter and number of particles of propping agent per unit area of the fracture. The high overburden loads and the sphericity of the particles impose very high stresses on formations. To place a value on this phenomenon, a test procedure was devised for establishing the pressure required to embed a simulated fracture prop in a given formation.[7]

Test Procedure

In one method[8] of determining the embedment pressure for a rock, a high speed steel ballpoint 0.05 in. in diameter is used. This is mounted on the upper platen of a hydraulic testing machine that loads the specimen hydraulically (see Fig. 6.3). The preferred size of rock specimen is a piece of 3½-in.-diameter core, 6 or more inches long—the additional length being desirable so

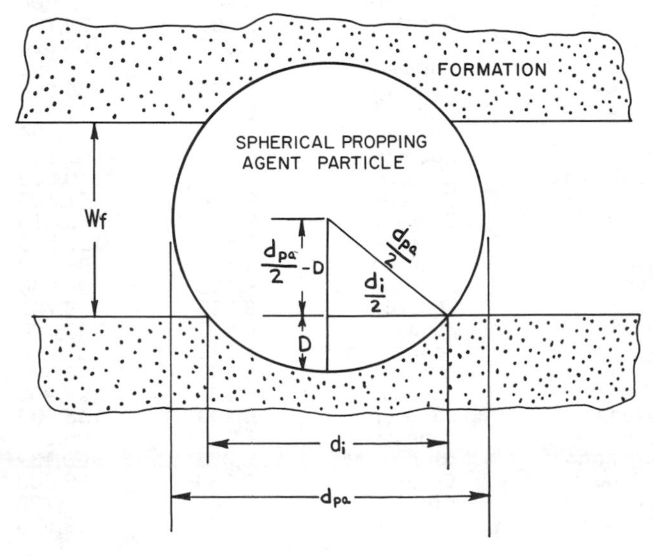

Fig. 6.2 Geometry of embedment.[1]

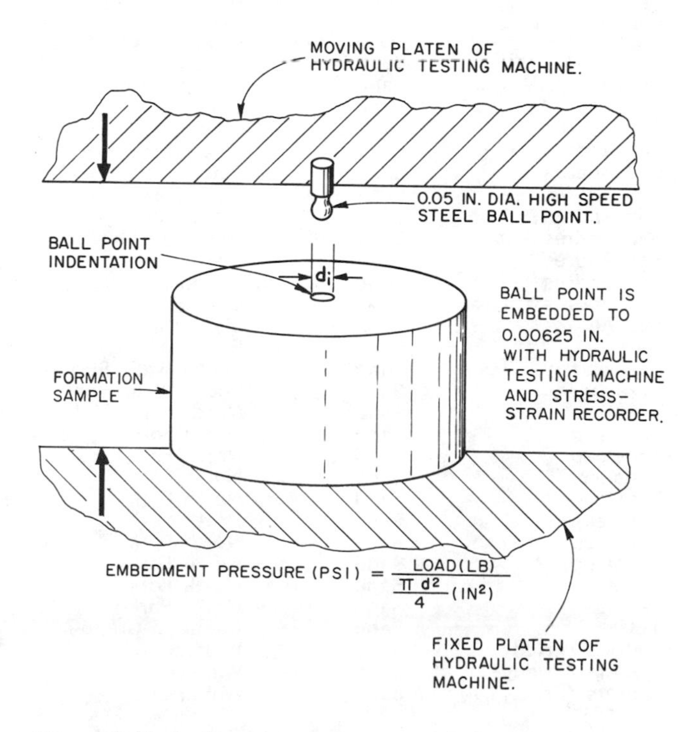

Fig. 6.3 Embedment pressure test technique.[8]

TABLE 6.2 — EMBEDMENT PRESSURE VALUES FOR VARIOUS FORMATIONS

Formation Name	Formation Type	Field Name	State or Country	County	Approximate Depth (ft)	Embedment Pressure (psi)
Annona Chalk	Carbonate	Pine Island	Louisiana	Caddo*	1,522	49,700
Atoka (Red Oak)	Sand	Red Oak	Oklahoma	Latimer	7,500	402,000
Atoka (Red Oak)	Sand	Red Oak	Oklahoma	Latimer	7,485	407,000
Atoka (Red Oak)	Sand	Red Oak	Oklahoma	Latimer	7,539	200,780
Basal Glauconite	Sand	Lobstick	Canada	Alberta**	5,900	229,700
Bromide	Sand	Wildcat	Oklahoma	Bryan	6,904	246,100
Bromide	Sand	Wildcat	Oklahoma	Bryan	6,918	173,500
Bromide	Sand	Wildcat	Oklahoma	Bryan	6,924	176,200
Bromide	Sand	Wildcat	Oklahoma	Bryan	6,915	187,400
Camerina	Sand	Big Lake	Louisiana	Camaron*	13,265	73,900
Cardium	Sand	Pembina	Canada	Alberta**	5,175	124,500
Cardium	Sand	Pembina	Canada	Alberta**	5,175	51,800
Cardium	Sand	Pembina	Canada	Alberta**	5,175	111,800
Cardium	Sand	Pembina	Canada	Alberta**	5,175	55,200
Cardium	Sand	Pembina	Canada	Alberta**	5,175	51,100
Cardium	Sand	Pembina	Canada	Alberta**	5,175	98,700
Cardium	Sand	Pembina	Canada	Alberta**	5,175	106,800
Cardium	Sand	Pembina	Canada	Alberta**	5,176	81,880
Cardium	Sand	Pembina	Canada	Alberta**	5,194	82,000
Cardium	Sand	Pembina	Canada	Alberta**	5,195	98,170
Cardium	Sand	Pembina	Canada	Alberta**	5,196	137,900
Cardium	Sand	Pembina	Canada	Alberta**	5,197	91,390
Cardium	Sand	Pembina	Canada	Alberta**	5,197	112,200
Crossfield	Lime	Crossfield	Canada	Alberta**	8,000	125,010
Fort Union	Sand	Wildcat	Wyoming	Fremont	10,000	38,200
Glauconite	Sand	Pembina Lobstick	Canada	Alberta**	5,948	85,700
Graneros	Sand	Jicarilla	N. Mexico	Rio Arriba	6,915	214,700
Grayburg	Lime	Hobbs	N. Mexico	Lea	4,100	312,600
Grayburg	Lime	Hobbs	N. Mexico	Lea	4,100	309,800
Grayburg	Lime	Hobbs	N. Mexico	Lea	4,100	333,200
Grayburg	Lime	Hobbs	N. Mexico	Lea	4,100	259,300
Grayburg	Lime	Hobbs	N. Mexico	Lea	4,100	350,300
Grayburg	Lime	Midland Farms	Texas	Andrews	4,800	138,180
Grayburg	Lime	Midland Farms	Texas	Andrews	4,800	112,600
Grayburg	Lime	Midland Farms	Texas	Andrews	4,800	148,300
Mesa Verde	Sand	Beaver Creek	Wyoming	Fremont	3,570	253,300
Mesa Verde	Sand	Beaver Creek	Wyoming	Fremont	3,570	252,100
Mesa Verde	Sand	Big Piney	Wyoming	Sublett	3,185	61,633
Mississippian Lime	Lime	Wildcat	Canada	Alberta**	11,000	279,900
Mississippian Lime	Lime	Wildcat	Canada	Alberta**	11,000	433,600
Mississippian Lime	Lime	Wildcat	Canada	Alberta**	11,000	190,600
Mississippian Lime	Lime	Wildcat	Canada	Alberta**	11,000	129,700
Mississippian Lime	Lime	Wildcat	Canada	Alberta**	11,000	111,700
Paluxy	Sand	Collins	Mississippi	Covington	12,417	57,600
Paluxy	Sand	Collins	Mississippi	Covington	12,419	73,600
Paluxy	Sand	Collins	Mississippi	Covington	12,532	62,160
Paluxy	Sand	Collins	Mississippi	Covington	12,535	45,500
Paluxy	Sand	Collins	Mississippi	Covington	12,536	104,600
Phosphoria	Dolomite	Beaver Creek	Wyoming	Fremont	10,513	341,000
Rodessa	Sand	Collins	Mississippi	Covington	13,762	328,300
Rodessa	Sand	Collins	Mississippi	Covington	13,764	388,900
Rodessa	Sand	Collins	Mississippi	Covington	13,767	117,600
San Andres	Lime	Slaughter	Texas	Cochran	5,000	231,300
San Andres	Lime	Slaughter	Texas	Cochran	5,000	193,130
San Andres	Lime	Slaughter	Texas	Cochran	5,000	102,800
San Andres	Lime	Slaughter	Texas	Cochran	5,000	139,700
San Andres	Lime	Slaughter	Texas	Cochran	5,000	266,000
San Andres	Lime	Devonian	Texas	Gaines	5,450	251,500
San Andres	Lime	Devonian	Texas	Gaines	5,450	176,600
San Andres	Lime	Devonian	Texas	Gaines	5,450	161,900
San Andres	Lime	Devonian	Texas	Gaines	5,450	130,000
San Andres	Lime	Slaughter	Texas	Cochran	5,058	301,820
Spiro	Sand	Red Oak	Oklahoma	Latimer	12,324	450,000
Sprayberry	Sand	South Wells Spraberry	Texas	Dawson	9,200	161,420
Sprayberry	Sand	South Wells Spraberry	Texas	Dawson	9,271	166,610
Tensleep	Sand	North Fork	Wyoming	Johnson	6,540	196,600
Tensleep	Sand	North Fork	Wyoming	Johnson	6,550	93,800
Tensleep	Sand	North Fork	Wyoming	Johnson	6,563	98,600
Tensleep	Sand	North Fork	Wyoming	Johnson	6,565	113,700
Tensleep	Sand	North Fork	Wyoming	Johnson	6,500	136,300
Tensleep	Sand	North Fork	Wyoming	Johnson	6,642	321,700
Tensleep	Sand	North Fork	Wyoming	Johnson	6,700	156,300
Tensleep	Sand	North Fork	Wyoming	Johnson	6,700	199,900
Tensleep	Sand	North Fork	Wyoming	Johnson	6,700	135,900
Tensleep	Sand	North Fork	Wyoming	Johnson	6,700	106,100
Tuff	Sand	Canadon Grande	Argentina	Chubut***	6,085	20,132
Walters (Priddy)	Sand	Walters	Oklahoma	Cotton	2,180	30,900
Walters (Priddy)	Sand	Walters	Oklahoma	Cotton	2,189	17,700
Walters (Priddy)	Sand	Walters	Oklahoma	Cotton	2,202	13,500

*Parish
**Province
***Town

that fracture capacity tests may also be run. However, if this size is not available, a smaller piece of rock may be substituted. Experience has indicated that smaller pieces will break more readily when loaded with the steel ballpoint. The rock cores should be taken, if possible, from the formation and the well that are to be hydraulically fractured. The rock specimen with two parallel faces, end face up, is placed on the lower platen of the tester. With the hydraulic system in operation, the ballpoint is brought into contact with the rock specimen, and the steel ballpoint is embedded to a depth of 0.0125 in. A strain recorder is used to denote the test results. When the desired embedment is reached, the load is recorded. At least three indentations are made on the test specimen about ½ in. or more apart. The final step in the test is to inspect the indentation in the rock core specimen under a microscope and measure its diameter in the plane of the rock surface. This diameter is used to calculate a "projected" area, which is then divided into the load to obtain the embedment pressure. This relationship is

$$\text{Embedment pressure (psi)} = \frac{4W_p}{\pi d_i^2} \quad . \quad . \quad . \quad (6.20)$$

Table 6.2 shows the embedment pressures for 22 different formations. Another method of obtaining values for rock embedment strength is presented in a paper by McGlothlin and Huitt.[19] Their procedure involves crushing proppant particles between plates of the same penetration hardness. The load required to crush the particle if a standard contact area exists at the point of crushing is calculated.

Embedment pressures observed in laboratory tests have varied from 13,000 to 527,000 psi. At the low end of this range, a hard, brittle proppant will embed in the core and create a larger area of contact for supporting the load, while at the high end of the range a hard proppant will more nearly retain a point contact. Experiments with sand — a hard, brittle proppant — have shown that at the same overburden pressure it will shatter when it comes in contact with high-embedment-pressure rock and will reduce the fracture capacity to near zero. Experiments with very high strength proppants such as steel shot maintain almost a point contact on rocks with higher embedment pressure. Sand normally will support the load and will yield reasonably high fracture capacity while in contact with a low-embedment-pressure rock.

Propping agents such as nutshells or plastic embed in the core a negligible amount and improve their load-carrying capacity by flattening slightly to increase the area of contact with the formation. This tendency to flatten under loading, however, makes the concentration and spacing of such proppants critical. If the concentration is too high and the spacing is not uniform, the particles will flatten under high overburden load and may touch each other and reduce or close the flow channels in the fracture. In instances where effective overburden does not exceed the combined strength of the supporting proppant, the conductivity of the fracture may be high.

In cases where the overburden is great enough to flatten nutshells excessively, a stronger material such as high strength glass, aluminum or steel should be used. Metallic propping materials also flatten under the influence of high overburden loads, but because of their higher compressive strength, they will withstand greater loads before the individual particles touch and block the flow channels. A concentration high enough to provide the required strength to support the overburden, and a distribution of the proppant particles uniform enough to insure support of the entire fracture area are required to obtain the highest fracture capacity.

The effect of embedment pressure on fracture capacity for various types, grades and sizes of proppants is presented later in this chapter.

6.4 Fracture Conductivity

The productivity of a well following fracturing is directly proportional to both the fracture-flow-capacity: formation-flow-capacity contrast (flow capacity of the fracture in md-ft divided by the flow capacity of the formation in md-ft) and the extent of the fractured area. The fracture-flow-capacity:formation-flow-capacity contrast, in turn, is directly proportional to the fracture flow capacity, since the formation thickness and permeability are fixed.

To illustrate the effect of this ratio of fracture flow capacity to formation flow capacity, refer to Fig. 6.4, in which it is plotted vs production increase from fracturing, with fracture penetration as a parameter. For example, a fracture-flow-capacity:formation-flow-capacity contrast of 10 and a fracture penetration of 40 percent will yield a fivefold production increase. Increasing the contrast to 100 yields an eightfold increase in well production with the same fracture penetration. Fig. 6.4 shows that very little additional production increase is achieved by fracture:formation-capacity contrasts higher than 100.[9,10]

The fracture's fluid carrying capacity is controlled by the fracture width, the proppant distribution, and the proppant concentration. The post-fracture width is controlled by the size of the prop that was placed during fracturing operations, the characteristics of the formation, the proppant strength, and the well depth. The second factor, distribution of the propping agent, is not easily controlled. Concentration of proppant, however, can be controlled by the use of pelletized spacer materials.[11]

To insure proper concentration and distribution, the propping agent may be diluted with another material that is similar in particle size, shape, and density, and that is soluble in well fluids. The diluent occupies space in the fracture and prevents the proppant from being deposited in a full monolayer. After fracturing, the spacer material is dissolved, leaving the proppant in a sparse distribution with wide channels throughout the fracture so that hydrocarbons can flow into the well.

Test Procedure

The fracture capacity effected by a given propping material can be determined in the laboratory by sandwiching the proppant between field core sections that are subjected to a simulated overburden pressure and temperature. With this method, the core sections (3½ in. in diameter and 2 in. long) are mounted in steel cups, using a controlled-melting-point alloy to confine the core so that about ¼ in. of the smooth core faces protrudes above the top edge of the cup. A 3/16-in.-diameter hole is drilled in the upper core half from the outside circumference of the core to the center and a tubing fitting is attached at the outside (see Fig. 6.5).

Overburden pressure on the two core halves, with propping material in place, is simulated by means of a hydraulic jack. Temperature is controlled by placing the two core halves in a heated box.

The fracture capacity is determined by flowing nitrogen gas from the hole in the center of the upper core half through the simulated fracture containing the propping material. The system used to measure the nitrogen is shown in Fig. 6.6. When a spacer is used with the proppant, the spacer is dissolved by flowing kerosene or water through the fracture before the nitrogen is introduced.

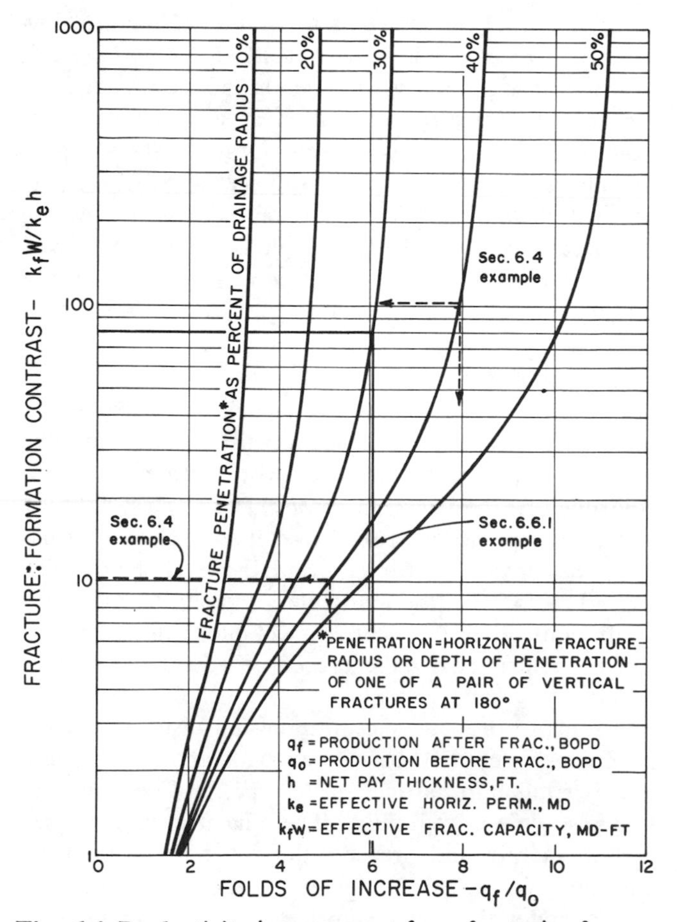

Fig. 6.4 Productivity improvement from fracturing.[9]

For quick, short-term tests, the propping material is placed on the core face in a desired layer fraction. The emplaced proppant is then subjected to overburden pressure at selected increments and a nitrogen flow measurement is taken after each increment has been reached. This type of test is completed within a few hours and serves as a method of screening propping materials.

To study the effect of time, the propping material is placed under the desired overburden load and left for 30 days. Nitrogen flow readings are taken at least once a week to measure the change in fracture capacity. These are referred to as long-term tests.

Effect of Packed Fracture

The design of hydraulic fracturing treatments requires the calculation not only of the areal extent of the fracture, but also of the effective flow capacity of the induced fracture system. Experimental work on the effect of proppant concentration on fracture capacity has demonstrated that partial monolayers of proppants, when they are used in high enough concentrations and when they are strong enough to support the overburden, afford high flow conductivity fractures.

Uniform or partial monolayer distribution of the propping agent is difficult to obtain. Proppant spacers and high pump rates are required throughout the entire fracture. Even under the most favorable circumstances, islands of proppants tend to form in horizontal fractures. These islands may be one layer thick or they may be several layers thick depending on rate of displacement, location in the fracture, and the treating procedure employed. Frequently, fracturing operations are halted to encourage deposition of the propping agent in the vicinity of the well. This is to insure maximum fracture width in that critical area.

Studies of proppant placement in vertical fractures have shown that the propping material tends to settle to the bottom of the fractures. The bottom portion of the fractured zone becomes packed with multilayers of propping material, while the top portion of the fracture contains little or no proppant. The number of layers of particles will depend upon the size, shape and concentration of the particles in the fluid, the width of the fracture during deposition of the proppant, and the rate at which the fracturing fluid is flowing over the fill.

When the fracturing process is complete, the fracture tends to close. The final width of the packed fracture depends upon its width at the end of the treatment and upon the effective overburden load. The conductivity of the fracture system is determined by this final fracture width and by the effective permeability of the propping material.

Experimental work was conducted in shallow wells with simulated horizontal and vertical fracture systems. These tests showed that fracture width during placement of the proppant was directly related to injection rate and to fluid-loss characteristics of the fracturing

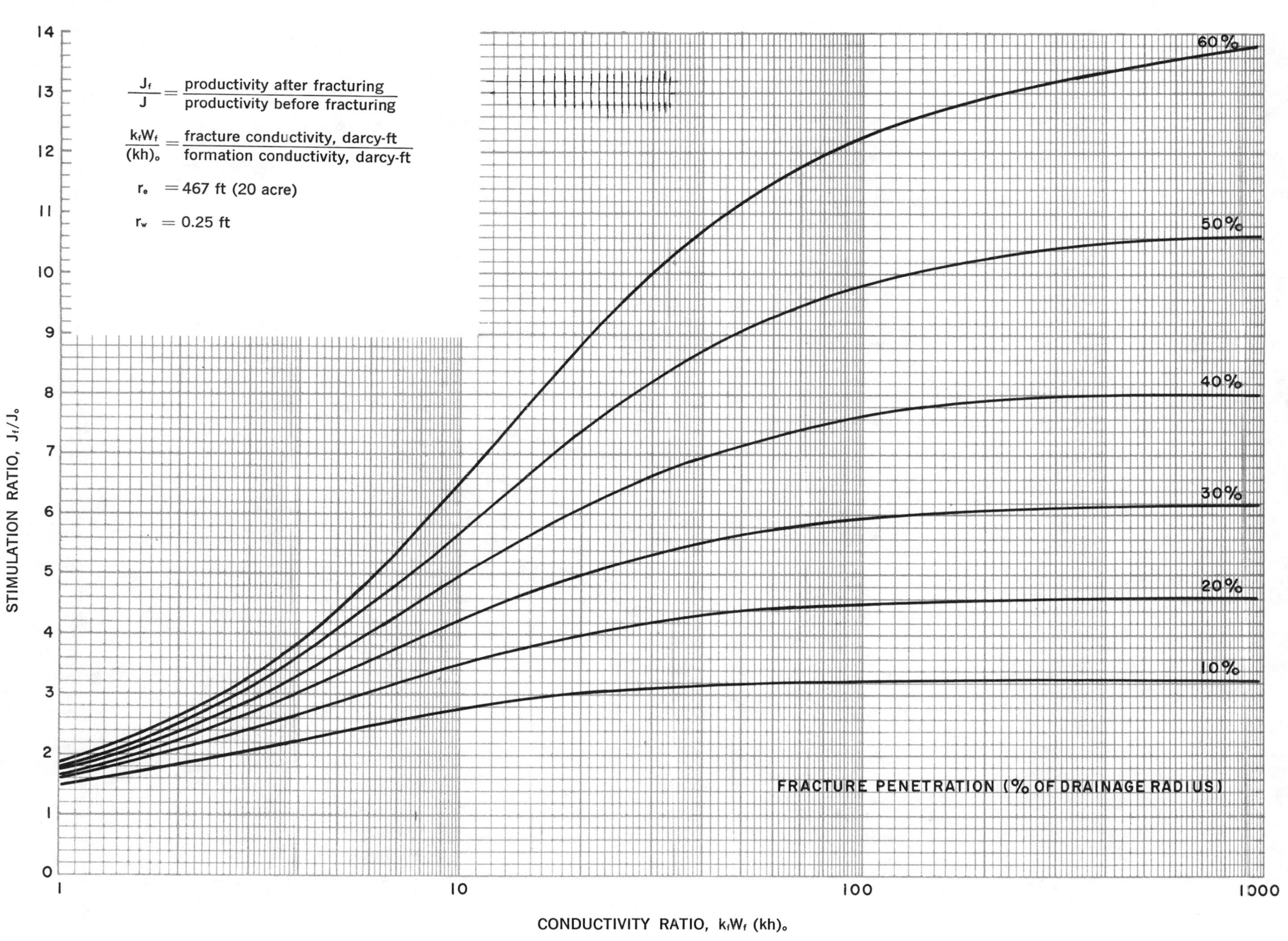

Fig. 6.4A Estimated production increase after fracturing (horizontal fractures).

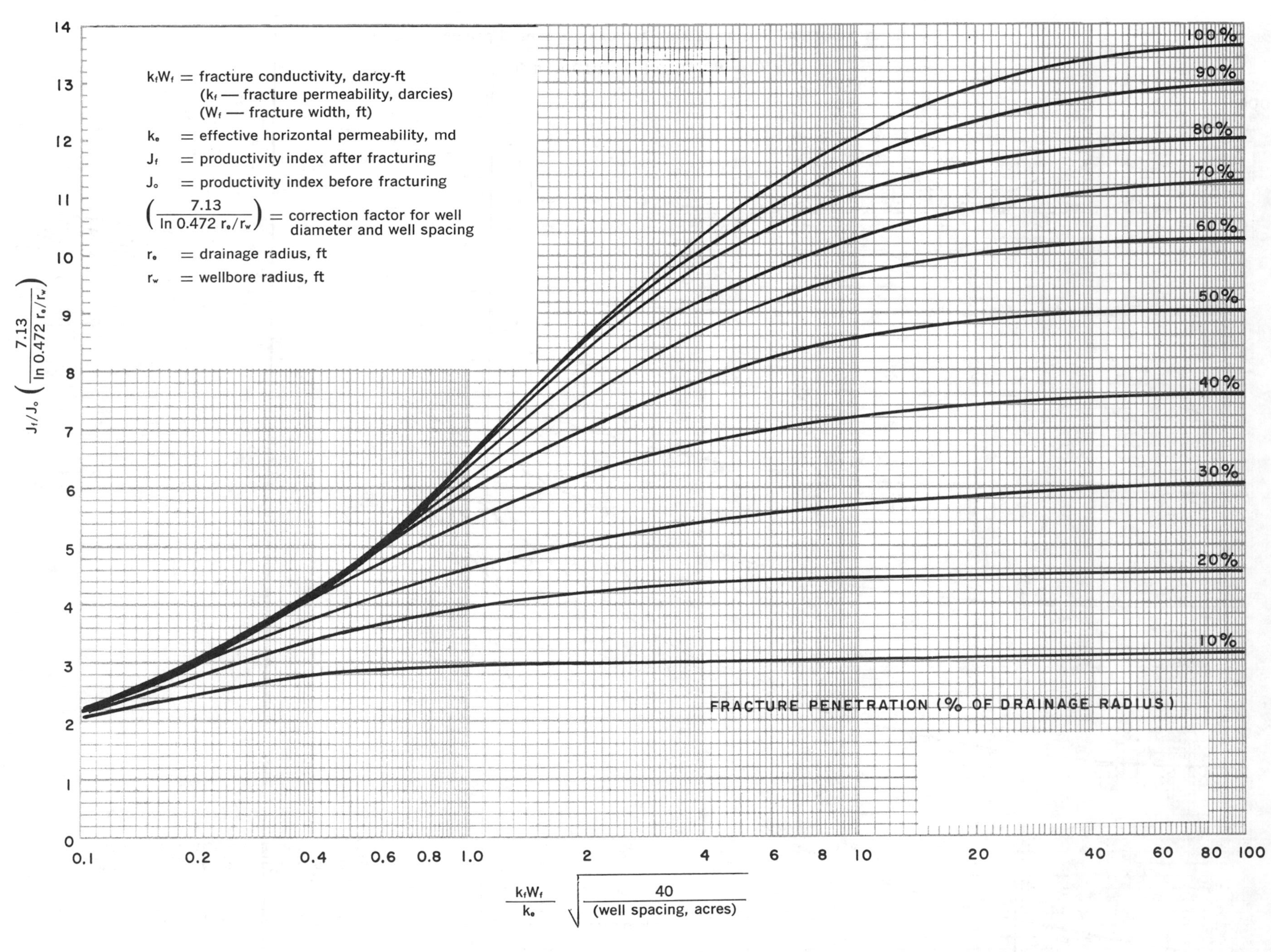

Fig. 6.4B Estimated production increase after fracturing (vertical fractures). After Dyes *et al.*[10]

agent; i.e., high rates and low fluid loss or high viscosity are required for wide fractures. The same high displacement rates and low fluid loss or high viscosity greatly enhance maximum distribution of the proppant. However, a high concentration of propping agent (greater than 3 lb/gal) in the fracturing fluid tends to allow the proppant to settle in multiple layers in the fracture. Typical results of such tests with a simulated horizontal fracture are shown in Fig. 6.7. These tests suggest that the layers, or packs, of propping agent in a horizontal fracture probably will not exceed three, even under the most favorable conditions. The number of layers of proppant in a vertical fracture system is governed by the fluid viscosity, pad volume pumped, the prespaced distance between the surfaces of the fracture rather than the proppant concentration in the fracturing fluid, and the injection rate.

To study the effect of multiple layers of proppant on fracture conductivity, a series of long-term (30-day) tests was run. The results of these tests are presented on Table 6.3, which indicates that as long as the proppant does not break or excessively deform, fracture flow capacity will continue to increase; and, at the three or four layer level, the flow conductivity or capacity will generally reach or exceed that of a partial monolayer.

Field tests have demonstrated that the multiple layer packed fracture in the vicinity of the wellbore is an effective device not only for increasing well production, but also for preventing the inflow of proppant into the well during post-fracture production.

It should be noted that very large proppants will accomplish the same objective as multiple layers of small proppants. Field practice shows, however, that it is much easier to pump the smaller size propping agent and effect a screenout than to handle the larger material. The size of the perforations through which the propping agent is pumped becomes a critical factor as the diameter of the proppant increases. Another factor that has restricted the use of the special, larger diameter prop is the difficulty the oil field service companies have had in obtaining large-sized proppants of the required sphericity and purity.

Effect of Partial Monolayer Spacing

Laboratory tests of fracture capacity have demonstrated that greater flow is obtained by placing high strength proppants in a partial rather than a full mono-

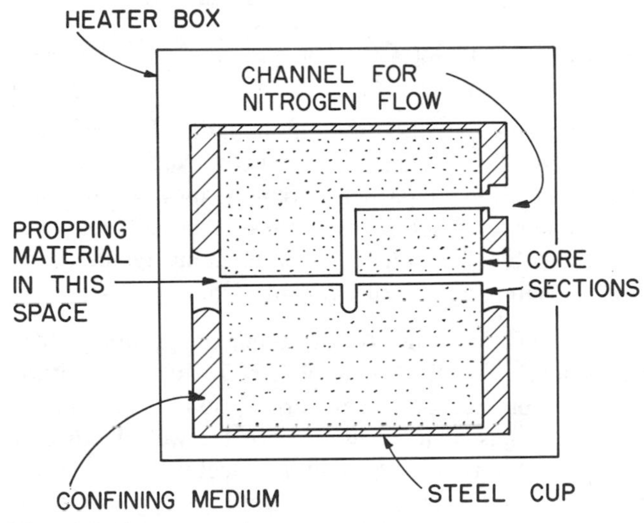

Fig. 6.5 Core mounting assembly.

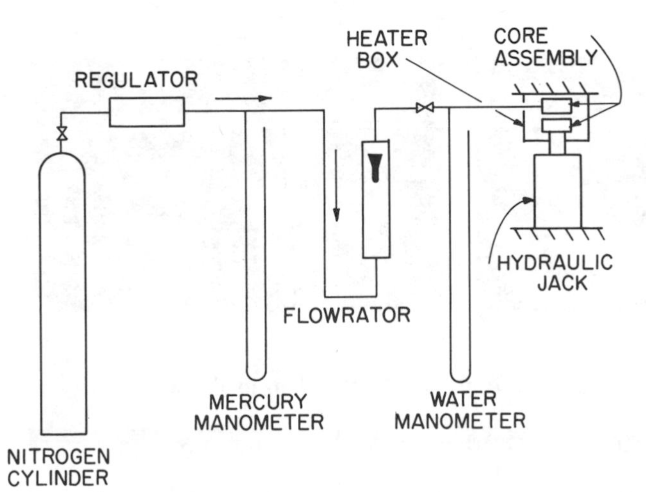

Fig. 6.6 Nitrogen flow system for measuring fracture capacity.

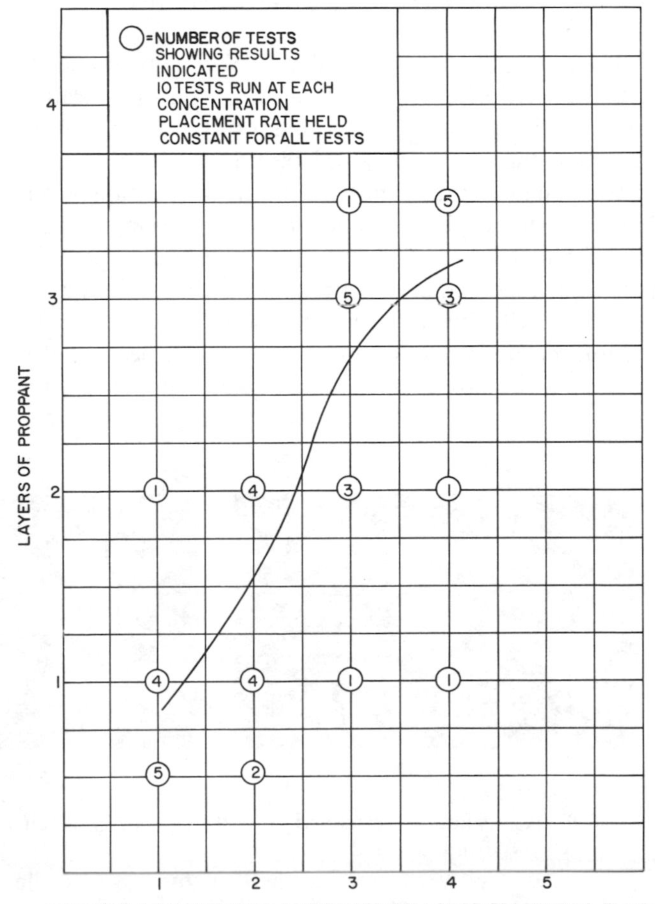

Fig. 6.7 Simulated horizontal fracture with fracture surfaces spaced five proppant diameters apart.

TABLE 6.3 — LONG TERM, MULTILAYER FRACTURE CAPACITY TESTS
(Formation Embedment Pressure — 150,000 to 250,000 psi)

Well Depth (ft)	Long Term Fracture Capacity (md-ft), Monolayers of Proppant ¼	½	1	2	3	4	Type of Proppant and Mesh Size (U.S. Standard Sieve)	
2,000	—	2,500	2,000	2,300	3,700	4,100	Sand	(−20+40)
4,000	—	1,250	750	1,100	1,800	2,000	Sand	(−20+40)
4,000	—	26,000	14,200	25,500	31,000	33,000	Glass beads	(−12+20)
6,000	—	17,200	8,100	11,000	15,000	15,500	Glass beads	(−12+20)
8,000	—	9,500	6,100	9,000	10,000	11,000	Glass beads	(−12+20)
10,000	—	5,100	4,100	5,600	7,800	8,500	Glass beads	(−12+20)
10,000	25,000	—	3,700	5,800	7,400	8,000	Metallic	(−12+16)
12,000	19,000	—	3,200	4,200	6,000	7,200	Metallic	(−12+16)
14,000	18,100	—	1,200	1,800	2,200	3,000	Metallic	(−12+16)

layer in the fracture. This is particularly evident with a deformable proppant such as nutshells or plastics. If, during fracturing, the deformable prop is packed in the fracture in a full monolayer or in multiple layers under conditions in which the weight of the overburden exceeds its strength, the proppant will be pressed into a tightly packed layer when the full load is applied at the completion of fracturing operations (see Photograph A on Fig. 6.8). Such a tightly packed layer of proppant results in less fluid conductivity than is obtained with a partial layer, such as that shown in Photograph C on Fig. 6.8. Nutshells, for example, that are uniformly spaced in a ¼ monolayer in a 7,000-ft well will have a fracture capacity three times greater than that of a full monolayer.

Table 6.4 shows the increase in fracture capacity obtained by using ¼ monolayer of nutshells rather than a full layer to prop formations at various depths. The ¼ monolayer is obtained by using only 25 percent of the weight of proppant required for a full monolayer. However, the optimum concentration of proppant may vary from 0.1 to 0.5 monolayer, depending upon the particular type and size of proppant used, the formation and the well depth.[8] Concentrations lower than 0.2 monolayer are generally not recommended because it is difficult to use so small a quantity of proppant in the field.

TABLE 6.4 — LONG TERM FRACTURE CAPACITY WITH ROUNDED NUTSHELLS

Well Depth (ft)	Long Term Fracture Capacity* U.S. Std. −12+20 mesh: ¼ monolayer 0.06 lb/sq ft (md-ft)	U.S. Std. −12+20 mesh: Full monolayer 0.24 lb/sq ft (md-ft)	U.S. Std. −8+12 mesh: ¼ monolayer 0.08 lb/sq ft (md-ft)	U.S. Std. −8+12 mesh: Full monolayer 0.33 lb/sq ft (md-ft)
1,750	28,000	9,300	64,600	18,900
4,500	13,800	5,000	37,400	10,900
7,000	8,600	2,600	21,600	6,200
9,400	5,600	1,700	16,500	3,700
11,600	3,600	1,200	11,200	2,300

*"Long term" tests are tests conducted over a period of 30 days on cores having 199,900 psi embedment pressure.

Deformable metallic propping materials have characteristics similar to those of nutshells except that, because of their high particle strength, they will usually produce higher fracture capacities in deep wells than are obtainable with other material. Some typical fracture capacities are given in Table 6.5.

The nondeforming, brittle type of propping materials also generally yield higher fracture capacities, provided that the optimum partial monolayer concentration is adjusted to conform with the well depth and the type of formation.[8] Sand is effective in a partial monolayer concentration to a depth of approximately 5,000 ft. A full monolayer concentration is generally

A

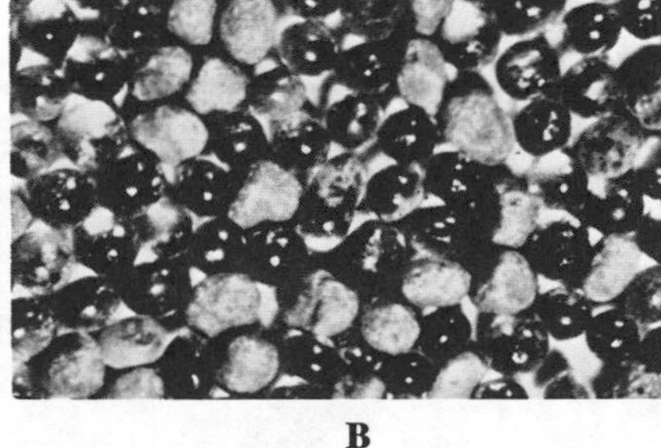

B

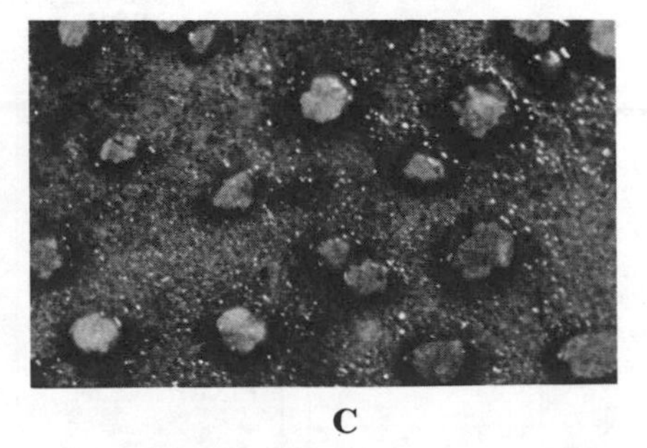

C

Fig. 6.8 **A.** View of restricted fracture's capacity when nutshells alone are used. Overburden load has been applied.

B. View of resin-nutshell mixture in place in the fracture. Overburden load has not been applied, hence "spherical" shape of nutshells.

C. Stage two of resin-spacer technique. The resin has dissolved and load has been applied. Although nutshells are now deformed, unobstructed passages remain for flow of formation fluid through fracture.

usable to 7,000 ft. Below 7,000 ft, strong glass beads can be used to advantage in properly designed partial monolayers. Typical fracture capacities for these materials are given in Tables 6.6 and 6.7.

These data demonstrate the desirability of partial monolayer concentrations of proppants and establish the need for a spacer material for fracture propping agents to insure optimum proppant distribution.

Propping Agent Spacers

The specific gravities of the various proppants are

Sand	2.7
Aluminum	2.7
Glass Beads	2.7
Nutshells	1.4
Most Plastics	1.1

Because the spacer material should have the same specific gravity as the proppant, three different specific gravities are needed for spacer materials. The spacer should be transportable and essentially insoluble in the fracturing fluid used (water-base or oil-base), yet it should have the quality of being easily removed from the fracture by injected solvents or by oil or water produced by the formation. The spacer must lend itself to prilling, must resist breakage while being pumped into the fracture, and must be capable of being stored and handled in the field. Three materials meeting these requirements have been developed.

Urea (NH_2CONH_2) is soluble in water and insoluble in oil, and therefore must be used with an oil-base fracturing fluid. It is dissolved from the fracture by produced water or by water pumped into the fracture at the end of the job. The specific gravity of urea is 1.3, approaching that of nutshells, and it can be readily prilled (spherical particle) and screened into mesh sizes suitable for use as a spacer. The breakage of particles during pumping has been found to be less than 10 percent. The solubility time of prilled urea in water or brine at 80°F is shown on Fig. 6.9.

A *hydrocarbon resin* that is highly soluble in naphthenic and aliphatic hydrocarbons has been developed. The resin is a low molecular weight copolymer of a vinyl aromatic and a cyclic diolefin that can be transported in a water-base fracturing fluid and dissolved from the fracture with produced crude oil. Its specific gravity is 1.1, making it suitable for use with nutshells or plastic props. A special process was developed for prilling this resin so that particles could be made with a roundness of greater than 0.9 Krumbein. Screening produces the desired size ranges that are compatible with proppants used in the field. The solubility time of 20 percent weight concentration of the prilled resin is 2 hours in 80°F kerosene, and 1 hour in diesel oil (see Fig. 6.10). Figs. 6.11 through 6.13 show the solubility characteristics of this resin in three types of crude oil. Solubility rates similar to that in kerosene were obtained in all of the crude oils tested. The breakage of this resin in field pumping operations is about 2 percent.

Sodium bisulfate has a specific gravity of 2.7, matching that of sand, glass beads, and aluminum. The rapid

TABLE 6.5 — LONG TERM FRACTURE CAPACITY WITH A DEFORMABLE METALLIC PROPPANT

	Long Term Fracture Capacity*			
	U.S. Std. —12+16 mesh**		U.S. Std. —8+12 mesh†	
Well Depth (ft)	¼ monolayer 0.10 lb/sq ft (md-ft)	Full monolayer 0.42 lb/sq ft (md-ft)	¼ monolayer 0.16 lb/sq ft (md-ft)	Full monolayer 0.65 lb/sq ft (md-ft)
7,000	36,600	4,900	188,400	16,300
9,400	26,100	3,800	131,900	13,800
11,600	19,500	3,100	88,900	11,300
14,000	18,200	1,100	36,800	10,600

*30-day tests.
**196,600 psi embedment pressure core.
†252,100 psi embedment pressure core.

TABLE 6.6 — LONG TERM FRACTURE CAPACITY WITH STRONG GLASS BEADS

	Long Term Fracture Capacity*			
	U.S. Std. —12+20 mesh		U.S. Std. —8+12 mesh	
Well Depth (ft)	½ monolayer 0.19 lb/sq ft (md-ft)	Full monolayer 0.37 lb/sq ft (md-ft)	½ monolayer 0.23 lb/sq ft (md-ft)	Full monolayer 0.46 lb/sq ft (md-ft)
4,000	26,000	14,200	34,000	23,000
7,000	16,500	7,800	25,000	16,900
9,400	9,300	5,400	13,100	9,100
11,600	4,600	3,500	6,000	5,300

*30-day tests conducted on cores having 135,900 psi embedment pressure.

TABLE 6.7 — LONG TERM FRACTURE CAPACITY WITH SAND

	Long Term Fracture Capacity*			
	U.S. Std. —20+40 mesh		U.S. Std. —10+20 mesh	
Well Depth (ft)	½ monolayer 0.1 lb/sq ft (md-ft)	Full monolayer 0.19 lb/sq ft (md-ft)	½ monolayer 0.17 lb/sq ft (md-ft)	Full monolayer 0.35 lb/sq ft (md-ft)
2,000	2,600	2,100	3,000	2,000
5,000	1,100	800	800	700
7,000	—	140	—	165

*30-day tests conducted on cores having 192,600 psi embedment pressure.

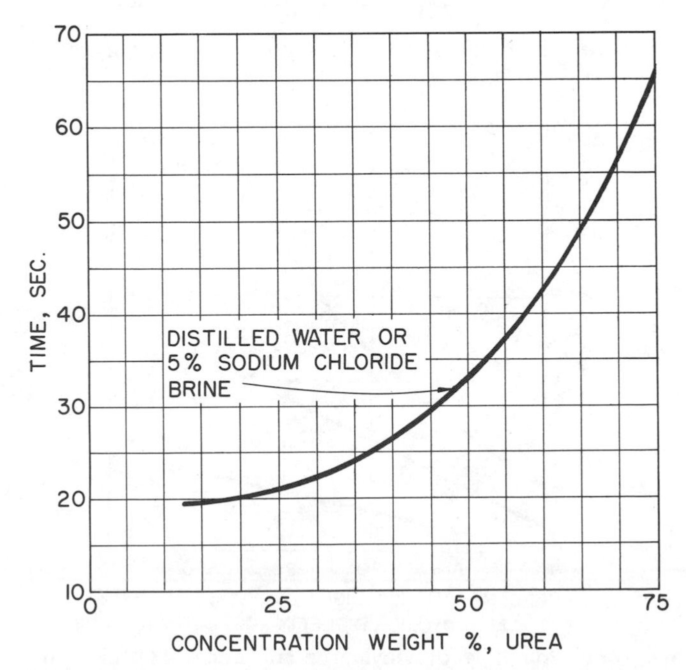

Fig. 6.9 Approximate solubility of prilled urea at 80°F.

solubility of sodium bisulfate in water means that it must be used in an oil-base fracturing fluid. Water produced from the formation or water injected after fracturing removes the sodium bisulfate spacer. The globular particles can be screened to the desired sizes, and particle breakage during pumping and handling is less than 10 percent. The rate of solution of sodium bisulfate in water or brine at 80°F is shown on Fig. 6.14. This material is not generally recommended for use in carbonate reservoirs because calcium sulfate scale may form.

Specifications for the commonly used propping agents and spacers are tabulated in Tables 6.8 through 6.10.

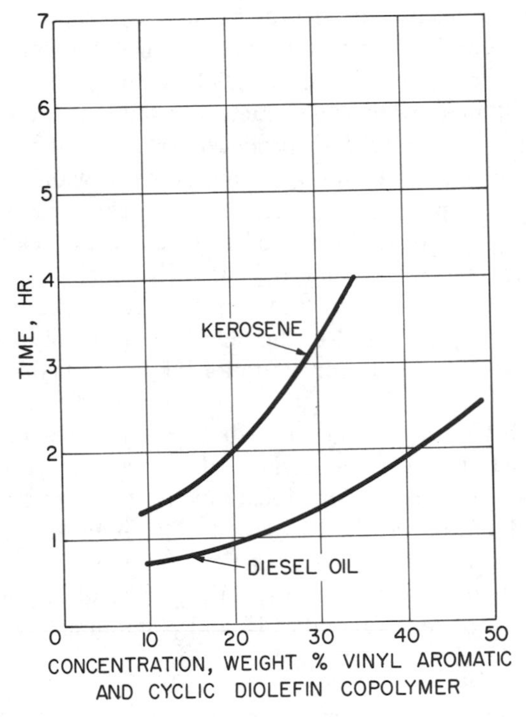

Fig. 6.10 Solubility of vinyl aromatic and cyclic diolefin copolymer at 80°F.

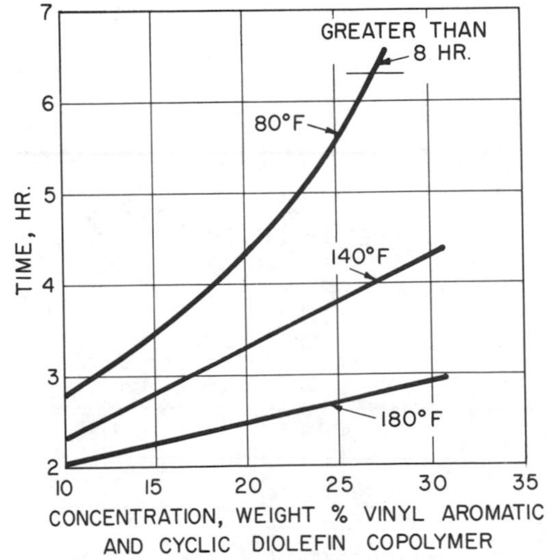

Fig. 6.11 Solubility of vinyl aromatic and cyclic diolefin copolymer in asphaltic crude.

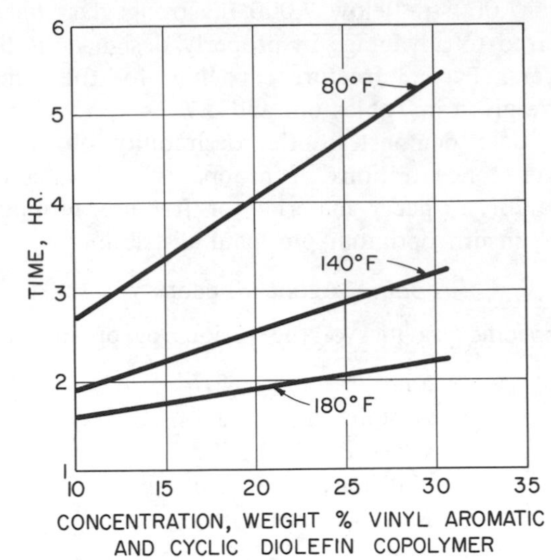

Fig. 6.12 Solubility of vinyl aromatic and cyclic diolefin copolymer in naphthenic crude.

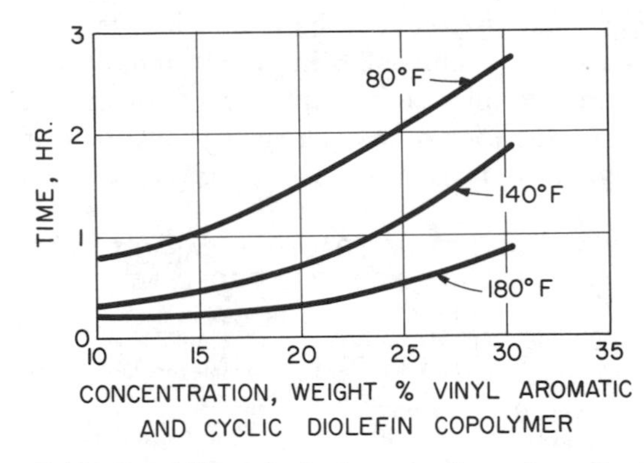

Fig. 6.13 Solubility of vinyl aromatic and cyclic diolefin copolymer in paraffinic crude.

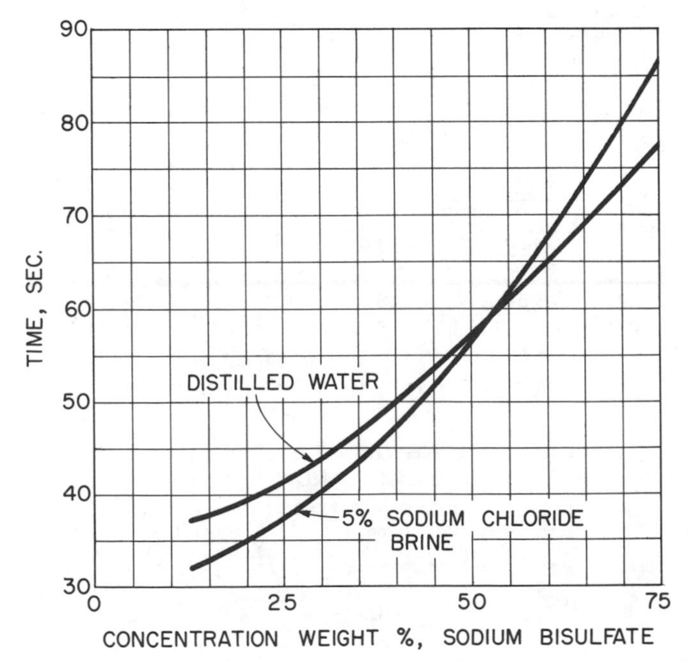

Fig. 6.14 Approximate solubility of globular sodium bisulfate at 80°F.

Note the close match in density and particle size distribution. Laboratory and field experience has indicated the need for matching these properties in proppant and spacer.

Spacer Field Test Results

To illustrate results of field usage of spaced proppant, Tables 6.11 and 6.12 present data from 35 treatments using hydrocarbon resin and eight treatments using prilled urea to space nutshell proppant. For comparative purposes, production data from 16 offset wells are presented.

An average of 8,600 lb of hydrocarbon resin spacer was used per treatment, at a weight ratio of 3 lb of resin to each pound of nutshell for most of the treatments. Due to the density difference between the nutshells and the resin, the 3:1 weight ratio results in a volume ratio of approximately 4:1, so that a 0.2 monolayer of nutshell propping material could be achieved.

Thirty-one wells fractured using resin spacer had an average potential of 261 BOPD, which is 44 BOPD higher than the average potential for 16 offset wells that were conventionally fractured. On 10 wells treated with

TABLE 6.8 — NUTSHELL, PLASTIC PROPPING AGENT, AND HYDROCARBON SPACER SPECIFICATIONS

	Hydrocarbon Resin Spacer	Nutshells	Plastic Proppants
Grain Size	—8+12 nominal, 100% —6+16 with 95% —8+12 mesh U.S. Std. Sieve	—8+12 nominal, 100% —6+20 with 95% —8+12 mesh U.S. Std. Sieve	—8+12 nominal, 100% —6+20 with 95% —8+12 mesh U.S. Std. Sieve
	—12+20 nominal, 100% —10+20 with 95% —12+20 mesh U.S. Std. Sieve	—12+20 nominal, 100% —10+40 with 95% —12+20 mesh U.S. Std. Sieve	—12+20 nominal, 100% —10+40 with 95% —12+20 mesh U.S. Std. Sieve
Average Krumbein Roundness	0.8 or higher	0.7 or higher	0.8 or higher
Minimum Oil Solubility	0.04 gm/ml of kerosene at 80° F	None	None
Specific Gravity	1.1	1.3 to 1.4	1.1
Particle Breakage	No more than 10% particle breakage when pumped in standard field fracturing pump at a concentration of 1 lb/gal	None	None

TABLE 6.9 — ALUMINUM, GLASS BEAD AND SAND PROPPING AGENT, AND SODIUM BISULFATE SPACER SPECIFICATIONS

	Sodium Bisulfate Spacer	Aluminum	Glass Beads	Sand
Grain Size	—8+12 nominal, 100% —6+16 with 95% —8+12 mesh U.S. Std. Sieve	—8+12 nominal, 100% —6+16 with 95% —8+12 mesh U.S. Std. Sieve	—8+12 nominal, 100% —6+16 with 95% —8+12 mesh U.S. Std. Sieve	—10+20 nominal, 100% —8+40 with 90% —10+20 mesh U.S. Std. Sieve
	—12+16 nominal, 100% —10+30 with 95% —12+16 mesh U.S. Std. Sieve	—12+16 nominal, 100% —10+30 with 95% —12+16 mesh U.S. Std. Sieve	—12+16 nominal, 100% —10+30 with 95% —12+16 mesh U.S. Std. Sieve	—20+40 nominal, 100% —16+60 with 90% —20+40 mesh U.S. Std. Sieve
	—20+40 nominal, 100% —18+60 with 95% —20+40 mesh U.S. Std. Sieve	—20+40 nominal, 100% —18+60 with 95% —20+40 mesh U.S. Std. Sieve	—20+40 nominal, 100% —18+60 with 95% —20+40 mesh U.S. Std. Sieve	
Average Krumbein Roundness	0.8 or higher	0.8 or higher	0.8 or higher	0.7 or higher
Minimum Water Solubility	0.25 gm/ml water at 80° F	None	None	None
Specific Gravity	2.7	2.7	2.7	2.7
Particle Breakage	No more than 10% particle breakage when pumped in standard field fracturing pump at a concentration of 1 lb/gal	None	None	None

TABLE 6.10 — NUTSHELL, PLASTIC PROPPING AGENT AND UREA SPACER SPECIFICATIONS

	Urea Spacer	Nutshells	Plastic Proppants
Grain Size	−8+12 nominal, 100% −6+16 with 95% −8+12 mesh U.S. Std. Sieve	−8+12 nominal, 100% −6+20 with 95% −8+12 mesh U.S. Std. Sieve	−8+12 nominal, 100% −6+20 with 95% −8+12 mesh U.S. Std. Sieve
	−12+20 nominal, 100% −10+20 with 95% −12+20 mesh U.S. Std. Sieve	−12+20 nominal, 100% −10+40 with 95% −12+20 mesh U.S. Std. Sieve	−12+20 nominal, 100% −10+40 with 95% −12+20 mesh U.S. Std. Sieve
Average Krumbein Roundness	0.8 or higher	0.7 or higher	0.8 or higher
Minimum Water Solubility	0.25 gm/ml water at 80° F	None	None
Specific Gravity	1.3	1.3 to 1.4	1.1
Particle Breakage	No more than 10% particle breakage when pumped in standard field fracturing pump at a concentration of 1 lb/gal	None	None

resin spacer where pretreatment well productivity data were available, the average production increased from 93 BOPD to 209 BOPD, for an average increase of 116 BOPD. The average potential was 291 BOPD for 21 wells fractured on completion using the resin-spaced nutshell technique; this was 51 BOPD higher than the 240 BOPD potential achieved with the conventional treatment of 13 new offset wells.

Table 6.12 lists the pertinent treating and production data available from eight wells in which prilled urea was used to space nutshells. An average of 11,250 lb of urea was used per treatment.

The increase in production as a result of the eight urea-spaced nutshell fracturing treatments averaged 50 BOPD. On seven wells where production data were available before treatment, the average production increase was from 21 BOPD to 77 BOPD.

6.5 Correlation of Propping Agent-Formation Characteristics

The importance of the influence of fracture capacity on productivity and the need for insuring proppant distribution have been established. Not to be overlooked, however, is an understanding of how fracture flow capacity is affected by such formation characteristics as embedment pressure, and by the type, grade, and size of the proppants. Each of the commonly used propping agents will be discussed. Plots of embedment pressure vs fracture capacity are presented on Figs. 6.15 through 6.24. The data derived from a series of coordinated embedment-pressure and fracture-capacity tests using cores from many different formations and using several different propping materials were correlated and a series of propping material selection charts (Figs. 6.15 through 6.24) was prepared.

Sand

Fracture-capacity flow tests were conducted over a period of 30 days, using sand as the proppant to establish a relationship between formation embedment pressure and proppant size. A summary of these tests is plotted on Fig. 6.15.

Sand is a brittle propping material that will embed in the formation if the formation embedment pressure is low. If the matrix has high embedment pressure, the sand either will support the overburden load with little embedment or will be crushed if the load-bearing capacity of the sand is exceeded. For example, Fig. 6.20 shows that −20+40 mesh sand used in wells 3,000 to 7,000 ft deep with embedment pressure greater than 150,000 psi, will effect very low ($<$ 500 md-ft) fracture capacities in single or partial layers. As illustrated by the data, sand is severely crushed when the formation embedment pressure is greater than 150,000 psi at a well depth of 7,000 ft. A full monolayer of sand will generally produce higher fracture capacities in wells 7,000 ft deep than will partial layers.

A well washed silica sand composed of particles having an average minimum Krumbein roundness of 0.7 and containing not more than 1 percent silt is most desirable. Silica sands may vary somewhat in crushing strength, but the strength variation of most well rounded, pure silica sands is not great enough to affect their use as proppants. However, sand crushing strength will be reduced with lower roundness or purity. Experimental work has shown that a narrow particle size distribution of sand propping material, such as −16+18 mesh (D) shown on Fig. 6.20, will produce higher fracture capacities than will the coarsely graded, commercially available particle size distributions (C). When the −16+18 mesh specially graded sand (0.3 lb/sq ft) is used as a proppant in the example 7,000-ft-deep well, fracture

TABLE 6.11 — FIELD TEST RESULTS, HYDROCARBON RESIN SPACER TREATMENTS

Location	Formation	Depth (ft)	Fracturing Fluid	Volume (gal)	Average Injection Rate (bbl/min)	Pounds of Propping Agent and Spacer Material						BOPD Before Treatment	Potential BOPD After Treatment
						20/40 Sand	10/20 Sand	12/20 Nutshells	12/20 Spacer	8/12 Nutshells	8/12 Spacer		
WYOMING — Field A													
Well 1	Muddy sand	4,800	Gelled water	16,400	16	5,000	5,000	—	—	300	900	New well	154
Well 2	Muddy sand	4,700	Gelled water	20,000	18	5,000	5,000	—	—	1,000	3,000	New well	175
Well 3	Muddy sand	4,900	Gelled water	20,000	20	5,000	5,000	—	—	1,000	3,000	New well	128
Well 4	Muddy sand	4,800	Gelled water	20,000	20	5,000	5,000	—	—	1,000	3,000	New well	243
Well 5	Muddy sand	4,700	Gelled water	14,000	20.5	2,500	3,500	—	—	2,000	6,000	New well	254
Well 6	Muddy sand	4,700	Gelled water	16,500	23.5	—	3,500	—	—	2,000	6,000	New well	21
Well 7	Muddy sand	4,700	Gelled water	20,000	23	2,500	3,500	—	—	2,000	5,000	New well	374
Offset Well 1	Muddy sand	4,800	Gelled water	11,000	18	5,000	5,000	—	—	—	—	New well	121
Well 8	Dakota sand	4,800	Gelled water	20,000	22	5,000	5,000	—	—	1,000	3,000	New well	382
Well 9	Dakota sand	4,800	Gelled water	20,000	16.5	5,000	5,000	—	—	1,000	3,000	New well	103
Well 10	Dakota sand	4,800	Gelled water	11,500	35.5	2,500	—	—	—	2,000	6,000	New well	135
WYOMING — Field B													
Well 11	"P" sand	2,700	Gelled water	6,000	22	—	—	—	—	1,000	2,500	38	75
Offset Well 2	"P" sand	2,700	Diesel oil	5,000	17.5	—	5,000	—	—	375	—	38	54
WYOMING — Field C													
Well 12	Tensleep sand	6,700	Gelled water	15,000	15	—	7,600	—	—	1,300	2,000	New well	320
Offset Well 3	Tensleep sand	6,900	Gelled water	31,000	13	28,000	7,000	—	—	—	—	New well	388
WYOMING — Field D													
Wells 13, 14 and 15	1st Wall Creek	900	Gelled water	10,000	40	—	—	—	—	2,450	7,350	140	300
Wells 16, 17 and 18	2nd Wall Creek	1,700	Gelled water	10,000	40	—	—	—	—	2,450	7,350	140	300
NEBRASKA — Field E													
Well 19	"J" sand	6,700	Gelled water	8,400	23	—	—	2,400	6,950	—	—	Flood not yet effective	
TEXAS — Field F													
Well 20	Fusselman	12,000	Gelled acid	50,000	70.4	5,000	—	—	—	4,500	13,500	New well	413
Offset Well 4	Fusselman	12,000	Gelled acid	50,000	72.9	37,500	—	—	—	—	—	New well	362
Offset Well 5	Fusselman	12,000	Gelled acid	50,000	N.A.	37,500	—	—	—	—	—	New well	152
Offset Well 6	Fusselman	12,000	Gelled acid	50,000	71	37,500	—	—	—	—	—	New well	266
Well 21	Devonian	11,000	Gelled acid	50,000	29.5	5,000	—	—	—	2,500	7,000	New well	276
Well 22	Devonian	11,000	Gelled acid	50,000	30	5,000	—	—	—	2,500	6,500	New well	329
Offset Well 7	Devonian	11,000	Gelled acid	50,000	N.A.	27,500	7,500	—	—	—	—	New well	158
Offset Well 8	Devonian	11,000	15% acid	19,000	2.8	Acid treatment, no propping agent						New well	269
Offset Well 9	Devonian	11,000	Emulsified acid	35,000	11	8,750	—	2,750	—	—	—	New well	246
Offset Well 10	Devonian	11,000	15% acid	10,000	3.6	Acid treatment, no propping agent						New well	215
Offset Well 11	Devonian	11,000	Emulsified acid	50,000	40.5	26,250	20,000	—	—	—	—	New well	292
TEXAS — Field G													
Well 23	Morrow	6,000	Water	25,000	17	—	—	3,300	9,680	—	—	31	120
Offset Well 12	Morrow	8,000	Water	20,000	24.7	15,000	—	1,000	—	—	—	64	258
OKLAHOMA — Field H													
Well 24	Upper McLish	7,700	Water	55,000	32	2,500	—	5,000	16,500	—	—	New well	449
Well 25	Upper McLish	7,500	Water	47,000	41.3	2,500	—	3,000	9,000	3,750	9,900	New well	696
Well 26	Basal Bromide	7,700	Water	66,000	37	5,000	—	4,000	4,800	5,000	9,080	New well	390
Well 27	Basal McLish	7,800	Water	47,000	29.5	2,500	—	3,000	9,000	3,750	11,250	New well	—
Well 28	2nd Bromide	8,900	Gelled water	80,000	26.3	7,500	—	7,300	27,400	—	—	New well	—
Well 29	Upper McLish	7,500	Gelled water	40,000	N.A.	—	—	3,000	9,000	—	—	New well	413
Well 30	2nd Bromide	8,000	Gelled water	40,000	N.A.	—	—	3,000	9,000	—	—	New well	509
Offset Well 13	Upper McLish	7,600	Salt water	33,700	34.6	7,500	—	3,000	—	—	—	New well	376
Offset Well 14	2nd Bromide	8,300	Salt water	33,700	18.1	7,500	—	3,000	—	—	—	New well	228
OKLAHOMA — Field I													
Well 31	Humphreys	6,300	Salt water	40,000	33	7,500	—	—	—	3,000	9,000	1	43
Offset Well 15	Humphreys	6,100	Salt water	30,000	43	4,000	—	5,000	—	—	—	12.5	37
Well 32	Humphreys	5,300	Salt water	42,000	47	11,000	—	3,200	9,600	1,800	5,400	24	54
Well 33	Humphreys	4,900	Salt water	18,500	23.4	—	4,000	—	—	2,500	7,500	4	Shut In
CANADA (ALBERTA) — Field J													
Well 34	Cardium	5,500	Water	23,000	33	1,000	—	1,840	—	2,000	11,520	New well	85
Well 35	Cardium	5,500	Water	10,920	23	—	—	3,670	11,010	—	—	New well	150
Several Offset Wells	Cardium	5,500	Water	N.A.	50	—	65,000	—	—	—	—	New well	45

TABLE 6.12 — FIELD TEST RESULTS, UREA SPACER TREATMENTS

Location	Formation	Depth (ft)	Fracturing Fluid	Volume (gal)	Average Injection Rate (bbl/min)	Pounds of Propping Agent and Spacer Material						BOPD Before Treatment	Potential BOPD After Treatment
						20/40 Sand	10/20 Sand	12/20 Nutshells	12/20 Spacer	8/12 Nutshells	8/12 Spacer		
WYOMING — Field D													
Well 36	1st Wall Creek	900	Crude	12,500	37.5	—	—	—	—	6,300	18,000	60	95
Well 37	1st Wall Creek	900	Crude	12,500	37.5	—	—	—	—	6,300	18,000	5	46
Well 38	1st Wall Creek	900	Crude	12,500	37.5	—	—	—	—	6,300	18,000	18	88
TEXAS — Field K													
Well 39	Grayburg	4,200	Crude	20,000	50	20,000	5,000	1,000	3,000	—	—	8	39
Well 40	Grayburg	4,200	Crude	20,000	N.A.	—	30,000	2,000	6,000	—	—	10	77
TEXAS — Field L													
Well 41	Grayburg	4,200	Crude	20,000	N.A.	—	30,000	2,000	6,000	—	—	16	132
CANADA (ALBERTA) — Field J													
Well 42	Cardium	5,500	Crude	10,890	19	—	—	3,670	10,000	—	—	30	80
Well 43	Cardium	5,500	Crude	10,890	21	—	—	3,670	11,000	—	—	N.A. (58 BOPD increase)	N.A.

capacity does not decline so rapidly as when the widely graded sands are used.

To be meaningful, fracture capacity tests must be conducted over a period of time long enough to reflect plastic deformation of either the proppant or the formation. A study of the reduction observed in long term fracture capacity indicates that the final fracture capacity averages about 60 percent of the initial fracture capacity.

In general, the $-8+12$, the $-10+20$ and the $-20+40$ U.S. Standard Sieve size sands are unsuitable for use in single layers in wells deeper than 7,000 ft with formations having greater than 150,000 psi embedment pressure. The $-16+18$ mesh, specially graded sand at depths less than 7,000 ft produces fracture capacities from 1,000 to 3,000 md-ft throughout the entire embedment pressure range. (The specially graded $-16+18$ mesh sand can be obtained only on special order.) A sand that is closely graded in size will better resist crushing and will yield a higher fracture capacity than will widely graded sand.

It has been found that sand of all sizes will produce high fracture flow capacity if placed in multilayers in fractures. Wells as deep as 12,000 ft have been suc-

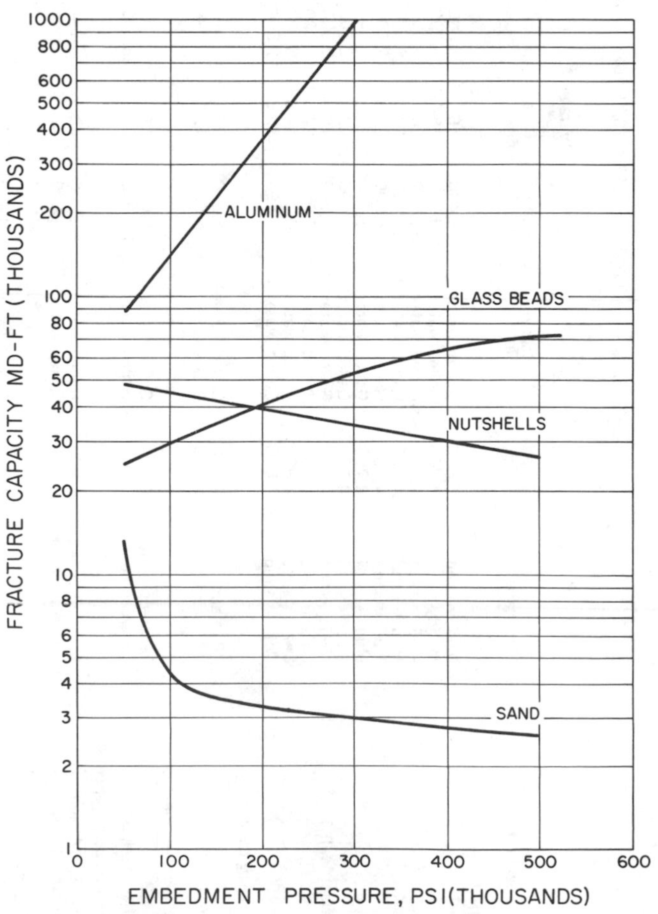

Fig. 6.15 Generalized selection chart, maximum fracture capacity obtained — well depth less than 4,000 ft. (Based on proppant performance at 4,000 ft.)

cessfully fractured by placing sand proppants in multi-layers using viscous oil based fracturing fluids.

Huitt and McGlothlin[12] showed that crushing of single layers of sand occurred in fractures where excessive overburden loads were imposed on formations with high embedment pressure. They also demonstrated that coating the sand proppant with a plastic restraining material effected fracture flow capacities 10 to over 100 times those observed with the same size sand uncoated. Examination of the coated grains of propping sand (after it was subjected to loading) showed that the sand grains were crushed, but that the coating held the crushed fragments in place.

Malleable Metals

Malleable metals such as aluminum particles exhibit interesting properties that tend to overcome propping limitations. As high loads are applied to sparsely placed particles, they deform slightly rather than shatter. The deformation results in a greater bearing area against the formation face, which reduces the average stress under the particle and, therefore, reduces particle penetration into the formation. In general, metals have sufficient strength to permit propping with a sparse array of particles, leading to very high conductivities.

The average stress on a malleable proppant is not constant as the particle deforms. This is because, as a result of work-hardening, the compressive strength of many metals such as aluminum* increases as the strain increases; and as the diameter of the sphere becomes large the friction between the ductile material and the formation fracture faces will cause considerable distortion of the yielding material. The increased distortion increases the resistance to deformation.

Field and laboratory tests indicate that aluminum proppant material is suitable for all depths, but is recommended primarily for applications where sand fails, meaning that aluminum should be used at depths greater than about 6,000 ft. The concentration at which aluminum performs best is from 0.1 to 0.25 monolayer, which in the −8+12 mesh size is a concentration of 0.07 to 0.16 lb/sq ft. In the −12+16 and −14+16 mesh size aluminum, a concentration of 0.04 to 0.1 lb/sq ft performs best, and for the −20+25 mesh size aluminum, a concentration of 0.025 to 0.065 lb/sq ft is ideal. Fracture capacity test data for aluminum are shown on Fig. 6.21.

*Aluminum proppant requires a special alloy to prevent interaction of the aluminum with the formation brines. This material is available on special order.

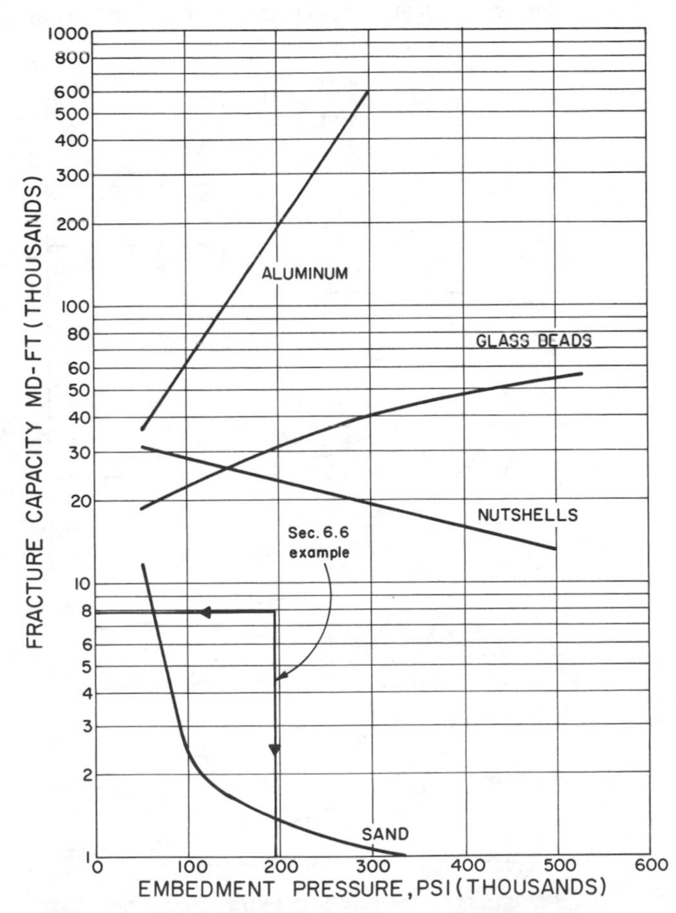

Fig. 6.16 Generalized selection chart, maximum fracture capacity obtainable — well depth 4,000 to 7,000 ft. (Based on proppant performance at 7,000 ft.)

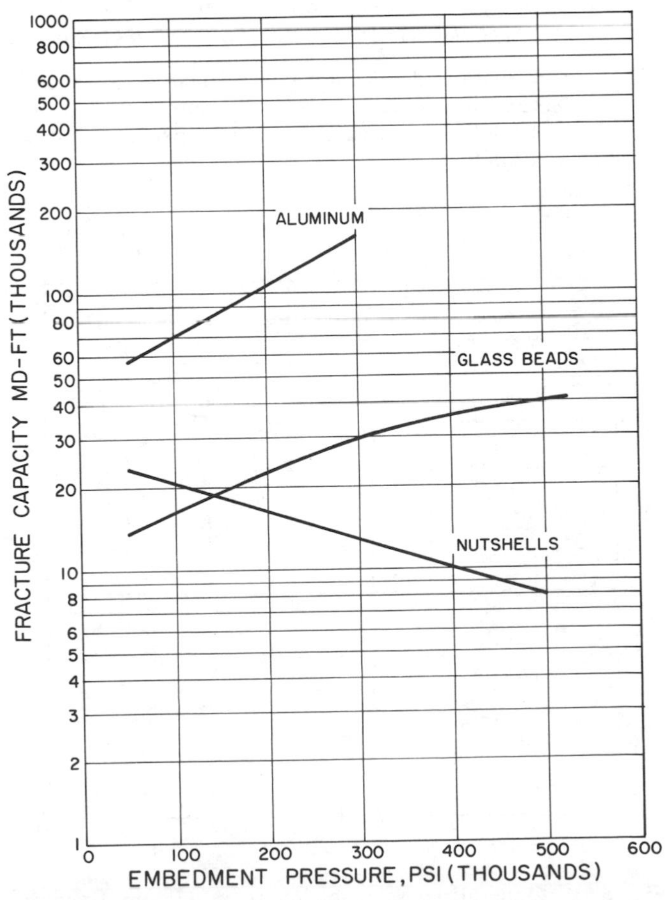

Fig. 6.17 Generalized selection chart, maximum fracture capacity obtainable — well depth 7,000 to 10,000 ft. (Based on proppant performance at 10,000 ft.)

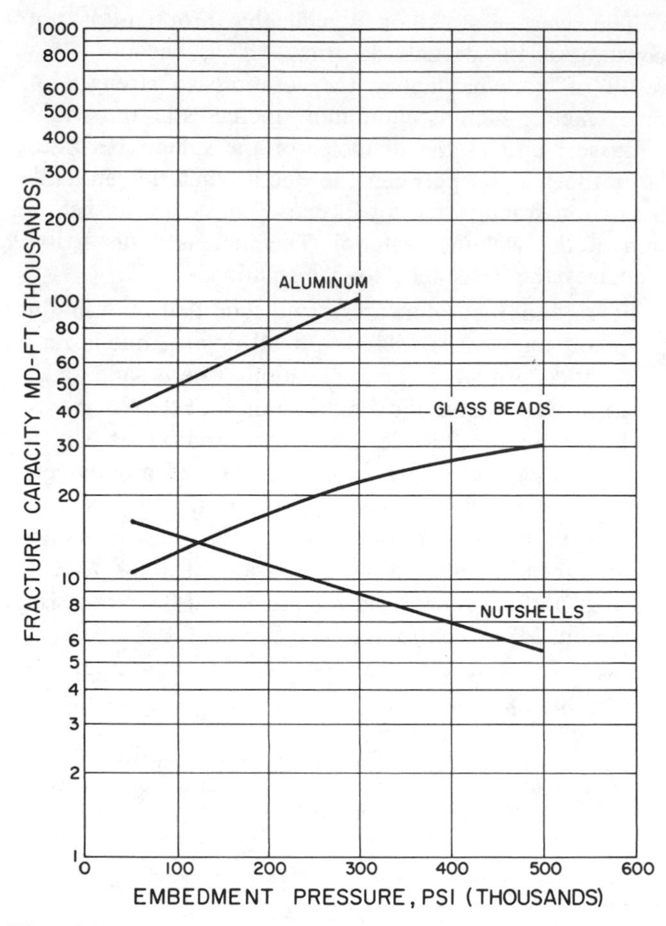

Fig. 6.18 Generalized selection chart, maximum fracture capacity obtained — well depth 10,000 to 12,000 ft. (Based on proppant performance at 12,000 ft.)

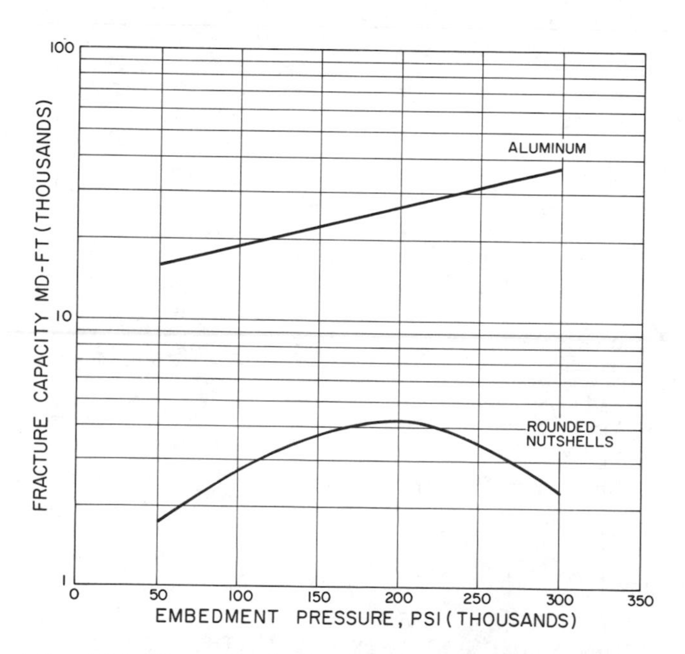

Fig. 6.19 Generalized selection chart, maximum fracture capacity obtainable — well depth 12,000 to 14,000 ft. (Based on proppant performance at 14,000 ft.)

Long range fracture capacity tests with aluminum indicate that the final fracture capacity will be about 75 percent of the initial fracture capacity.

Nutshells

Short range fracture capacity tests made on both angular and rounded nutshells indicate that the rounded nutshells are superior to the angular variety. In general, the fracture capacities obtained on short range tests are not so high as for aluminum, mostly because the compressive strength of the nutshells is low. However, in many instances where the formation production capacity is low, adequate fracture-flow-capacity: formation-flow-capacity contrast can be obtained with nutshells. The rounded nutshells in the $-8+12$ and the $-12+20$ mesh size are suitable as a propping material at all embedment pressures and at all overburden depths tested to 14,000 ft. A concentration of ¼ monolayer approaches an optimum condition. For the $-8+12$ and $-12+20$ mesh rounded nutshells, the optimum concentrations are 0.08 lb/sq ft and 0.06 lb/sq ft, respectively.

Long range fracture capacity tests conducted with angular nutshells indicate that the final capacity can be reduced to 40 percent of the initial capacity. Tests with the rounded nutshells have shown that the final fracture capacity can be expected to approach 65 percent of

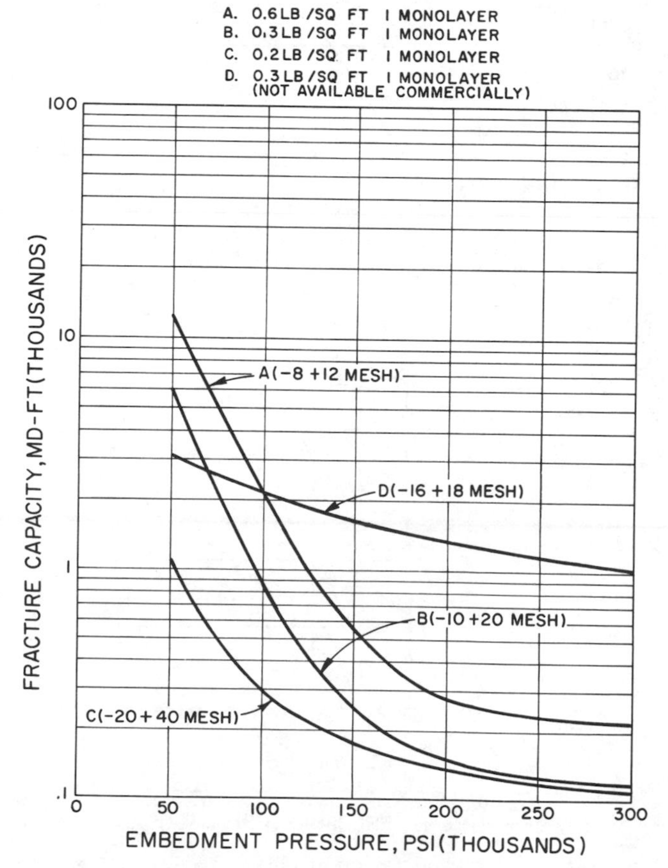

Fig. 6.20 Propping material selection curve for sand — well depth 3,000 to 7,000 ft. (Based on proppant performance at 7,000 ft.)

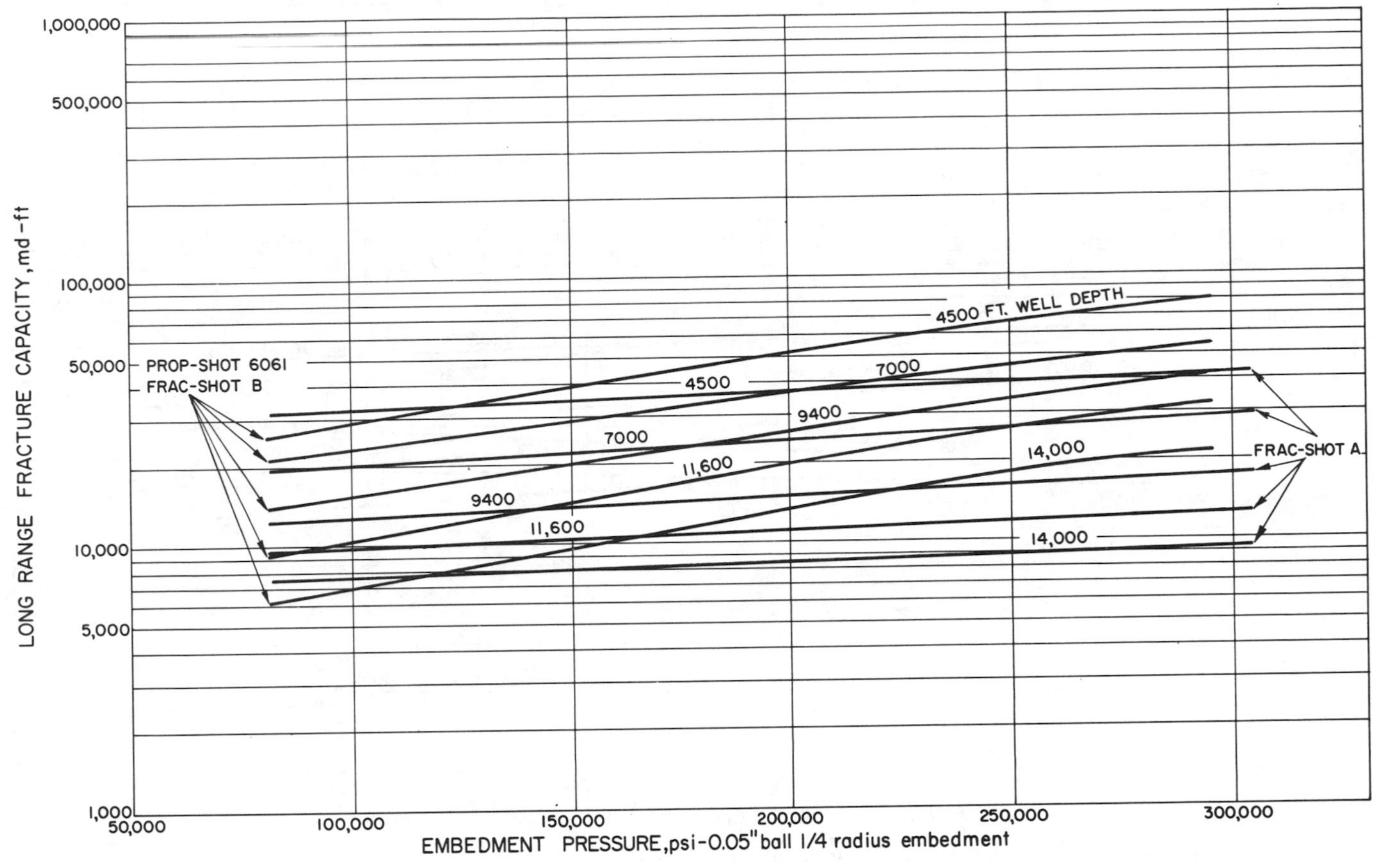

Fig. 6.21 Propping material selection curve for aluminum (—12+16 mesh U. S. Standard Sieve, 0.1 lb/sq ft.)

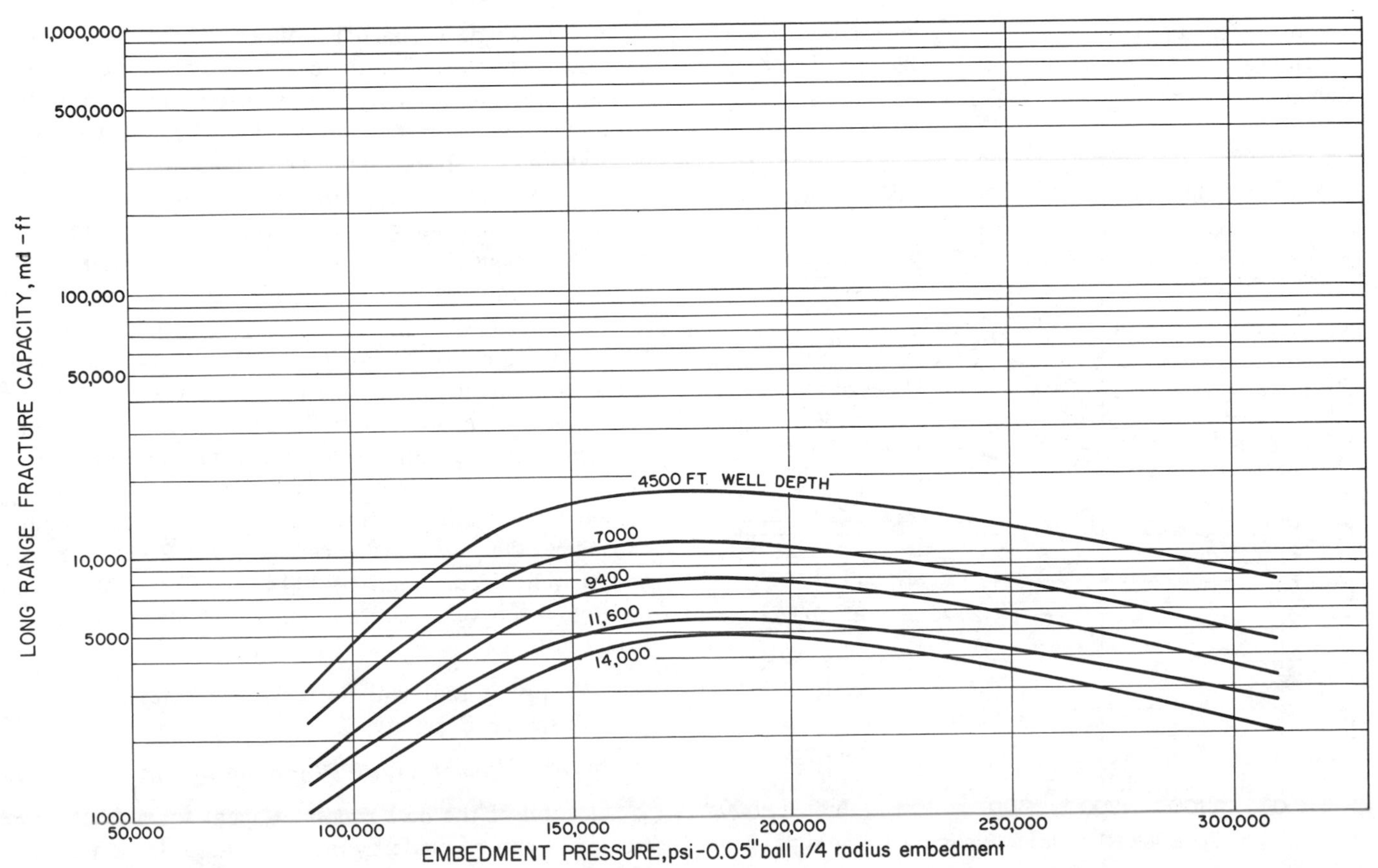

Fig. 6.22 Propping material selection curve for rounded nutshells (—12+20 mesh U. S. Standard Sieve, 0.06 lb/sq ft.)

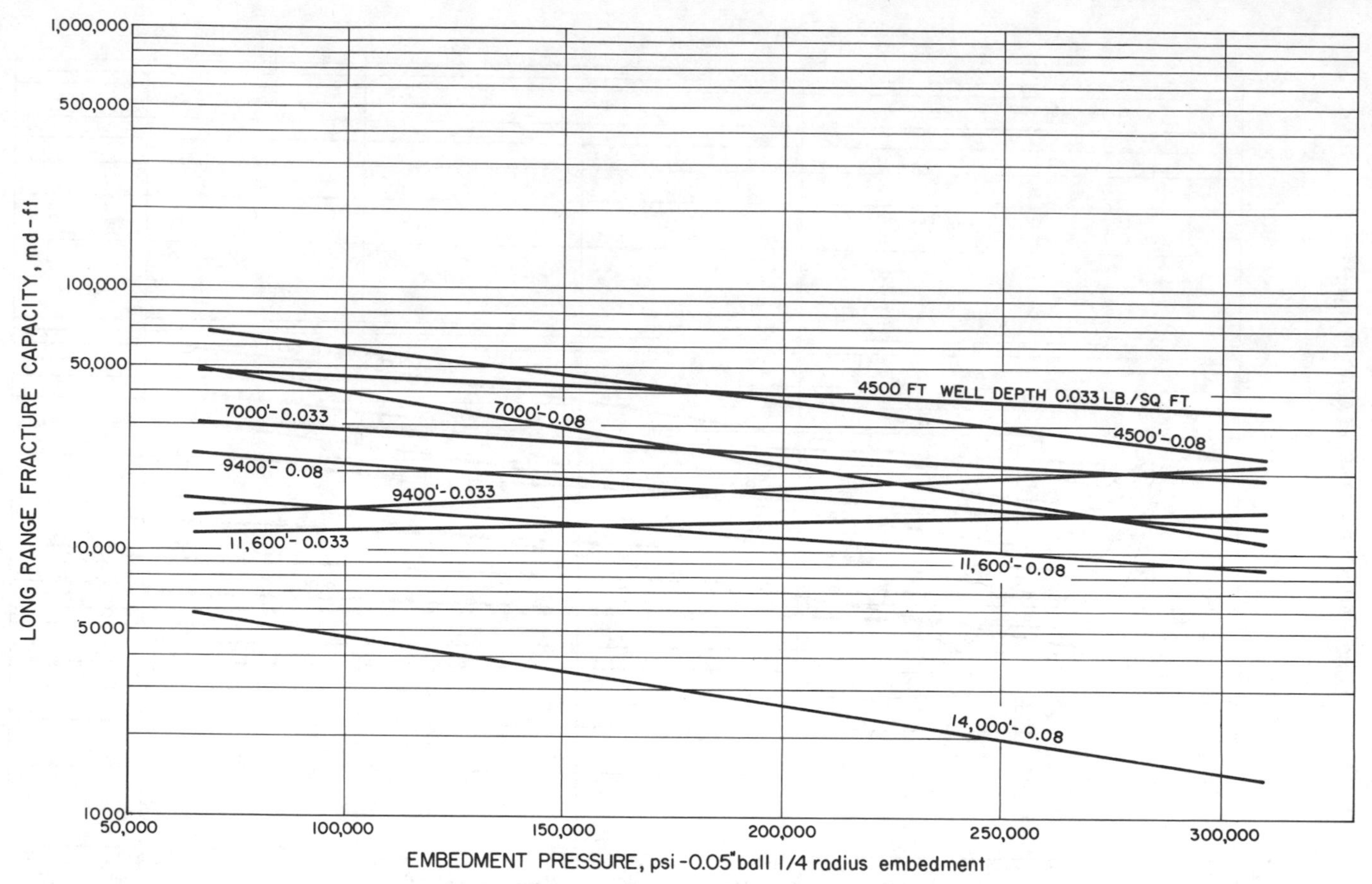

Fig. 6.23 Propping material selection curve for rounded nutshells (—8+12 mesh U. S. Standard Sieve.)

the initial fracture capacity. Long range fracture capacities for angular and rounded nutshells are shown on Figs. 6.22 and 6.23.

Glass Beads

High strength glass bead proppants are generally suitable in single or partial monolayers for all but the very deepest wells. This material produces an extremely high fracture capacity and is useful in deep wells with medium to hard formations. The glass particles are spherical and therefore have less tendency to bridge or screen out than do the more angular proppants. The optimum concentration to produce the highest fracture capacity depends upon well depth and formation hardness. A ½ monolayer of −8+12 mesh and −12+20 mesh glass beads, which produces very high fracture fluid carrying capacities under most conditions, requires concentrations of 0.23 lb/sq ft and 0.19 lb/sq ft, respectively. On soft formations (< 50,000 psi embedment pressure), glass proppants generally will embed more severely than the deformable proppants such as nutshells and, thus, will produce slightly lower fracture capacities. On very hard formations (> 250,000 psi), and in wells deeper than 14,000 ft, a full monolayer or multiple layers of glass are required to prevent the glass from crushing.

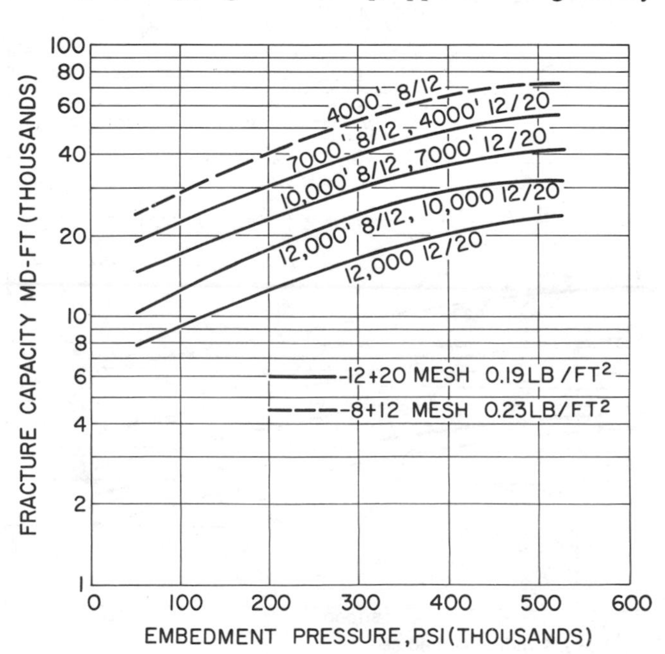

Fig. 6.24 Propping material selection curve for glass beads, 0.5 monolayer.

6.6 Propping Material Selection for Maximum Fracture Capacity

Selection of Propping Agent

Selection of the best propping agent for a fracturing treatment can be made by means of the selection charts presented on Figs. 6.15 through 6.24.

The *first* step in the selection of the best propping

agent for a fracturing treatment is to determine the fracture capacity required to achieve the desired productivity following fracturing. This fracture capacity depends upon the flow capacity of the producing formation and upon the penetration of the fracture into the reservoir. The generalized relationship between the fracture:formation-capacity ratio and the production increase ratio, with fracture penetration as a parameter, is shown on Fig. 6.4. The data for these plots were calculated from Darcy's equation[4] for radial flow of a homogeneous fluid moving through a formation with discontinuous radial variations in permeability. Specific data for horizontal and vertical fractures are presented on Figs. 6.4A and 6.4B. These curves show that a high fracture-flow-capacity:formation-flow-capacity contrast is required to obtain the maximum benefit from fracturing, and that, where fracture penetration is deep and formation flow capacities are high, fractures with high capacity are more advantageous.[9]

The *second* step in the selection of a propping agent is to determine the formation embedment pressure in the portion of the pay zone to be fractured. This may be done according to the procedures previously described. Table 6.2 lists some typical embedment pressures.

The *third* step is to select the type of propping agent required. This may be determined from Figs. 6.15 through 6.19 which are generalized propping agent selection curves. These curves present the maximum fracture capacity obtainable for sand, rounded nutshells, glass beads and aluminum alloy proppants for a well depth of 0 to 14,000 ft. The maximum fracture capacity that can be obtained for a given embedment pressure and propping agent must lie in the zone beneath the curve for the specific proppant.

The *fourth* and final step is to select the propping agent size and concentration that will produce the desired fracture capacity. This is accomplished by using the curves plotted on Figs. 6.20 through 6.24, which are propping material selection curves for sand, rounded walnut shells, glass beads and aluminum alloy.

Example Calculation

Assume the following well conditions:

Well depth	6,800 ft
Formation	San Andres
Embedment pressure (Table 6.2)	193,000 psi
Planned increase in well productivity	6
Formation capacity	160 md-ft
Planned fracture penetration	30 percent drainage radius

Fig. 6.4 (see example problem on this figure) shows that a 60:1 fracture:formation-capacity contrast is required to effect the preplanned sixfold increase in well production. Thus, the fracture capacity must be 160 md-ft (formation capacity) $\times$ 60-fold contrast, or 9,600 md-ft.

After the required fracture capacity and the formation embedment pressure have been determined, the type of propping agent can be selected. First the generalized selection chart (Fig. 6.16) is used. Since embedment pressure is 193,000 psi, and the required fracture capacity is 9,600 md-ft (see example problem on Fig. 6.16), sand would not yield the required fracture capacity, but rounded nutshells would.

Referring to the selection curves for rounded nutshells (Figs. 6.22 and 6.23), Fig. 6.22 ($-12+20$ mesh nutshells) should be used to determine the size and concentration of nutshells required. Fig. 6.22 (example problem) shows that a fracture capacity of 12,000 md-ft can be attained in a formation with an embedment pressure of 193,000 psi with $-12+20$ mesh rounded nutshells at a concentration of 0.06 lb/sq ft of fracture surface.

In another method of determining the conductivity of a propped fracture, developed by a major fracturing service company, the formation characteristics are broadly classified as "soft," "hard" and "very hard" rather than by a specific embedment pressure. Figs. 6.25 through 6.29 are plots of fracture conductivity vs proppant concentration, with overburden pressure as a parameter, 20-40 round sand in soft, hard, and very hard formations, and for two sizes of Ucar props in very hard formations. This method is satisfactory for estimating fracture conductivity vs type of proppant, but formation embedment characteristics should be more specifically defined for final fracture treatment design.

Determining the Quantity of Propping Material and Spacer

Propping materials should be placed in the fracture in concentrations that will vary with formation characteristics and well depth. Suggested partial monolayer concentrations of the various proppants and spacers per 1,000 sq ft of fracture area are tabulated in Table 6.13. These factors may be used to calculate the amounts of proppant and spacer needed for a treatment.

The optimum concentration of a proppant as related to well depth and type of formation is established from propping agent selection charts (Figs. 6.20 through 6.24).

Special well conditions of depth and formation hardness may require that the concentrations shown in Table 6.13 be changed to achieve optimum results. As a general rule, the volume of proppant and spacer should be equivalent to a full monolayer, the amount of proppant being determined as previously described and the spacer occupying the remaining volume.

Under some conditions, a multilayer of the proppant and spacer blend conceivably can be placed in the fracture. Even in this situation the probability of depositing a full monolayer or more of deformable proppant is lessened by the use of spacers. For example, if a double layer of proppant-spacer blend in a 1:7 vol-

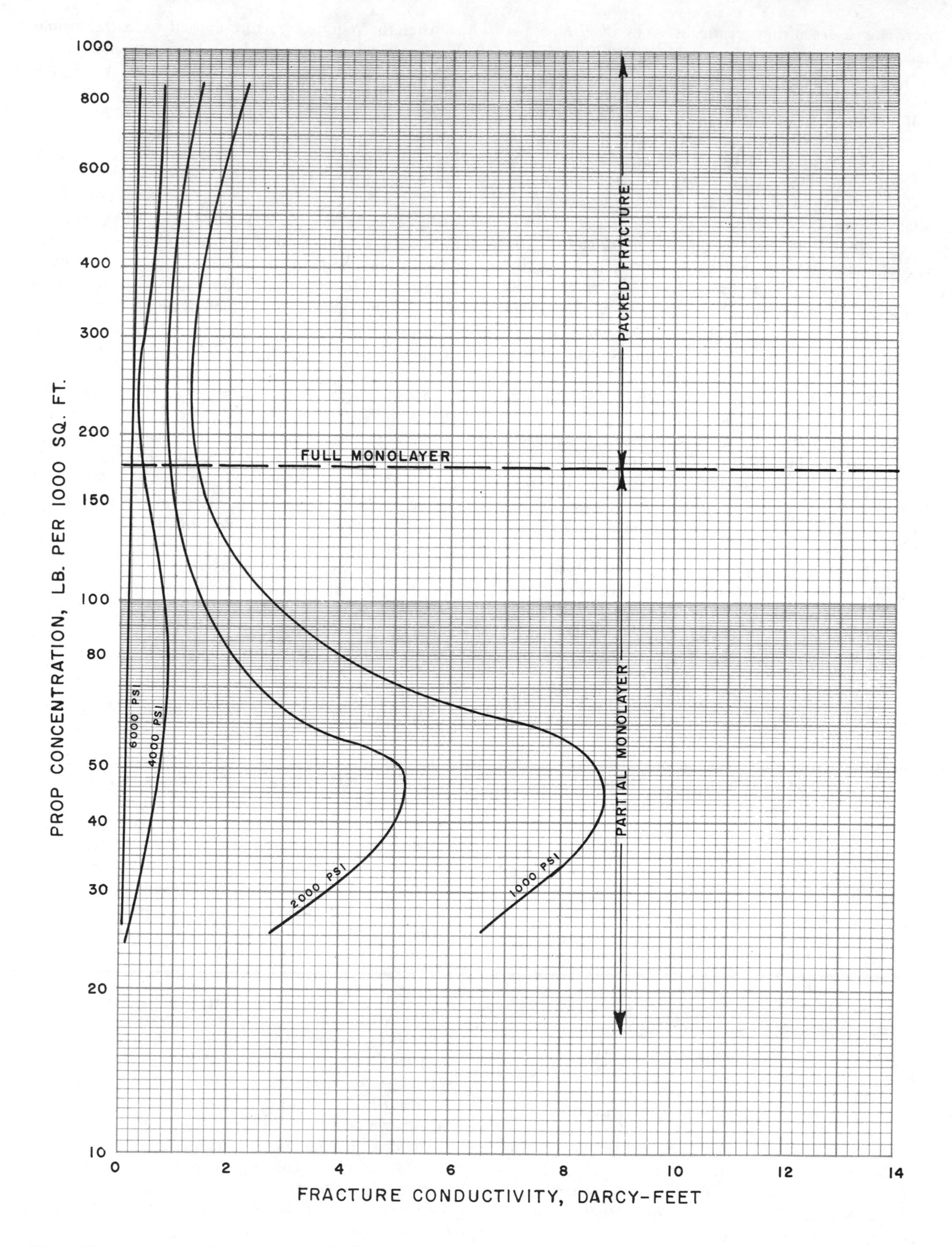

Fig. 6.25 Fracture conductivity vs load pressure — 20-40 round sand, soft formation.[20]

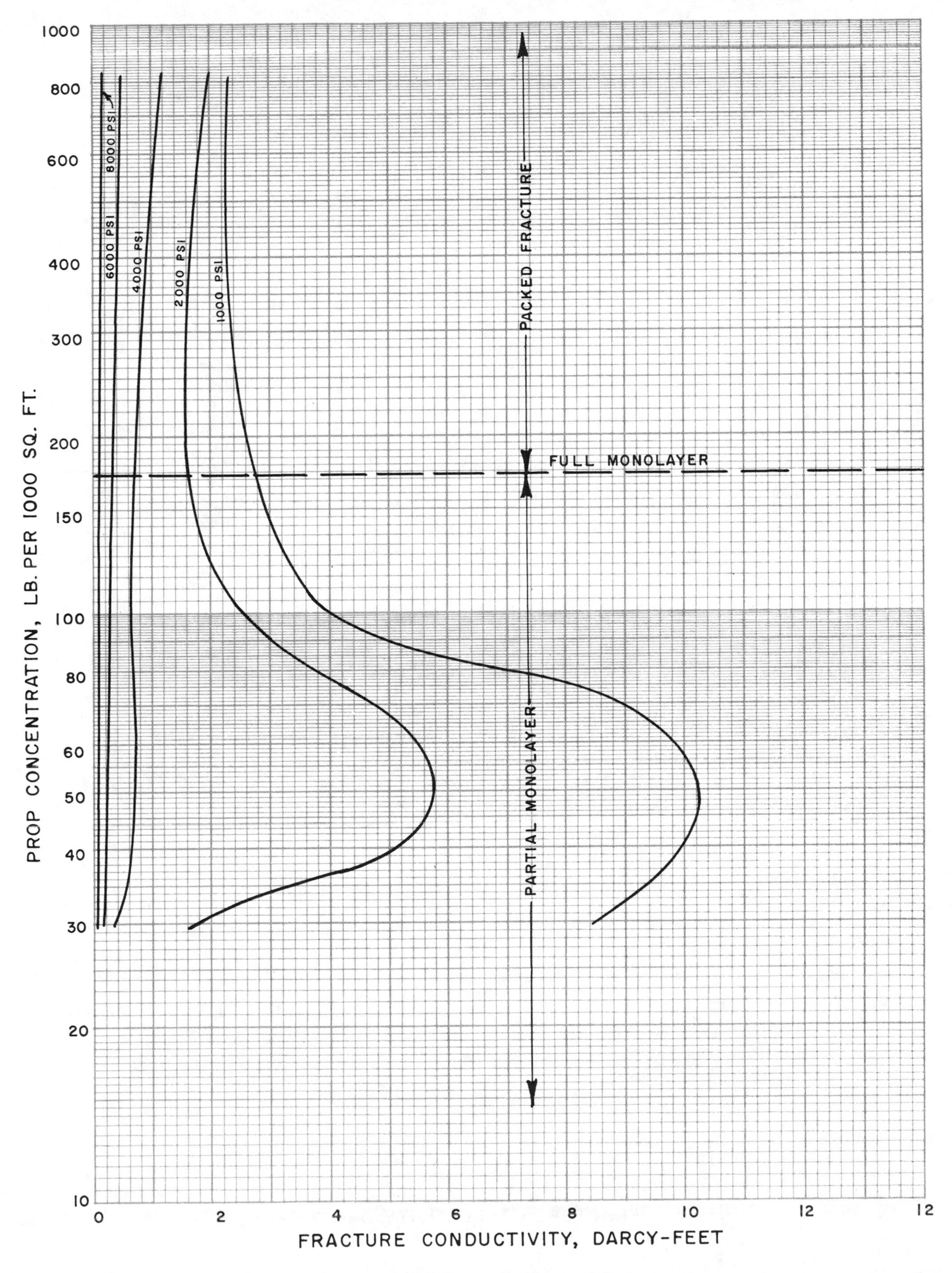

Fig. 6.26 Fracture conductivity vs load pressure — 20-40 round sand, hard formation.[20]

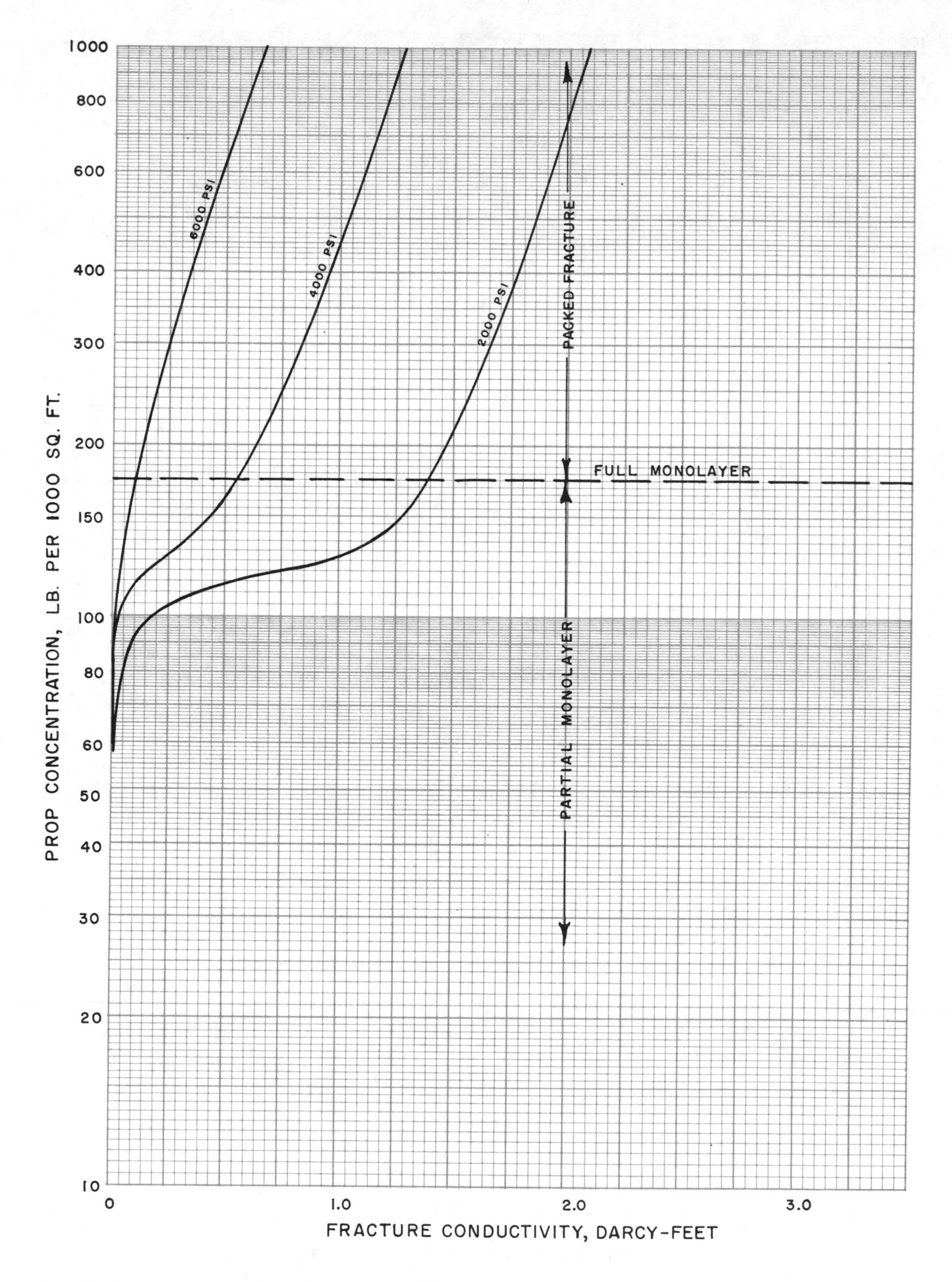

Fig. 6.27 Fracture conductivity vs load pressure — 20-40 round sand, very hard formation.[20]

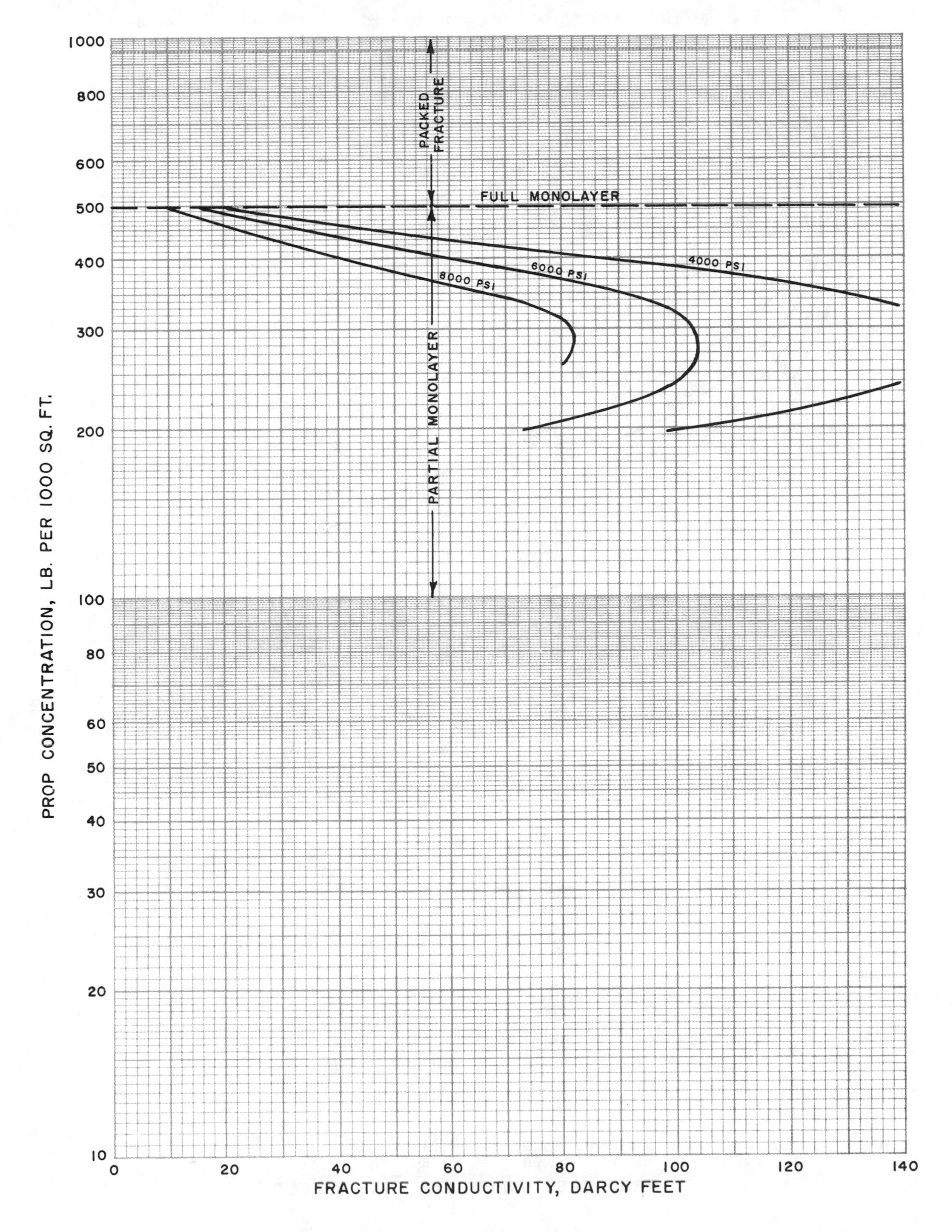

Fig. 6.28 Fracture conductivity vs load pressure — 8-12 UCAR props, very hard formation.[20]

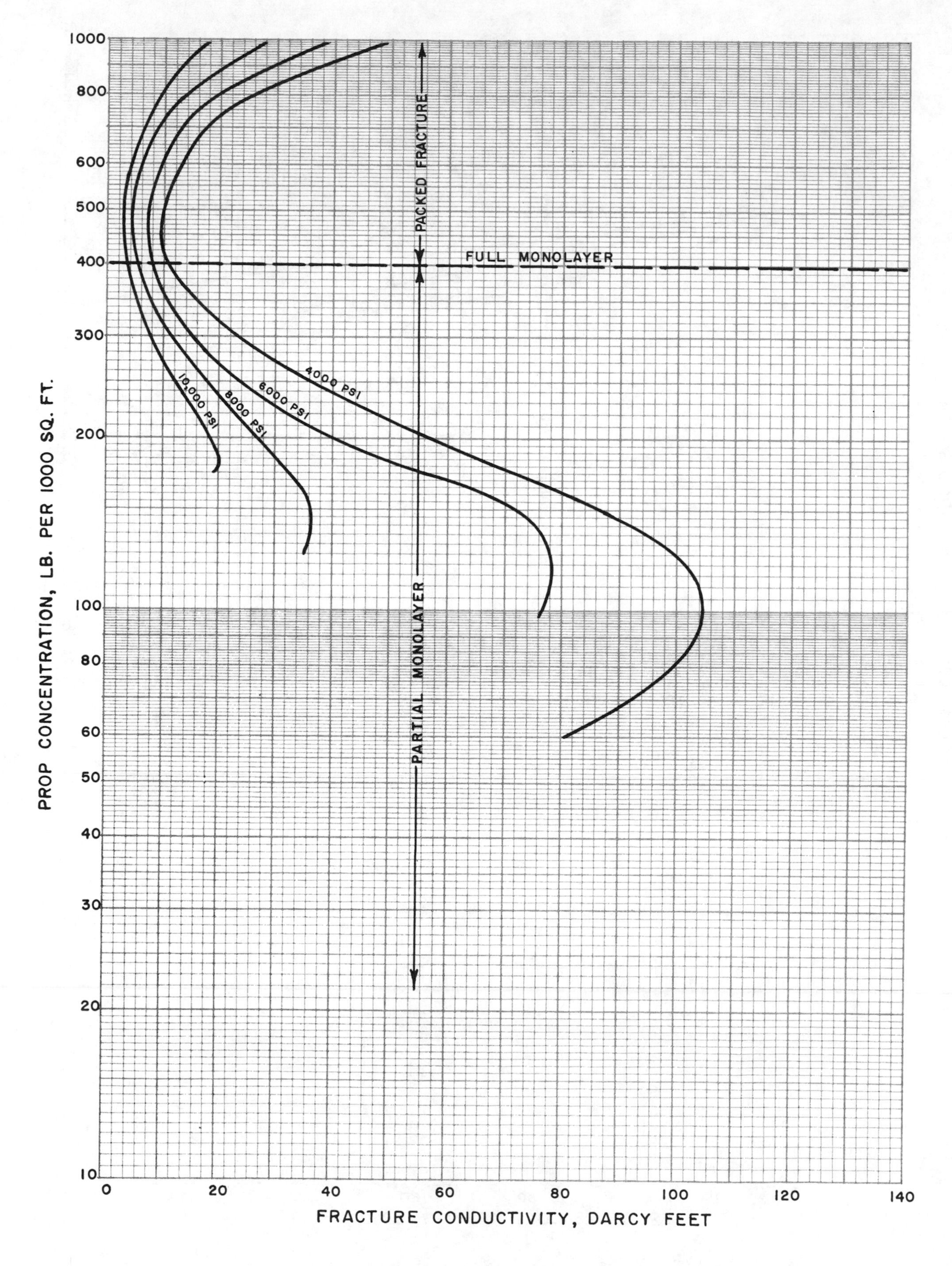

Fig. 6.29 Fracture conductivity vs load pressure — 12-20 UCAR props, very hard formation.[20]

TABLE 6.13 — QUANTITY OF PROPPANT AND SPACER REQUIRED PER 1,000 SQ FT OF FRACTURE AREA

Proppant	Proppant Concentration	Spacer Concentration
Nutshell (0.2 monolayer)	50 lb —12+20 mesh nutshells	150 lb —12+20 mesh resin
Nutshell (0.2 monolayer)	50 lb —12+20 mesh nutshells	185 lb —12+20 mesh urea
Aluminum (0.25 monolayer)	100 lb —12+16 mesh aluminum	300 lb —12+16 mesh sodium bisulfate
Aluminum (0.25 monolayer)	160 lb —8+12 mesh aluminum	480 lb —8+12 mesh sodium bisulfate
Glass beads (0.5 monolayer)	190 lb —12+20 mesh glass beads	190 lb —12+20 mesh sodium bisulfate
Glass beads (0.5 monolayer)	230 lb —8+12 mesh glass beads	230 lb —8+12 mesh sodium bisulfate
Sand (0.5 monolayer)	100 lb —20+40 mesh sand	100 lb —20+40 mesh sodium bisulfate
Sand (0.5 monolayer)	170 lb —10+20 mesh sand	170 lb —10+20 mesh sodium bisulfate

NOTE: Special well conditions of depth and formation hardness may require that these concentrations be changed to achieve optimum results.

ume ratio is placed in the fracture, the resulting concentration of proppant in the fracture is a 0.25 monolayer (0.125 monolayer/layer × 2 layers = 0.25 monolayer) after the spacer is dissolved. In a 7,000-ft well, a fracture propped with a double layer of −12+20 nutshells would have a 5,200-md-ft capacity. If the proppant is spaced with a soluble material in a ratio of 1:7, the capacity of the fracture would be the same as ¼ layer, or 8,600 md-ft. Similarly, a triple layer of proppant-spacer blend in a 1:11 volume ratio would result in a final fracture concentration of 0.25 monolayer.

In deep wells with low embedment pressure formations, multilayers of proppant will usually yield the best results.

6.7 High Strength Propping Agent Tail-In

Because the exclusive use of high strength proppants in hydraulic fracturing treatments would increase costs appreciably, Fig. 6.30 was prepared to demonstrate the post-fracturing productivity that may be obtained by using special proppant materials to tail in after a conventional fracturing treatment in a well where the sand propping agent yields low fracture capacities (200 md-ft).

These curves show the theoretical increase in productivity for wells of various formation capacities when sand and a high strength proppant are used in propping a horizontal fracture having a radius of 200 ft.

Curve I — The entire 200-ft fracture radius contains high strength propping agent (5,000 md-ft).

Curve II — 50 percent of the 200-ft fracture radius (25 percent of the total fracture area) adjacent to the wellbore is propped with high strength proppant (propped fracture capacity is 5,000 md-ft), and the remaining fracture is propped with sand (200 md-ft).

Curve III — 25 percent of the 200-ft fracture radius (6.25 percent of the fracture area) adjacent to the wellbore contains high strength proppant (propped fracture capacity is 5,000 md-ft) and the remaining fracture area is propped with sand (200 md-ft).

Curve IV — The entire area of the 200-ft fracture is propped with sand (fracture capacity is 200 md-ft throughout).

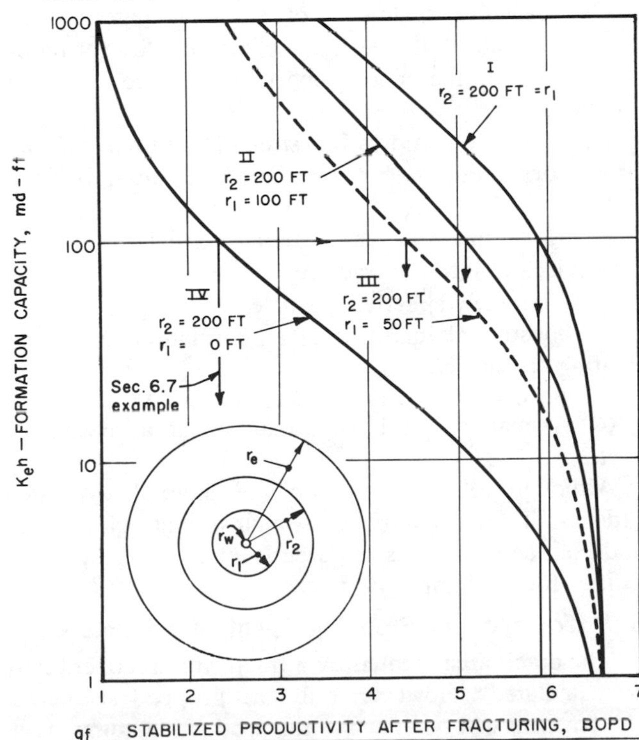

Fig. 6.30 Productivity improvement resulting from tailing-in with high-strength propping agent.

These data show that substantial improvement in post-fracturing production will occur if high strength proppant is substituted for sand in a 200-ft-radius fracture. For example, if the natural formation capacity is 100 md-ft, a sand-propped fracture will increase post-fracturing productivity 2.4 times pretreatment production. If the same fracture is propped with material yielding a capacity of 5,000 md-ft, the productivity increase will be 5.9.

If the treatment is conducted so that the inner 100 ft of the 200-ft-radius fracture is propped with high strength proppant, the productivity increase will be 5.1, whereas if only the 50-ft radius (6.25 percent of the area) adjacent to the well is propped with the strong prop, the productivity increase is 4.4. These data demonstrate that if a high capacity propping material is placed in the portion of the total fracture radius adjacent to the well the productivity is increased substantially more than if the well is completely sand propped. The cost of using a high strength proppant such as glass beads can be reduced appreciably without seriously decreasing productivity by first propping the fracture with sand and then tailing in with this more effective, more expensive material.

Tail-in treatments in vertical fractures will yield similar results if the tailed-in proppant can be placed adjacent to the well. Special techniques — use of very viscous or high-gel-strength fluids, high pumping rates, tapered density proppants and the like — will be required.

6.8 Desirable Propping Agent Characteristics

The preceding evaluation of propping materials has helped to establish the desirable characteristics for propping material, the most important of which are the following.

1. Sufficient compressive strength and malleability under formation loading to insure maximum fracture clearance.
2. Maximum size and narrow particle range for easy injection into the fracture.
3. Uniformly spherical particles.
4. Substantial inertness to all formation fluids and treating chemicals.
5. A specific gravity in the range of 0.8 to 3.0.
6. Availability in large quantities at a reasonable cost.

Although not a single one of the materials evaluated satisfies all these requirements, high strength glass beads and rounded nutshells apparently offer the best possibilities for field application.

6.9 Transport of Propping Agent in a Fracture

The crack area created by a fracturing treatment may be calculated.[9] However, only the propped section of the fracture will be effective for production stimulation. Many variables, such as injection rate, treatment volume, fluid viscosity, and proppant concentration will affect transport efficiency.

Nikuradse,[15] showed that in the region of viscous flow surface roughness of a fracture has no appreciable effect upon the resistance to flow. In the region of turbulent flow, however, surface roughness is an important factor.

Durand,[16] and Wahl and Campbell,[17] identified three types of solid transport in fluids according to the size of the particles in the mixture — homogeneous mixtures, intermediary mixtures, and heterogeneous mixtures. When the transported particles have diameters of less than about 20 microns, they will form essentially homogeneous mixtures with water. Even small materials, however, will tend to settle out under laminar flow conditions.

Wahl and Campbell[17] further concluded that mixtures containing solids over 50 microns in diameter do not achieve total homogeneity even under turbulent flow conditions. Particles from 50 microns to 0.2 mm in diameter can be moved in fully suspended flow at normal transport velocities, even though the concentration in the vertical plane is not uniform. Above 2 mm in diameter, solid materials are transported along the bottom of the conduit at a velocity substantially less than that of the liquid itself. When they are between 0.2 and 2 mm in diameter, the particles tend to be in a transition zone between heterogeneous suspended flow and deposit flow at normal hydraulic transport velocities. The sand used in fracturing usually falls in this size range.

Horizontal Fracture

The most significant quantity obtained from the experimental data developed by Wahl is the rate of advancement of the sand proppant in the fracture.[17] This quantity is equivalent to the rate of advance of the sand pack formed on the bottom of the fracture.

The rate of sand advance is important from two standpoints. First, the ratio of the advance of the proppant v to the bulk velocity v_B yields an efficiency factor for the sand transport mechanism. This factor, coupled with a knowledge of the fluid loss characteristics of the fracturing fluid, may be used to estimate the portion of the fracture area containing propping agent. Secondly, the rate of advance of the sand in the fracture, for a given flow rate and sand concentration, fixes the amount of sand distributed in the fracture. The areal extent and thickness of the sand pack are important quantities in fracture design.

Wahl and Campbell developed the following equation for determining the velocity or rate of advancement of proppants through the fracture.[17]

$$v_{pa} = 0.705 v_b \, 1.35 C_v 1.07 \mu_o - 0.143 \quad . \quad . \quad . \quad (6.21)$$

To evaluate proppant transport efficiency, Wahl and Campbell[17] stated that

$$\frac{v_{pa}}{v_b} = 0.705 v_b 0.35 C_v 1.07 \mu_o - 0.143 \quad . \quad . \quad . \quad .(6.22)$$

The effect of displacement rate on transport efficiency is shown on Fig. 6.31. This figure shows that flow rates with 7-cp oil have a marked effect on proppant displacement efficiency v_{pa}/v_b. If a radial flow pattern is assumed, the superficial bulk velocities in these tests would exist 17, 35 and 133 ft from the wellbore for a pump rate of 20 bbl/min. Fluid loss during a hydraulic fracturing treatment would cause these velocities to be attained nearer the wellbore.

Fig. 6.32 demonstrates the effect of viscosity on v_{pa}/v_b at a flow rate of 4 gal/min. Again, sand placement efficiency is plotted against sand concentration. High efficiencies are obtained at high sand concentrations with the more viscous oil.

Fig. 6.32 also shows that the sand placement efficiency is not directly proportional to the ratio of oil viscosities, but that high sand placement efficiencies may be obtained with high viscosity oils. High sand placement efficiencies are particularly critical in the placement of partial monolayers. At injection rates of about 4 gal/min, attempts to place a partial monolayer sand pack using a 7-cp oil probably would result in a limited multilayer sand pack.

Fig. 6.33 demonstrates the effect of liquid flow rate and oil viscosity on transport efficiency. The curves were obtained at a constant sand injection rate of 8 lb/min. As flow rate increases, sand concentration decreases; and more fluid is used to transport the identical amount of sand. When small amounts of liquid are

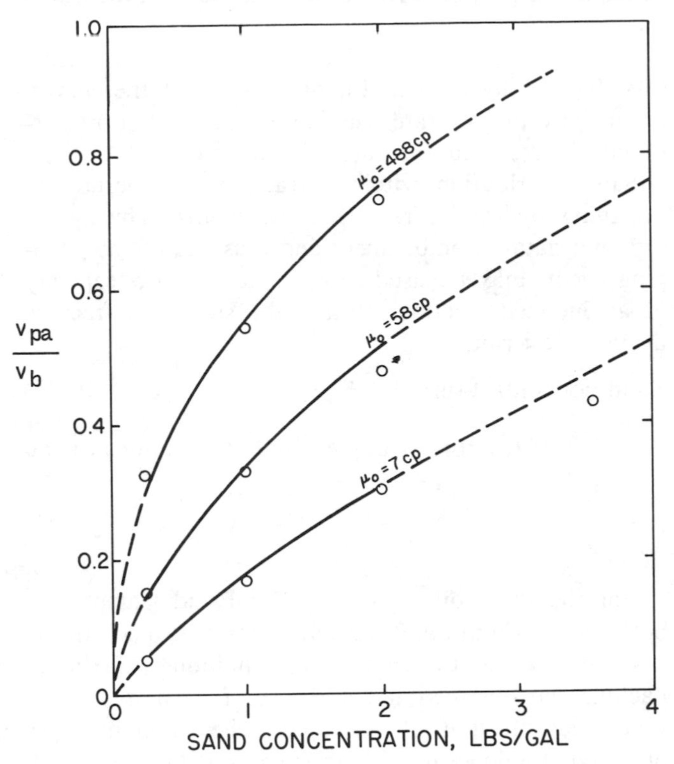

Fig. 6.32 Variation of sand placement efficiency with sand concentration and oil viscosity at a liquid flow rate of 4 gal/min. Fracture thickness ½ in., 20-40 mesh beads.[17]

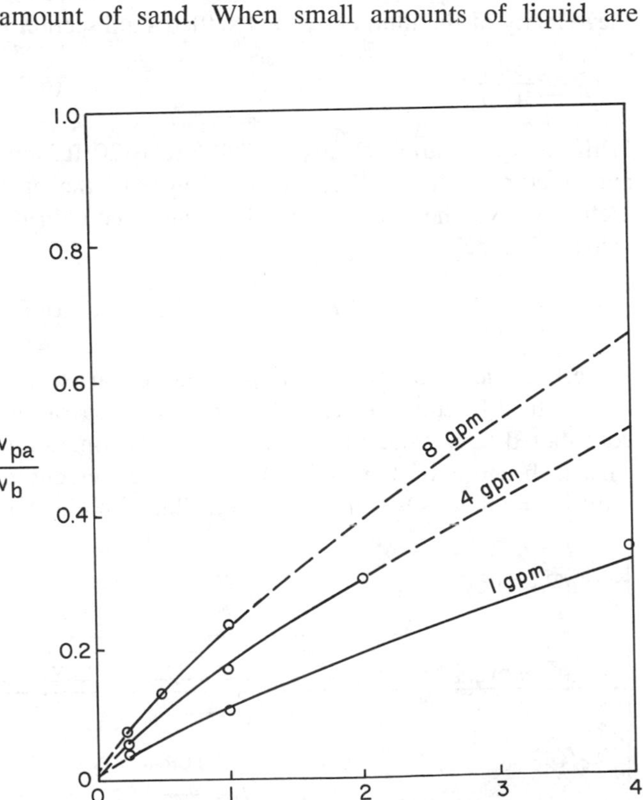

Fig. 6.31 Variation of sand placement efficiency with sand concentration and liquid flow rate for a 7-cp oil. Fracture thickness ¼ in., 20-40 mesh beads.[17]

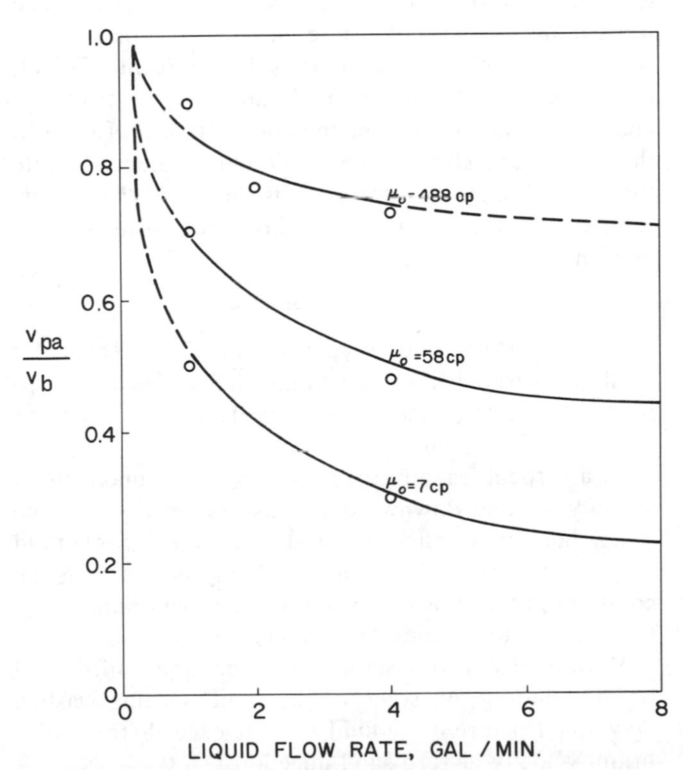

Fig. 6.33 Variation of sand placement efficiency with liquid flow rate and oil viscosity at a flow rate of 8 lb/min. Fracture thickness ¼ in., 20-40 mesh beads.[17]

added to the proppant fracturing fluid slurries, the transport efficiencies drop rapidly. However, as more liquid is added and the flow rate increases, the rate of decline of v_s/v_b becomes small. In this portion of the curves, an increase in flow rate will cause essentially a proportionate increase in the rate of sand advance.

Sand distribution within a fracture is important in designing hydraulic fracturing treatments. This quantity, and data on embedment and crushing of the propping agent, may be used to estimate the flow capacity of an induced fracture. Within the simulated fracture during a test run,

$$\text{Sand concentration} = \frac{C_v v_b}{v_{pa}}, \quad (6.23)$$

Fraction of fracture occupied by bulk volume of sand

$$= \frac{(1.58)C_v v_b}{v_{pa}} \quad (6.24)$$

For the three oils tested by Wahl and Campbell,[17] both the maximum and the minimum bead pack thickness occurred at the same flow conditions. Maximum packing was obtained at a flow rate of 1 gal/min and a sand concentration of 8 lb/gal. The minimum was obtained at a flow rate of 8 gal/min and ¼-lb/gal sand. As shown in Table 6.14, the maximums ranged from 0.875 to 0.529 and the minimums varied from 0.234 to 0.043.

In these tests, a single layer of beads occupies approximately 10 percent of the fracture thickness. Values less than 0.1 indicate a partial monolayer pack. The last column in the table represents the thickness of the sand pack when the transport is 100 percent efficient. These values represent a minimum possible thickness and are a function only of the concentration of sand in the introduced slurry. These minimum values indicate that even though the sand pack is thicker at the highest sand concentrations, the transport mechanism is more efficient.

TABLE 6.14 — FRACTION OF FRACTURE THICKNESS OCCUPIED BY TRANSPORTED SOLIDS

Q_i (gal/min)	C_v (lb/gal)	7-cp oil	58-cp oil	488-cp oil	Minimum Value
1	8	0.875	0.654	0.529	0.430
8	¼	0.234	0.092	0.043*	0.018
1	1	0.656	0.338	0.204	0.071
4	1	0.397	0.207	0.123	0.071
8	1	0.308	0.160	0.093*	0.071

*Not confirmed experimentally

Vertical Fracture

Kern, Perkins and Wyant[18] studied movement of sand in a vertical fracture and concluded that the proppant falls during placement regardless of the type of fracturing fluid used.

In a vertical fracture the sand has two components of velocity — one downward because of gravity and the other horizontal effected by the movement of fluid through the crack. If the fluid velocity is less than the equilibrium value, a bed of settled sand will build up at the bottom of the crack (Fig. 6.34).

Kern *et al.*[18] also observed that while the settled sand bed is building up, some of the sand is being washed down the fracture if the fluid velocity exceeds the equilibrium velocity at zero sand injection.

The rate at which the height of the settled sand bed grows depends upon the falling rate and concentration of the sand in the fracturing fluid. The growth rate of the bed is given approximately by

$$\frac{dh_s}{dt} = \frac{V_g C_s}{13}, \quad (6.25)$$

where 13 is the bulk density of sand in the settled bed in pounds per gallon.

The distance the settled sand bed extends from the wellbore while it is building depends on the horizontal velocity, the falling rate, and the height of the crack. For a crack of rectangular cross-section this distance is approximately

$$r_B = \frac{34\ i}{W_f\ V_g}. \quad (6.26)$$

Loss of fluid to the matrix will reduce r_B as calculated by Eq. 6.26. Also, leakoff will increase the rate of growth of the settled sand bed height because leakoff has the effect of increasing the concentration of sand in the fluid.

For a vertical fracture of rectangular cross-section, the velocity of the fluid above the settled sand section is

$$v_{fH} = \frac{34\ i}{W_f\ h_o} \quad (6.27)$$

With an equilibrium velocity of 7 ft/sec (420 ft/min) and fracture width of 0.25 in., the height of the open section above the settled sand bed when equilibrium velocity is reached is

$$h_o = \frac{34}{(1/4)(420)} \cong 0.3\ i \quad (6.28)$$

Kern *et al.*[18] observed that proppant in suspension in a vertical fracture at the end of fracturing operations contributed very little to the success of the operation. This is because of the very small volume percent of sand in the fluid. Overflush also contributes nothing to

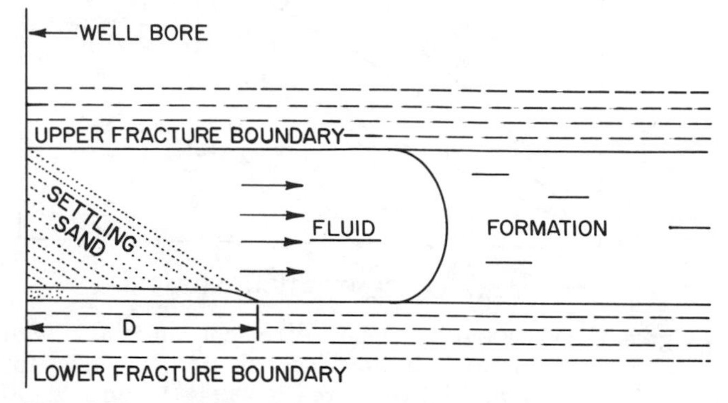

Fig. 6.34 Buildup of settled sand bed.[18]

post-fracturing production.

A very important observation in this study is that the sand nearest the wellbore is the sand injected during the first part of treatment; and the sand farthest from the wellbore is the sand injected during the last part of the treatment (see Fig. 6.35).

This means that the technique of tailing in with large sand in a vertical fracture using normal, nonviscous fracturing fluids has been carried out in reverse. If equilibrium velocity is attained, the large sand will be washed over the small sand. If it is not attained, the large sand will be distributed in a thin, long layer atop the settled bed of small sand. The large sand will not be ultimately nearest the wellbore, where it is needed unless special fracturing fluids and techniques are used.

When large sand is beneficial in vertical fracturing, it should either be injected *first,* rather than last, or be used all at once unless either a fracturing fluid with sufficient gel strength to suspend the sand is used or very high injection rates are used.

6.10 Summary

1. High fracture:formation-permeability contrasts are required to effect maximum productivity increases from a fracture.
2. The force required to embed a steel ball in a rock core is used as an index in classifying formations according to the ratio of overburden load to type, size, and amount of propping agent.
3. Partial monolayers (0.10 to 0.50), where applicable, effect very high fracture capacities.
4. Deformable, metallic propping materials provide higher fracture capacities in deep wells than are obtainable with other agents. This proppant should be considered for wells deeper than 12,000 ft.
5. A partial monolayer of deformable nutshell proppant generally will provide higher fracture capacities than are obtainable with a full layer of proppant. This material is normally used in the 5,000- to 12,000-ft range.
6. The nondeforming, brittle propping agents such as sand or glass beads generally yield higher fracture capacities in full layers than the deformable prop. The use of sand is limited to 5,000 ft in partial monolayer concentrations and to 7,000 ft in full monolayers. Strong glass beads may be used to depths of 12,000 ft or more.
7. Partial monolayers of proppants can be achieved by using oil-soluble or water-soluble spacers matched in density and particle size to the propping agent.
8. The rate of proppant advance in a horizontal fracture system and the placement efficiency can both be calculated.
9. The horizontal extent and vertical height of proppant fill in a vertical fracture can be calculated.
10. A bed of settled sand builds up in the bottom of a vertical fracture unless injection rate per foot of formation is very high or unless a high gel strength or very viscous fracturing fluid is used. Sand injected later in the treatment is washed over this settled sand bed. Since this settled bed is nearest the wellbore, it is the most important factor affecting fracturing results.
11. The area of a fracture and the required fracture: formation contrast (flow capacity) are dependent on the productivity increase desired, on the type of rock, and on the characteristics of the producing formation.

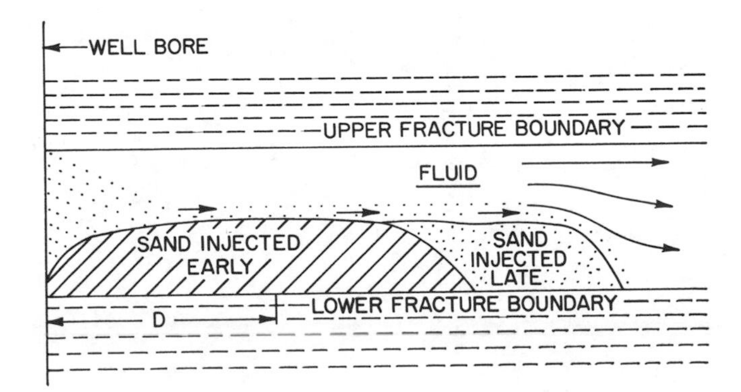

Fig. 6.35 Final position of sand injected late in treatment.[18]

References

1. Darin, S. R. and Huitt, J. L.: "Effect of a Partial Monolayer of Propping Agent on Fracture Flow Cacapacity", *Trans.,* AIME (1960) **219,** 31-37.
2. Lamb, Horace: *Hydrodynamics,* 6th ed., Dover Publications, Inc., New York (1945).
3. Darcy, H.: *Memoires a l'Academic des Sciences de l'Institute imperial de France* (1858) **15,** 141.
4. Muskat, M.: *Physical Principles of Oil Production,* McGraw-Hill Book Co., Inc., New York (1949).
5. Carman, P. C.: *Trans.,* Inst. Chem. Eng., London (1937) **15,** 150.
6. Wyllie, M. R. J. and Gregory, A. R.: "Fluid Flow Through Unconsolidated Porous Aggregates", *Ind. and Eng. Chem.* (1955) **XLVII,** 1379.
7. Huitt, J. L., McGlothin, B. B., Jr., and McDonald, J. F.: "The Propping of Fractures in Formations in Which Propping Sand Crushes", *Drill. and Prod. Prac.,* API (1959) 120.
8. Rixe, F. H., Fast, C. R. and Howard, G. C.: "Selection of Propping Agents for Hydraulic Fracturing", *Drill. and Prod. Prac.,* API (1963) 138.
9. Fast, C. R., Flickinger, D. H. and Howard, G. C.: "Effect of Fracture-Formation Flow Capacity Contrast on Well Productivity", *Drill. and Prod. Prac.,* API (1961) 145.
10. Dyes, A. B., Kemp, C. E. and Caudle, B. H.: "Effect of Fractures on Sweepout Pattern", *Trans.,* AIME (1958) **213,** 245-249.
11. Mallinger, M. A., Rixe, F. H. and Howard, G. C.: "Development and Use of Propping Agent Spacers to Increase Well Productivity", *Drill. and Prod. Prac.,* API (1964) 88.

12. Huitt, J. L. and McGlothin, B. B., Jr.: "The Propping of Fractures in Formations Susceptible to Propping-Sand Embedment", *Drill and Prod. Prac.,* API (1958) 115.

13. Nadai, A.: *Theory of Flow and Fracture of Solids,* 2nd ed., McGraw-Hill Book Co., Inc., New York (1950) **I,** 340-343.

14. Kern, L. R., Perkins, T. K. and Wyant, R. E.: "Propping Fractures with Aluminum Particles", *J. Pet. Tech.* (June, 1961) 583-589.

15. Nikuradse, J.: *Forsch, Gebiete Ingenieurw Forschungsheft* (Sept.-Oct., 1932); reprinted in *Pet. Eng.* (1940) **XI,** Nos. 6, 8, 9, 11 and 12.

16. Durand, R.: "Basic Relationships of the Transportation of Solids in Pipes—Experiment Research", *Proc.,* Minnesota Hydraulic Convention, Part I (1953).

17. Wahl, H. A. and Campbell, J. M.: "Sand Movement in Horizontal Fractures", paper SPE 564 presented at SPE Production Research Symposium, Norman, Okla., April 29-30, 1963.

18. Kern, L. R., Perkins, T. K. and Wyant, R. E.: "The Mechanics of Sand Movement in Fracturing", *Trans.,* AIME (1959) **216,** 403-405.

19. McGlothin, B. B., Jr., and Huitt, J. L.: "Relation of Formation Rock Strength to Propping Agent in Hydraulic Fracturing", *J. Pet. Tech.* (March, 1966) 377-384.

20. *Frac Guide Data Book,* Dowell Div. of Dow Chemical Co., Tulsa, Okla. (1965).

Chapter 7

Mechanics of Hydraulic Fracturing

7.1 Introduction

Hydraulic fracturing, although basically simple, is complicated by many important variable factors that significantly influence the results obtained and the over-all cost of the job.

Because this type of stimulation is performed frequently and with great success there often is not enough importance attached to these variables. However, operating experience clearly indicates that engineered treatment planning is essential if the best results and the greatest economic benefits are to be realized from the substantial expenditures made. Well preparation, pump truck rental, and additives constitute the major out-of-pocket expense items involved in most hydraulic fracturing treatments. As will be shown in this chapter, the interaction of these expense items must be effectively utilized for maximum treatment results and minimum total cost.

The purpose of this chapter is to summarize information regarding the numerous factors involved in hydraulic fracturing; these factors range from completing the well in preparation for fracturing to selecting a specific treating plan from the many materials and techniques available. A number of innovations applicable to hydraulic fracturing also are discussed.

7.2 Fracture Plane Inclination

It is important to predict correctly the inclination of a fracture in order to (1) determine realistically the probable increase in post-fracturing productivity, (2) determine whether multiple fracturing is feasible, and (3) prevent inadvertent fracturing into zones containing undesirable fluids.

Horizontal fractures are generally most effective where the following conditions exist.

1. Homogeneous formations are present.
2. The oil reservoir is underlaid by an active bottom-water drive, or is overlaid by a gas cap.
3. Gravity drainage is the major energy source and high capacity drainage channels are required.
4. Uniform injection of fluid into large areas of the reservoir is required.
5. Multiple fractures are required to drain massive pays.

A vertical fracture is generally most effective in the following situations.

1. Closely spaced horizontal stratification exists in the producing reservoir.
2. Fractures are desired at a given azimuth to facilitate drainage or injection.
3. Fluid injection is aided by vertical distribution in the producing interval.
4. Very deeply penetrating fractures are required.

The problem of effecting a vertical or horizontal fracture is discussed later in this chapter.

Although hydraulic fracturing treatments have been conducted since 1949, published field data are somewhat limited regarding the actual inclination of hydraulically created fractures as a function of depth, formation characteristics, and fracture treating gradient. Data presented in Chapter 1, Section 1, and on Figs. 1.3 and 1.4 show a vertical fracture in a 9,530-ft well and a horizontal fracture in a 2,635-ft well following squeeze cement jobs. Data from fracturing treatments in other fields follow.

Pine Island Field Data

In 1954, Pan American Petroleum Corp., in an attempt to determine the orientation of hydraulically induced fractures, conducted an experiment[1] in the Pine Island field near Shreveport, La. A well (163) was drilled in the center of a plot of approximately 10 acres, containing four corner wells. Fig. 7.1 presents the well pattern. Well 163 was an open-hole completion in the Annona Chalk with the casing set at 1,606 ft and with a total depth of 1,636 ft.

The fracturing treatment was carefully controlled, and included a low-rate formation breakdown using crude oil containing a radioactive tracer followed by a low-rate injection of a residual fuel oil fracturing fluid. This phase of the treatment was followed by the injection of 10,000 gal of viscous fuel oil containing an average of 1 lb of 20/40 sand per gal of oil at an average injection rate of 56 bbl/min.

During the treatment, surface and bottom-hole temperatures and pressures were measured in the fractured well, No. 163. The maximum bottom-hole treating pressure gradient was 0.67 psi/ft, and the average gradient was 0.61 psi/ft. During and after the treatment, samples were taken from offset Wells 33 and 112, which were 230 ft away, and Wells 67 and 83, which were 290 ft away. These samples were measured for viscosity and radioactivity.

During treatment, fracturing oil and sand broke through in Well 112. Recovery of radioactive oil from these wells indicated the proximity of Wells 33, 67, and 83 to the apparently horizontal fracture that was induced during this fracturing operation. Fig. 7.1 shows the location of the wells, the estimated areal extent of the sand-propped fracture, and the extent of the radioactive oil invasion.

Sacatosa Field Data

Reynolds *et al.*[2] in 1960 reported the results of a series of tests conducted in the Sacatosa field in Maverick County, Tex., to determine fracture areal extent and orientation. N. J. Chittim Well 37-1 was completed in the San Miguel No. 1 sand at a total depth of 1,438 ft. To aid in initiating a horizontal fracture, the well was perforated in a single plane with six shots in groups of three at 180°. The formation was broken down with lease crude oil at a surface pressure of 2,300 psi.

A mixture of 176,000 gal of diesel oil and lease crude oil treated with a fluid-loss agent was used as the treating fluid, and 270,000 lb of 20-40 mesh Poteet sand at a concentration of 1.5 lb/gal was added as the fracture propping agent. The injection rates during treatment averaged 30 bbl/min. At the end of the treatment an instantaneous shut-in pressure showed a treating pressure gradient of 1.6 psi/ft.

Following the fracturing treatment, 14 test holes were drilled in a radial pattern at various locations around the test well. By employing several different testing techniques, the presence of the fracture and some of its characteristics were determined. Fig. 7.2 is a location plot of the 14 test holes around the fractured well, No. 37-1. This plot shows the fracture to be horizontal. The fracture stayed within the San Miguel formation a horizontal distance of at least 250 ft.

Following are conclusions[2] drawn about the fracture in N. J. Chittim Well 37-1:

1. The fracture was horizontal.
2. The areal extent of the fracture was at least as large as that calculated from the Howard and Fast equation[3] — it extended more than 250 ft from the wellbore.
3. The fracture was not uniform in all directions or in thickness.
4. The perforations were only slightly eroded during the fracturing job.
5. Propping sand was not changed appreciably during the fracturing job.

Howard Glasscock Field Data

In 1961, Fraser and Pettitt[4] reported on a test conducted in the Howard Glasscock field to determine the orientation of induced fractures to be used in planning a waterflood project.

The area involved is in eastern Howard County, Tex. Production is from the 1,600-ft Queen sand of Permian age. This sand has an average porosity of

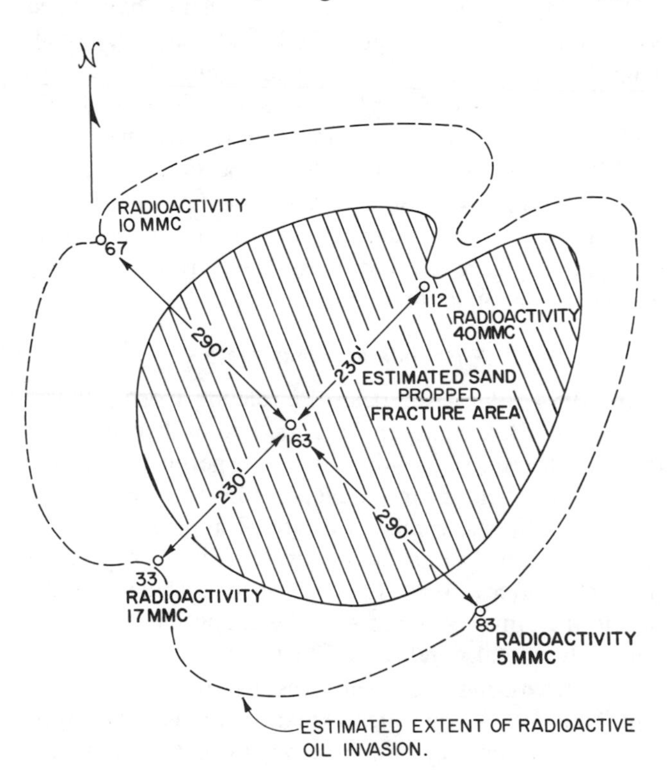

Fig. 7.1 Areal extent of fracture system, special tests, Pine Island field, La.[1]

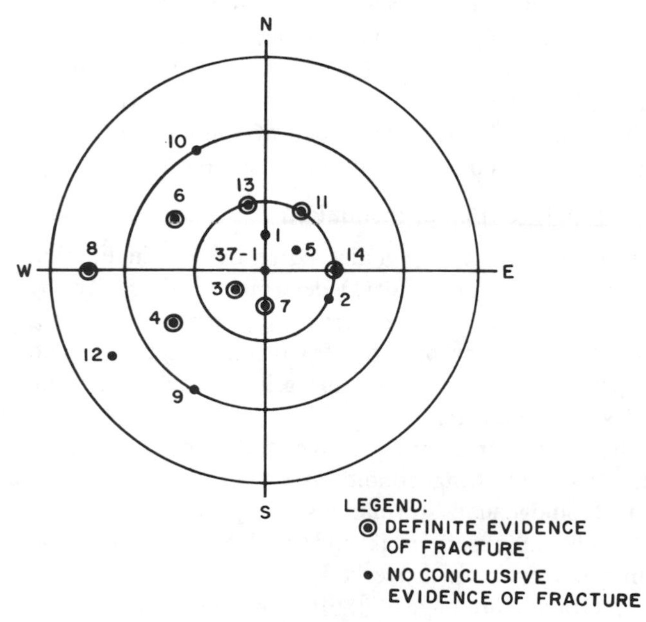

Fig. 7.2 Test well areal plot.[2]

23 percent, an average permeability of 57 md, an average water saturation of 35 percent, and an average gross thickness of 25 ft. The sand is very fine grained and contains considerable amounts of gypsum, silt, and clay. Geologically, the project is on the Eastern Platform of the Midland basin.

It was believed that if the orientation of the fracture system could be defined, the project area could be developed more economically by using fewer injection and producing wells. The problem, then, was to analyze data from previous treatments.

Sarah Hyman No. 13 was selected for the test and drilled to within about 15 ft of the top of the Queen sand at 1,607 ft using a 7⅞-in. bit. Then a 4¾-in. rathole was drilled to total depth. The rathole was filled with sand, and 4½-in. casing was cemented at the top. The rathole was cleaned out, and the pay zone was fractured with 20,000 gal of lease crude containing 30,000 lb of sand. After treatment, the instantaneous shut-in pressure was measured, and the fracture propagation gradient was calculated to be 0.75 psi/ft. It was concluded that a vertical fracture had been created; and the radius of this fracture was calculated to be 400 ft, assuming a fracture width of 0.25 in.[3]

To establish fracture orientation, an inflatable formation packer was used. A special rubber element was developed that possessed sufficient "memory" to bring an impression to the surface after a reasonable setting time. Since packer elements were available in 53-in. lengths, two of these were run in tandem. To determine the orientation of the fracture imprint, a standard directional survey was run after the packers were seated.

The packer assembly was set opposite the pay section, and an inflation pressure of 1,300 psi was applied. The packers were left seated for approximately 13 hours. Prior to unseating, the orientation measurement was taken, and then the packers were pulled from the hole.

Running the entire length of the packer elements was an impression of a vertical fracture with an orientation of North 73° East.

Allegheny Field Data

Anderson and Stahl[5] presented data in 1966 showing that vertical fractures were induced in three wells 1,600 to 1,700 ft deep in the Allegheny field in western New York.

To gain a better understanding of fractures that exist in a formation, various tests were run in the Richburg oil sand. This reservoir is part of a series of predominantly grey sandstones and shales with occasional brown sandstone layers; it is known as the Caneadea unit of the Canadaway group of Upper Devonian age. Average physical characteristics of 12 cored wells in this area were:

Thickness of gross sand formation — 13 ft
Thickness of net sand — 12 ft
Permeability — 1.7 md
Porosity — 11 percent
Oil saturation (before secondary recovery) — 44 percent
Water saturation (before secondary recovery) — 16 percent

Wells upon which tests were conducted are designated Wells 1, 2, and 3. Well 2 was located 440 ft north of Well 1, and Well 3 was located approximately 880 ft west of Well 1. Wells 1 and 2 were completed with cable tools; and the producing interval of Well 3 was rotary cored using fresh water as the drilling fluid to obtain oriented cores. Tests were divided into separate phases on each well as the fracturing program progressed. The tests consisted of measurements performed prior to treatment, during and after the formation breakdown operation, during fracture treatment, and after treatment.

To verify location and azimuth of the fracture created, and to obtain an impression of wellbore irregularities, a packer was used. This packer was lowered on tubing into the wellbore opposite the interval under study and inflated in stages to allow the impression material to conform to the wall of the wellbore. A magnetic compass was attached to the tool for orientation reference. Impressions made prior to treatment showed no existing fractures in the wells under study. Impressions made after both formation breakdown and treatment revealed fracture width and pattern at the wellbore.

Acoustic logs using a standard, single receiver sonde were made with various transmitter-receiver spacings. Gamma-ray logs were also made prior to treatment in order to provide background information with which to compare subsequent gamma-ray logs made after treatment with radioactive proppant. These logs were made with an instrument utilizing a scintillation-type detector located at the lower extremity of the sonde.

Surface and subsurface pressures were measured by calibrated transducers located on the treating manifold and a few inches from the lowest open portion of the well. All wells were fractured through 3-in. tubing with a packer set in open hole. Fluid used throughout the treatments and tests was formation crude. Because of the injection rates used (14 to 31 bbl/min), a friction reducing agent was added to the crude during the treatment of Well 3.

The impression packer tests conducted in each of the three wells showed that vertical fractures were formed. Fig. 7.3 shows a typical packer impression formed in these wells.

General Observations

The preceding data show that in the Pine Island and Sacatosa fields the induced fractures were horizontal, and in the Howard Glasscock and Allegheny fields they were vertical. It is our experience that geologic conditions and well depth in most areas will control the fracture plane inclination, whereas in other regions, the fractures in shallow wells may be oriented in either a vertical or a horizontal plane depending upon the stimulating technique used.

Additional correlation of theoretical and actual field results is needed before an accurate method of predicting fracture inclination in shallow wells in new areas can be developed. In areas where there has been appreciable fracturing, the inclination of the fracture system can be estimated with reasonable reliability by using post-fracturing well performance data.

To obtain background information on fracture inclination certain data may be collected and analyzed for areas where formation stimulation may be necessary. For example formation breakdown pressure and fracture treating gradients (psi/ft) may be observed on several wells completed in each formation in different fields and geographical areas. Reliable fracture treating gradients can be calculated using the instantaneous shut-in pressure technique described later in this chapter. These data may be correlated with formation characteristics, communication tests, production logs, and tracer surveys to determine fracture inclination.

These data will also be very useful for planning other phases of a fracture treatment and, in later secondary recovery operations, for establishing the maximum injection rate that may be used without extending existing formation fractures or without creating new ones.

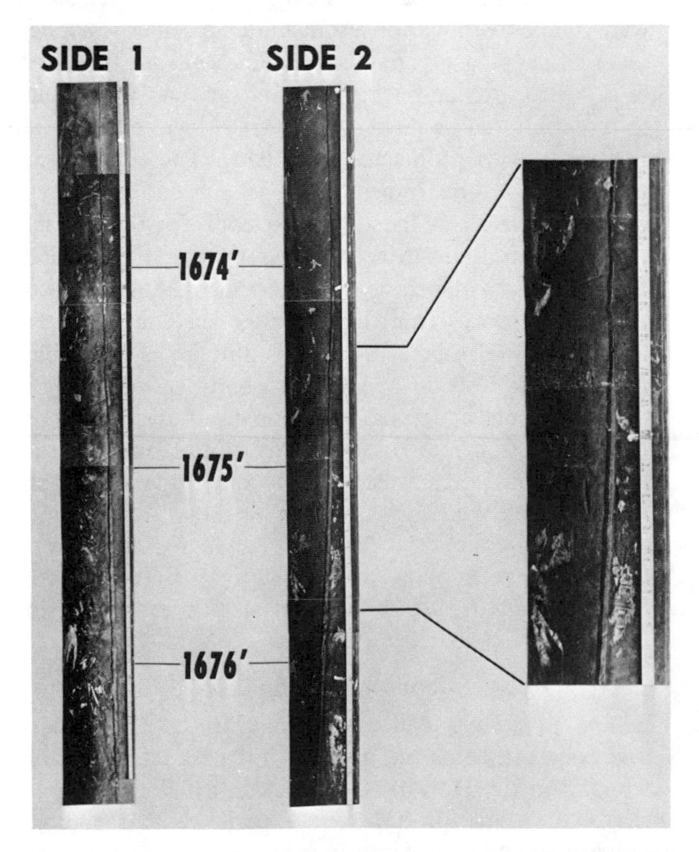

Fig. 7.3 Impression packer after treatment, Well No. 3.[5]

7.3 Fracture Placement and Orientation Methods

The depth and orientation at which a fracture is effected are often important in achieving maximum productivity and oil recovery. For example, in shallow reservoirs near depletion, an increased utilization of gravity drainage by placing a horizontal fracture at the base of the producing interval will extend the economic life of the well. Water or gas coning can often be reduced by the placement of horizontal fractures at appropriate elevations in the formation.

Although it is necessary to choose the right technique for selective or multiple fracturing, it is even more important to have an effective cement job between the intervals to be treated. Frequently, it was thought that selective stimulation had been accomplished in perforated completions, but subsequent tests showed that the intervals were in communication behind the casing. In these cases, considerable money was spent in treating the zones individually; yet the results were inferior to those that would have been obtained had the zones been treated indiscriminately. It is important, therefore, to evaluate realistically the feasibility of physically isolating, in both open- and cased-hole completions, the interval to be selectively stimulated.

Unsuccessful efforts to place horizontal fractures in individual sections of a zone may be erroneously attributed to failure of either the cement job or the down-hole mechanical equipment when, actually, the reservoir stresses in that particular field are such that only vertical fractures could be initiated. This is usually the case in deeper wells. However, if vertical or horizontal fractures are possible and zone isolation can be achieved, several techniques are available for selectively treating different zones. It should be emphasized that the fact that a fracture is started in a desired orientation does not insure that it will continue through the formation in the same direction. The fracture may alter its course away from the well, depending upon stresses and joints in the rock.

Sand-Liquid Jetting

The method that has been used most successfully to place horizontal fractures selectively involves jetting of sand and liquid at high rates to circumferentially notch the formation. In laboratory tests conducted in a 3-ft.-diameter concrete drum, the casing, cement, and formation were penetrated several inches after rotating a jetting tool for 30 minutes while pumping sand and water through the jet at 2,000 psi differential pressure. In shallow test wells in a uniform Oologah limestone formation near Tulsa, Okla., fracture placement has been repeatedly achieved in both cased hole and gauge open hole using this method. Also fracture initiating pressures were consistently lower with this technique than with any other method attempted either in open hole or through casing. This method has been successfully applied in the field by several different operators and service companies.

In attempts to orient vertical fractures in shallow test wells in the Oologah limestone it was found that the sand-liquid jetting method could sometimes be used to place vertical fractures in gauge open hole. However, because the formation apparently contained natural weaknesses or joints in the east-west direction, the initiated fracture often tended to go that way, regardless of the direction of the notch. As a result, special techniques were required to control the orientation of the fracture.

In order to confine the fracturing fluid and stresses on the formation to the notch, a plastic casing patch[6] reinforced with glass fiber was placed in the gauge open hole section where the fracture was desired. The patch and formation were then notched in a desired direction by using a sand-water jetting technique to cut notches in both sides of the hole simultaneously. A two-element packer was then set in the patch to straddle the notched zone, and the formation was fractured through the notch. With this method we were consistently able to initiate vertical fractures in any desired direction.

Shaped Charges

Many configurations of shaped explosive charges have been used in fracture placement tests. Circumferential cutting shaped charges have been used in attempts to place horizontal fractures. These charges, which were designed primarily for cutting casing to facilitate its recovery, are generally inefficient and rarely will cut more than 2 in. in a hard limestone or sandstone formation. Some limited success has been experienced with this type of charge in the placement of horizontal fractures in shallow test wells and in the field. However, due to the limited penetration and the large quantity of explosive required, it has been difficult to form secondary fractures outside the circumferential cut.

Vertical cutting shaped charges have been used experimentally to notch formations in placing vertical fractures. Although this type of shaped charge is inherently more efficient than a circumferential cutting charge, its penetration is appreciably less than that of a sand-water jetting tool. Experience in vertical placement has not been good with this tool in test wells. Although some success has been achieved, to make this method feasible it must be used with the plastic casing patch, described above.

High density perforating was evaluated many years ago[7] for the placement of horizontal fractures in conjunction with squeeze-cementing operations. As a result of this and other work, four-way, single-plane, shaped charge perforators — and, later, six-way[8] and eight-way guns — were developed for use in the placement of horizontal fractures in both open hole and casing. These shaped charges, which are much more efficient and which penetrate more deeply than the linear cutting charges, have been used in many wells in attempts to place horizontal fractures. Success has been reported[2,8] on many of these tests. Conventional shaped charges have been fired unidirectionally in an effort to place vertical fractures in both cased and open holes. Although there is no documented success with this method, it appears logical that this would be successful in areas where formation stresses favor vertical fractures.

Laboratory and field tests indicate, however, that success with any method of fracture placement depends primarily upon the tectonic stresses in the formation and upon the zonal isolation achieved.

Mechanical and Penetrating Fluid Methods

Mechanical methods of notching formations to enhance horizontal fracture placement have been tested with some success in both the laboratory and the field.[7,9] The primary disadvantages of these methods are the high costs involved and the limited penetration of the formation that can be achieved.

Another approach[10] to fracture placement is the use of a nonpenetrating fluid to restrict fluid egress from the well, along with the use of a penetrating fluid placed at a selected elevation to effect a fracture. The key to fracture placement — vertical or horizontal — is to restrict fluid egress to the desired plane and elevation and, if possible, to weaken the formation in the desired direction. Again, the penetrating fluid technique can be successful only if the in-situ formation stresses are favorable to the type of fracture desired.

7.4 Multiple Fracturing

Selective stimulation of each zone in an open-hole completion or of each perforated interval is often desirable to obtain maximum post-treatment well productivity and to deplete each zone uniformly as the well is produced. Although there are several techniques for this, the objective is not always attained. The limitations and relative advantages of the more popular treating procedures and other proposed methods are presented in the following sections.

The small fluid volumes and viscous gels employed in early fracturing dictated the use of tubing and packers. The latter were conventional single types set above the pay interval, with fluid entering the formation below, and creating only one fracture. It soon became obvious that some technique was needed for treating wells where more than one zone was present so that each zone could be isolated and stimulated individually. Tools were developed containing dual packers arranged for variable spacing, thus permitting the straddling of a producing interval for treatment between the two packer elements. In this manner several sets of perforations could be treated by merely moving the assembly from one interval to another. Tools like this were also built for use in open hole.

Diverting Agents

In wells with open-hole diameters too large for

formation packers, it was not possible to isolate the treatment to specific sections of the open hole. Other wells were difficult to treat because poor casing cement, either around the shoe joint or between perforated sections, made effective isolation and treatment of a given section impossible. It was to combat these problems and, where possible, to obtain multiple fractures when the use of packers was impracticable that the multiple-fracturing method[11] using diverting agents was developed.

In shallow well tests and later in the field it was found that a single fracture started and extended by hydraulic fracturing may be plugged at the face on the well wall by the introduction of a bridging material that will exclude further penetration of the fracturing fluid into the crack. It is then possible to increase the hydraulic pressure in the hole so that another fracture will occur at some other elevation where the fluid seal on the well wall is not so complete, or where the rock is markedly weaker than at other depths. Each successive fracture thus formed and not temporarily plugged can be extended with a fracturing fluid containing no bridging material. This process could be repeated, yielding additional fractures.

Evaluation work on bridging materials indicated that the granular-type additive is more effective than either the fibrous or lamellated type in restricting the pressure to the wellbore. This is due primarily to the fact that with the granular material the large particles bridge at the opening of the fracture, and the smaller particles, in turn, bridge on the larger particles, and thus a seal is formed at the surface of the fracture.[12]

The ability of the granular-type bridging material to seal cracks, fractures, and small holes proved, in a large number of laboratory-simulated well tests, to be an important factor in restricting the hydraulic fracturing operation to the productive zone of the well. This characteristic was shown to be particularly important in the isolation of the treatment in wells with poor primary cement jobs or with cement that is cracked and shattered during casing perforating operations.

The most commonly used oil-soluble, temporary bridging material is ground naphthalene, and the most commonly used water soluble bridging agent is graded rock salt. These materials are generally used in an oil- or water-base gel or emulsion.

The diverting mixture is normally pumped between stages of a selective stimulation treatment to exclude the zone previously treated and thereby divert to another untreated interval the stimulating fluid that follows. Many of the mixtures also incorporate a breaker that is catalyzed by the bottom-hole temperature to reduce the viscosity of the injected gel so it may be produced following the treatment. The diverting material is convenient to use. It is routinely furnished by the service company, thereby minimizing advance planning, and normally no special down-hole equipment is required.

The principal disadvantage of this selective technique is the uncertainty associated with its use. It is probable that the surface pressure will increase when the diverting material is pumped to bottom. This may be interpreted as a breakdown of a different zone, whereas the stimulating fluid injected after the diverting material may continue to enter the previously treated zone.

Field tests have shown that multiple fractures can be obtained by first sealing intrinsic or previously produced fractures with temporary, solid bridging materials carried in gels, emulsions, or thickened acids, and then creating additional fractures with an appropriate fracturing fluid. The most satisfactory fracture-plugging agents are temporary bridging materials that possess adequate, effective strength initially, and then are slowly and completely soluble in produced well fluids.

Multiple-fracturing operations are particularly adaptable to wells when: (1) poor primary cement job prevents isolation of the zone to be treated; (2) there are open-hole sections, which prevent the use of packers; (3) they are needed in long open-hole or perforated casing sections to drain the reservoir effectively; and (4) productive sections have such low effective vertical permeability that they cannot drain the reservoir effectively.

Perforation Ball Sealers

The principle of the perforation sealing process[13] is similar to that involved in a ball and seat valve. The sealing elements are rubber-coated nylon or aluminum balls that are carried by the flow of treating fluid into the well, down the treating string and, finally, against the casing perforations through which the treating fluid is passing. When a sealing element comes in contact with a perforation, the element restricts the flow of fluid, thus creating a pressure differential across the perforation. This pressure differential causes the element to seat and seal tightly against the perforation, thus preventing further flow through the perforation. The element remains sealed against the perforation as long as the pressure in the well is kept higher than that in the surrounding formation.

In a fracturing treatment where two or more perforated intervals are exposed, the treating fluid is injected into the well and enters the perforations opposite the most permeable zone. Balls are injected into the flush fluid following the fracturing treatment of the first zone. Since this perforated interval readily takes fluid, these balls follow the fluid stream to the perforations where they provide a seal due to the differential pressure holding them in place.

While the pressure on the well is maintained, the second fracturing stage immediately follows the first stage flush. Since the first perforations are sealed, pressure builds up and the formation opposite the second perforated interval is broken down and the fracturing fluid injected. This process is repeated as many times as is necessary to complete the treatment of all zones.

Upon completion of the fracturing treatment, pressure on the wellhead is released, and the differential pressure from the formation toward the wellbore causes the balls to be released from the perforated holes. The specific gravity of the balls causes them to fall to the bottom of the well.

If the presence of the rubber balls on bottom is objectionable, they may be recovered with a bailer, or, if the well is capable of flowing, they can be washed from the well by a high rate of flow. In most wells, the balls are left on bottom. To be practical and effective in sealing off casing perforations in multi-stage fracturing treatments, the balls used must meet the following conditions.[14]

1. They must be of such a size and density as to be retained in the fluid stream so that each ball will be directed to and held on a perforation.
2. They must be tough enough so they will not be extruded through the perforation under the pressure differentials actually encountered in the field.
3. They must not not only adequately seal the perforation during treatments, but also free themselves from the perforation when the pressure differential is decreased.
4. They should be of such specific gravity that they will settle to the bottom of the well when released.
5. They must be drillable.

The efficiency of the sealers [15] is primarily influenced by: (1) the velocity of the balls falling down the pipe, and (2) the velocity of the fluid entering the perforations. To divert the sealer to the perforation, the inertial force of the ball must be overcome by the drag force created by the fluid velocity through the perforation.

The force tending to remove the ball from the perforation is created by the fluid drag upon the exposed part of the ball. The force tending to hold the ball on a perforation is proportional to both the area of the perforation and the pressure differential across the ball (on the perforation). In order to dislodge the ball from the perforation, the fluid drag on the ball must exceed the vertical vector of the holding force.

The differential pressure that holds the ball on the perforation will be at a minimum the instant of sealing and will increase thereafter because of the subsequent bleed-off of fracture fluid to the formation. Under normal conditions, the change in holding force caused by this bleed-off is nominal and is neglected.

Equations have been derived[15] for calculating the velocity of the ball, its inertial force, the drag force tending to seat the ball, the force tending to unseat the ball, the holding force, and the unseating force.

The primary disadvantage of the ball-sealer selective fracturing technique is that if several intervals are exposed simultaneously there is no assurance that each zone is positively treated and excluded. For positive treatment, each zone should be individually perforated and treated and then enough ball sealers should be injected to achieve a positive shutoff. (Generally, two to three balls per open perforation are necessary to obtain a complete shutoff.) As the ball sealers are hydraulically maintained on the first zone, the next zone is perforated. This second zone is then fractured and the process is repeated as many times as necessary to complete the selective well treatment.

Limited Entry

Shortly after perforation ball sealers were introduced, it was recognized that casing perforations have a limited capacity to take fracturing fluid.[16,17] It was also recognized that too many perforations would be detrimental to a selective fracturing job when several perforated zones were to be treated with ball sealers. Some methods were presented of designing fracturing treatments to achieve multiple fracturing in two or more perforated intervals.[18,19] This was accomplished by using a limited number of perforations in a well and injecting the fracturing fluid at a high enough rate to enter all of the open perforations.

To treat more than one perforated interval, the bottom-hole treating pressure must be raised above the fracture initiation pressure of each successive zone to be treated. This can be accomplished by limiting the number and diameter of the perforations in the casing. As shown on Fig. 7.4, the perforation friction pressure varies directly with the rate at which the fracturing fluid is pumped through the perforation. Therefore, by increasing the injection rate, the perforation friction will be increased. In other words, the perforations are acting as individual bottom-hole chokes. They create an increase in the bottom-hole casing

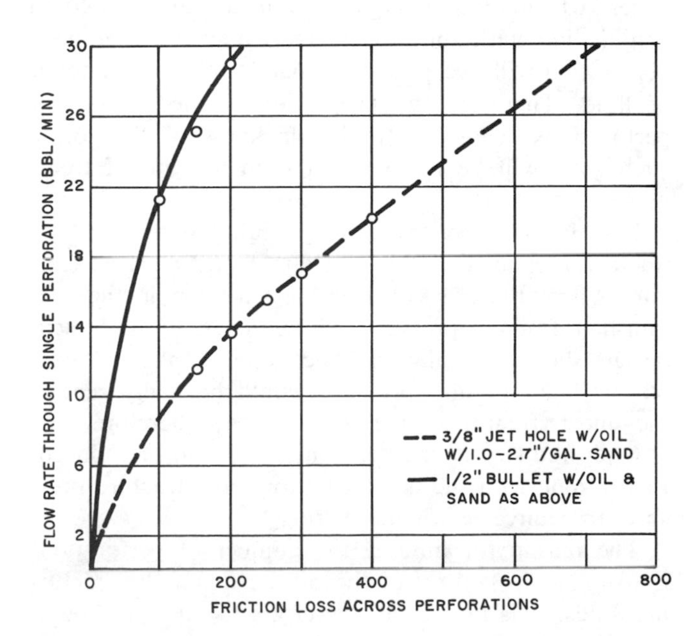

Fig. 7.4 Flow rate vs friction loss, laboratory measured (Halliburton).

pressure as the injection rate is increased. The accompanying increase in pressure in the casing will then initiate a fracture in the next zone.

The process of breaking down each successive zone occurs rapidly, since maximum pressure and rates are established early in the treatment. With an adequate injection rate at the surface, this process can be continued until either all of the perforated zones are being fracture treated or the maximum permissible pressure on the casing is reached.

According to Lagrone and Rasmussen,[19] the best results are obtained by maintaining perforation friction at a maximum during treatment. This assures that, within the permissible casing pressure limitations, all of the perforated interval that will accept fluid will be treated. It is recognized that all the perforations could be treated simultaneously at a lower injection rate. This would not be true, however, if the bottom-hole fracture pressure of the individual zones varies significantly. Therefore, to assure that all zones are being treated, an injection rate that will give a maximum permissible casing pressure is necessary.

Small-diameter perforations are preferred in limited entry treatments to increase perforation friction and to lower the hydraulic horsepower requirements. Fig. 7.4 shows that, for the same perforation friction, approximately twice as much fluid can be injected through a ½-in. hole as through a ⅜-in. hole. Therefore, by using the small perforations, less hydraulic horsepower is required to deliver an injection rate adequate to maintain maximum perforation friction. Therefore, ⅜-in. perforations are generally used for limited entry treatments.

Limited entry treatments can be designed so that the desired amounts of fluids will be injected into each interval. This is an important advantage where thick zones, which require larger treatments, are treated in conjunction with thin zones. It is assumed that each perforation will accept approximately the same amount of fluid. Therefore, by proportioning the number of perforations according to the thickness of the zones, each zone will be given the desired amount of treatment.

For this method to be successful, the bottom-hole fracture pressures of the individual zones must be similar. Where there is considerable variation in the bottom-hole fracture pressures of the zones, the treatment design should be altered. The zone with the lowest bottom-hole fracture pressure would normally receive the most treatment per perforation; therefore, the number and size of the perforations should be reduced. In the zone with the highest bottom-hole fracture pressure, the converse would be true.

The reason for limiting the number of perforations is to maintain control of the placement of the fracturing fluids. It is important, therefore, to know the number of perforations to use for a desired injection rate to obtain maximum perforation friction.

The equation for perforation friction is:

$$p_{pF} = p_{ti} - p_{ISI} - p_F \quad \text{(7.1)}$$

This equation was derived by substitution in the following equations:

$$p_{bf} = p_{ti} + p_h - p_F - p_{pF}, \quad \text{(7.2)}$$

$$p_{bf} = p_{ISI} + p_h \quad \text{(7.3)}$$

A limited entry treatment is designed by trial and error. First, to treat all of the pay interval and properly proportion the treatment, a minimum number of perforations are chosen. Second, an injection rate is determined for those perforations that will maintain maximum perforation friction (within casing pressure limitations). If the calculated injection rate is considered unreasonable (either too high or too low), the number and placement of the perforations should be altered.

The limited entry technique provides field data that can be used to determine the number of intervals that were treated. If this analysis indicates that all zones are not being treated, the completion design can be altered.

The three requirements for determining the number of perforations accepting fluid are (1) accurate injection rates, (2) accurate surface injection pressures, and (3) an instantaneous shut-in pressure (ISIP) at the beginning of the job.

The instantaneous shut-in pressure should be measured at the start of the treatment so that the actual number of perforations accepting fluid during treatment can be calculated. ISIP can be defined as that static pressure required to hold a fracture open. Fig. 7.5 is a treatment pressure chart. To get an ISIP while pumping into the formation at fracture pressure, pumps are stopped instantaneously. The recorded surface pressure falls abruptly to a stabilized pressure and then bleeds off slowly into the formation. The abrupt stabilized pressure point is a measurement of the ISIP. Note that the ISIP at the start of the treatment illustrated in Fig. 7.5 is 2,400 psi and increases to 3,000 psi at the end of the treatment. This is not a freak occurrence. The ISIP increases during most treatments.

The major difficulty that has been encountered in

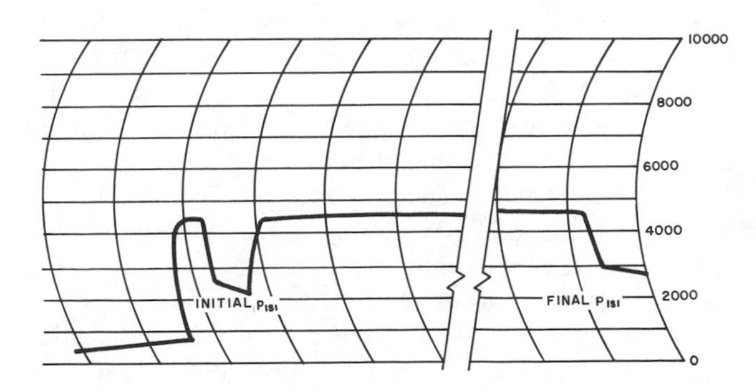

Fig. 7.5 Pressure chart.[19]

limited entry treatments has been insuring that all holes are open prior to the fracture treatment. Seldom are all of the perforations able to accept fluids without first being acidized. If the well is perforated with a solids-free fluid, and if a differential pressure from the formation to the well is maintained during perforating, formation breakdown and perforation plugging difficulties are usually eliminated.

Limited entry fracture treatments have been performed with injection pressures and rates, and treating-fluid types and volumes similar to those of conventional treatments. Sometimes it is undesirable or impossible to have injection rates high enough to insure treatment of all the perforations. In this case, ball sealers can be used as a diverting agent. Because of the higher injection rates per hole and the greater separation between perforations, ball-sealer action is usually very effective in limited entry treatments.

Individual perforations sometimes sand out during treatment. A decrease in injection rate is indicative of the time and number of perforations affected when sand-out occurs. A continuous-rate recorder is necessary for observing the decrease in the number of perforations taking fluid. It is also helpful in determining the proper number of ball sealers to "drop" during a job.

The limited entry technique is effective in treating multiple zones and thick pays. In many instances, even with higher hydraulic horsepower cost, this method results in lower treating costs, particularly where dual horizons can be treated simultaneously.

Bridge Plug and Sand Plugback Methods

Bridge Plugs. There are many types of bridge plugs that are available for use in multiple or selective fracturing of cased wells. Permanent (drillable) bridge plugs have been used since the early days of fracturing for selective stimulation of multiple zones. The procedure is to perforate the bottom zone in the well and fracture it either down the casing or through the tubing equipped with a packer set above the perforations. Following the treatment, a drillable bridge plug is set above the perforations. The next zone is then perforated and fractured as was the first zone. This procedure is repeated as many times as is desired. After all the fracturing treatments are completed, drilling tools are run in the well and the bridge plugs are removed or knocked to bottom. The principal disadvantages of this procedure are that (1) a drilling rig is required to remove the bridge plug; (2) it is slower than some of the other methods; and (3) communication with the previously fractured zones below generally cannot be detected during the treatment.

In the middle 1950's the bridge plug method of fracturing was greatly improved by the development and introduction of the retrievable bridge plug. The operation of this tool is simple and positive, utilizing pressure differential from either above or below to achieve and maintain a packoff. It is run in on tubing, released, and then retrieved by means of a retrieving head. This tool, when used in conjunction with a retrievable squeeze packer, makes a straddle packer with the distances between packers unlimited (see Fig. 7.6). It may also be used alone for down-the-casing stimulation treatments in multiple-perforated intervals. As with the drillable plugs, the retrievable plug may be used when all zones in the casing are perforated prior to treatment. It may also be used by starting at the bottom and successively perforating and treating each zone. To avoid having to pull the tubing to perforate each zone, a through-tubing, debris-free perforator can be run through a full-opening squeeze packer. To insure full perforation hole size and penetration, a mechanical or magnetic positioning device should be used on the perforating gun. The principal advantages of using a retrievable bridge plug are that it is much quicker and therefore less costly than using drillable plugs, and that no drilling rig is required.

Sand Plugback. The sand plugback procedure, which can be very effective, can also be the least costly of all mechanical methods of isolating zones. With this technique, the lowest zone is first perforated, stimulated, and then excluded by a sand column, which is placed with the flush fluid. The process is then repeated on each zone until all intervals are selectively stimulated. The sand is reversed circulated from the well while the pump trucks used for fracturing are still on location.

The quantity of sand to place following each stage can be closely approximated, since the interval between zones will be known and the proper volume of

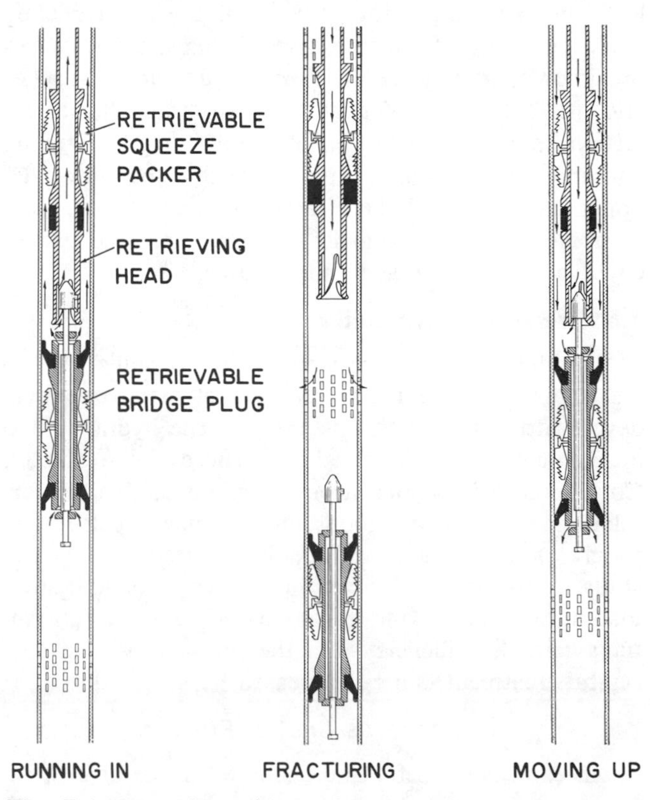

Fig. 7.6 Fracturing with retrievable bridge plug (Baker).

fracture propping sand can be added to the fluid pumped when flushing operations are started. A small quantity of loss-of-circulation preventive material, added to the last sack of sand introduced into the well, will make the sand plug impermeable even though the sand column is only a few feet high. It is advisable to verify that the desired fillup has been obtained before the next zone is perforated. This may be done by pumping into the casing. A surface pressure buildup with negligible injection of fluid is a positive indication that the previously stimulated zone has been excluded.

If treating pressures require that fracturing fluid be pumped down tubing that is isolated from the casing by a packer, a short tail pipe should be run below the packer to facilitate reversing the sand from the well. A through-tubing perforator may be used to perforate each zone. These guns are available with a mechanical or magnetic device to position the gun adjacent to one side of the casing before firing so that consistent hole size and penetration are achieved. The number of perforations should be correlated with the expected injection rate and should be grouped within a narrow vertical interval to minimize the possibility of channeling behind the casing.

A variation of this technique is to initially perforate all zones and set a packer above the lower zone, treat the zone, and then exclude it with a sand plug. The packer is then reset above the next zone and the process is repeated until all zones are individually treated. The sand plug is then reversed from the well. A suitable annulus pressure (for example, 500 psi less than the breakdown pressure for the lower zones) should be maintained to minimize the possibility of channeling while breaking down the zone being treated. After breakdown, it may be desirable to unseat the packer and manifold the tubing and annulus for a higher injection rate. Disadvantages of manifolding the tubing and annulus with this technique are that the perforated upper zones could be fractured prematurely; and pumping the stimulating fluid past an unseated packer could be objectionable in some cases.

7.5 Treatment Hydraulics

The high success ratio achieved in hydraulic fracturing is probably the major reason that many operators devote little thought and planning to the hydraulic and mechanical factors involved in a successful treatment. Considerable economic incentives are available for correlating the tubular goods arrangement, perforating program, pump rate, and additives used. By careful study of all factors involved in the design of well stimulation treatments, the costs can be reduced substantially and the efficiency and the probability of a successful treatment can be increased.[20]

Instantaneous Shut-In Pressure

An extremely useful and accurate means of evaluating several important aspects of treatment design is to observe the surface pressure when pumping operations are suddenly halted after a formation fracture has been extended a distance from the wellbore. This instantaneous shut-in pressure can be effectively utilized to determine (1) accurate breakdown and fracture treating gradients, (2) the effectiveness of friction reducing additives under field conditions, (3) the horsepower requirements for fracturing, (4) the approximate number of perforations taking fluid, and (5) the effectiveness of the mechanical devices being used to reduce friction horsepower.

For many formation evaluation techniques, primary data must be plotted or further analyzed before any meaningful results are obtained. ISIP data, however, can be used immediately in conducting the treatment or in diagnosing the over-all treatment efficiency. The relationship of the instantaneous shut-in pressure to other elements constituting the total surface pressure required to inject fluids into a formation is

$$p_{ti} = p_{bf} + p_F + p_{pF} - p_h \quad . \quad . \quad . \quad . \quad (7.2)$$

It should be noted that if pumping operations are suddenly stopped, the two friction terms of Eq. 7.2, p_F and p_{pF}, drop out and the surface pressure, p_{ti}, is then the difference between the bottom-hole fracturing pressure, p_{bf}, and the hydrostatic pressure, p_h. The surface pressure observed at the moment pumping operations are suddenly discontinued is termed the instantaneous shut-in pressure (ISIP, or, in the equations, p_{ISI}), and Eq. 7.2 can now be expressed as follows:

$$p_{ISI} = p_{bf} - p_h \quad . \quad . \quad . \quad . \quad . \quad . \quad . \quad . \quad (7.3)$$

Eq. 7.3 shows that the ISIP is simply the surface pressure required to inject fluids into a fracture system under dynamic conditions if there were no pipe or perforation friction involved. The ISIP can be a very valuable and useful technique in planning and diagnosing stimulation treatments.

Effect of Tubular Goods

A well stimulation treatment should be planned concurrently with the drilling and completion program if optimum results and lowest total cost are to be achieved. This is particularly true with regard to the tubular goods and perforating program, which often significantly affect total treating costs.

The production casing size is usually selected on the basis of many considerations other than hydraulic requirements for stimulating the well. Fig. 7.7 graphically illustrates that the horsepower required to overcome tubular goods friction can constitute the major portion of the total horsepower used in some treatments. In this example, the well was completed open hole and perforation friction is not a factor. Thus, the total horsepower consisted of only two components: (1) the horsepower required to inject fluids into the formation, and (2) the additional horsepower to overcome friction for the various combinations of tubular goods depicted.

It should be noted that the horsepower required to inject the fracturing fluid into the formation varies directly with the pump rate and cannot be controlled by the operator. The horsepower lost to tubular goods friction increases exponentially with the pump rate; it can be controlled to a considerable degree, however, by using large size tubing, by treating down the casing, or by manifolding the tubing and casing and using friction loss additives.

In most areas, the cost of pumping equipment will average (as of 1969) approximately $1/hhp for a 4-hour period. On the basis of this cost figure and horsepower data from Fig. 7.7 for a pump rate of 15 bbl/min, the cost of tubular goods friction horsepower varied from $95, if the job was conducted down 5½-in. casing, to $2,700 if the fluid was pumped through 2⅞-in. tubing. In both cases, the same horsepower was delivered to the formation, yet the cost per horsepower delivered at the formation face varied from $1.16 when pumped down the 5½-in. casing to $5.62 when the fluid was pumped through tubing.

The curves on Fig. 7.7 show that the friction horsepower can be a major portion of the total used in some treatments, particularly with smaller sized tubular goods and higher pump rates. In other cases, such as when pumping down casing at moderate rates or injecting through larger diameter casing at high rates, the total friction horsepower is negligible and expending money to further reduce the friction horsepower would not be worthwhile. On the other hand, if it is possible to pump down both the tubing and tubing-casing annulus, either before or after the formation is broken down, then hydraulic horsepower and, thus, cost can be reduced considerably.

Cross-Over Valve Technique

The potential reduction in horsepower cost that will result from treating down manifolded tubing and casing instead of down tubing alone often is not obtained in wells where the surface pressure required to initiate the fracture exceeds the working pressure of the casing. However, pressures recorded during well stimulation treatments indicate that after the fracture is initiated, the surface pressure often declines and is within the working pressure of the casing throughout the remainder of the job.

Fig. 7.8 shows a surface pressure and injection rate record for a stimulation treatment where a water-base fluid was used in a typical 7,000-ft. well. The surface pressure required to initiate the fracture in this well was approximately 5,500 psi; then it dropped to approximately 5,100 psi at a pump rate of 5 bbl/min. When injection was stopped suddenly, the pressure dropped to 2,600 psi. This instantaneous shut-in pressure (marked ISIP on Fig. 7.8) is the surface pressure (exclusive of friction losses) required to inject into the formation after breakdown. The 5,100 psi injection pressure while pumping at 5 bbl/min through tubing includes 2,500 psi friction loss (5,100 minus 2,600 ISIP).

These pressure-injection rate relationships ultimately resulted in the development of the cross-over valve and

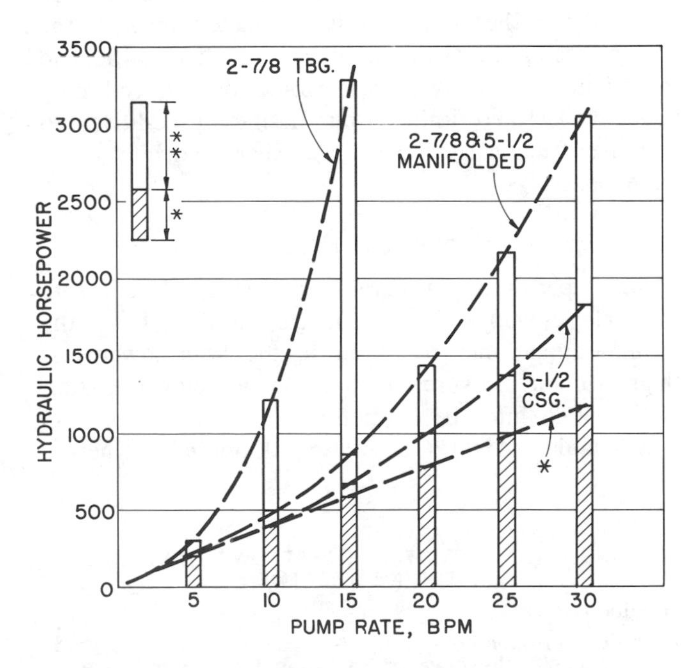

Fig. 7.7 Horsepower distribution during a fracturing treatment.[20]

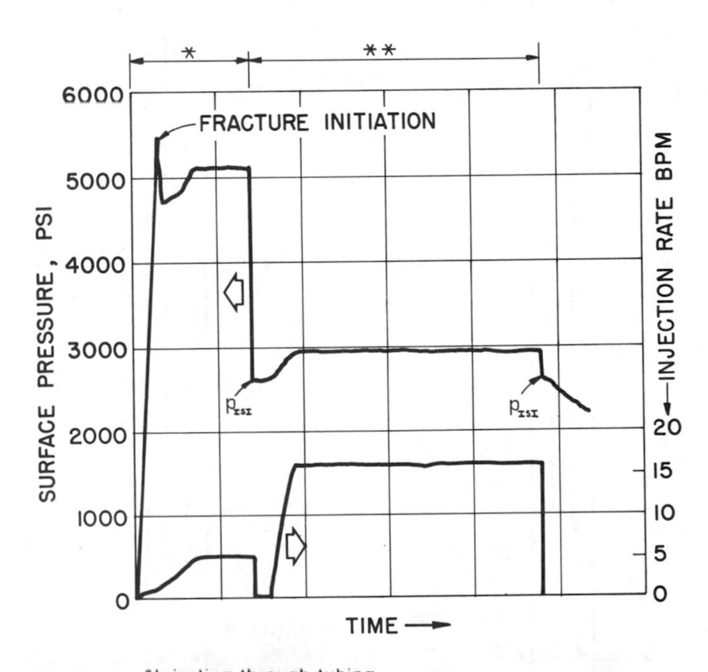

Fig. 7.8 Surface pressure and injection rate while fracturing.[20]

packer arrangement, which is shown in Fig. 7.9. The valve permits the casing to be isolated when the fracture is being initiated, and then during the remainder of the treatment permits simultaneous injection through the tubing and annulus. The cross-over tool is installed as an integral part of the tubing string and is pressure-controlled from the surface. It can be opened as often as necessary during well stimulation by simply applying pressure to the annulus, and it can be closed at any time by discontinuing injection down the annulus. Injection down the tubing can be continued, uninterrupted during operation of the valve.

An example of the advantage of using a cross-over tool is shown on Fig. 7.8. Noted that, after the formation was broken down through the tubing containing a packer and a cross-over valve, it was possible to increase the injection rate into the formation through the manifold tubing and the tubing-casing annulus to 16 bbl/min, yet the surface pressure required was only 3,000 psi. This compares with a surface pressure of 5,100 psi required to inject only 5 bbl/min through the tubing.

Field tests of the cross-over valve have been very successful. The injection rate down the manifold tubing and casing is typically two to four times higher than the rate attained down tubing, and these rate increases are achieved with decreased injection pressure. The savings from using the cross-over valve are achieved by substantially decreasing the horsepower lost to friction.

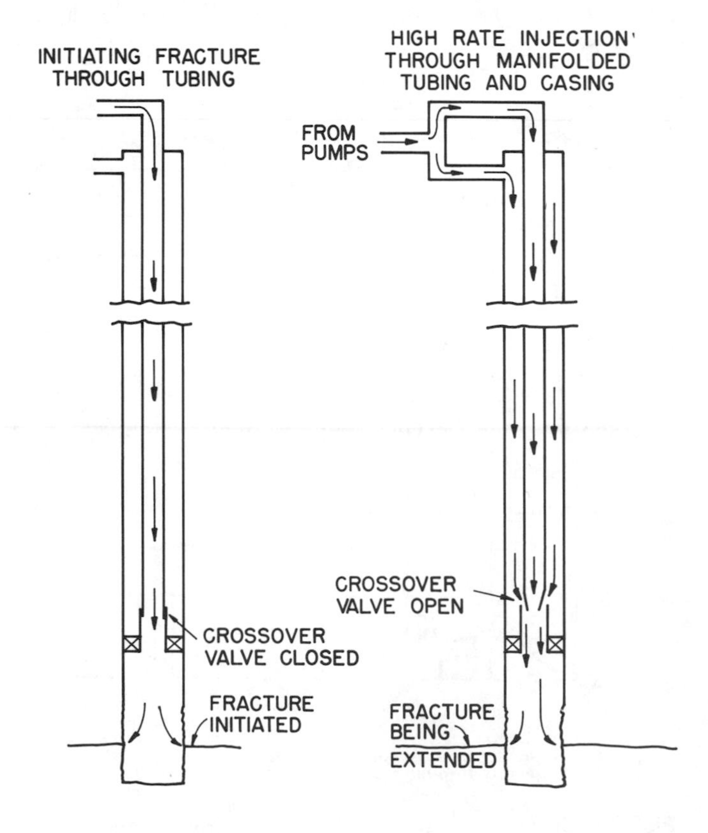

Fig. 7.9 Cross-over valve technique.[20]

Friction Reducing Additives

The economics of using a friction reducing additive in the stimulating fluid will depend upon well depth, efficiency and cost of the additive, quantity of fluid to be used, injection rate required, size of tubular goods used, allowable casing working pressure, and cost of the pumping equipment. For example, when the amount of horsepower lost to friction is small during the pumping of liquid through casing or a manifold small tubing-large casing combination, an additional 50- to 75-percent reduction in the friction loss will not materially reduce the total horsepower required.

On the other hand, if the horsepower lost to friction is high, as would be the case with tubing or manifolded tubing-small casing, the use of friction loss additives will often reduce the total well treating costs. Using the cost figure of $1/hhp and horsepower data from Fig. 7.7, the cost of the hydraulic horsepower (hhp) lost to friction for a pump rate of 15 bbl/min for different tubular goods arrangements is summarized in Table 7.1.

The data in Table 7.1 show that the cost of friction horsepower in these common tubular goods arrangements can vary from $2,700 to $95 even though the pump rate is constant at 15 bbl/min in each case.

Using the 2⅞-in. tubing example in Table 7.1, an additive that decreases the pumping friction loss by 50 percent would reduce the horsepower cost $1,350. If 20,000 gal of fluid were treated at a cost of 2¢/gal for the additive, a net savings of $950 would be realized. However, the friction additive should not be purchased if the treatment is conducted through the 5½-in. casing, since the additive would cost $400 and the friction horsepower cost would be reduced only about $50. Indiscriminate purchasing of additives to minimize friction can result in an increased total treating expense.

Perforating Program

Many perforating programs are designed to insure adequate openings for oil or gas to flow from the formation, and not to minimize the horsepower loss (high friction pressure) involved in pumping the stimulating fluid into the formation. It is not unusual to inject liquids at rates of 20 to 60 bbl/min, whereas

TABLE 7.1 — FRICTION HORSEPOWER COST IN FRACTURING TREATMENTS

Conditions

7,000-ft well, open-hole completion; treating pressure gradient, 0.7 psi/ft; fresh-water fracturing fluid; pump rate 15 bbl/min

Tubular Goods Description	Friction hhp Cost	Percent of Total Pump Cost
2⅞-in. tubing	$2,700	82
Manifolded 2⅞-in. tubing and 5½-in. casing	270	32
5½-in. casing	95	14

in most prorated areas, oil is seldom produced at rates in excess of 1/12 bbl/min or 120 B/D. It is important, therefore, that the opening to the formation be designed to accept the stimulating fluid without excessive horsepowers being lost to friction. If the casing perforating program (shaped charges, bullets, abrasive or mechanical notching, etc.) is primarily designed for the planned fracture injection rate, it will usually be adequate for post-fracturing producing conditions.

Hole Size. The hole size produced by a perforator and the spacing of the shots are extremely important when fluids are to be injected through the perforations. Hole size is particularly important when the perforated well is to be fractured.

Fig. 7.10 is a plot of pressure required to pump water containing 1½ lb of sand per gallon through perforations of various sizes. Since the pressure loss across perforations is directly proportional to the specific gravity of the fluid mixture being pumped and is essentially independent of the viscosity of the mixture, the pressure losses for any other system could easily be determined by dividing the pressure loss plotted at any particular rate by 1.1 (specific gravity of the water-sand mixture) and multiplying by the specific gravity of the mixture being considered.

The pressure loss vs pump rate data shown on Fig. 7.10 were calculated for a perforation where the cement behind the casing is enlarged to a diameter greater than the perforation and where the perforation is not enlarged by sand and fluid erosion. If the pressure losses and the volume of fracturing fluid handled are known, the hydraulic horsepower and pump truck cost can be calculated. The minimum cost condition occurs when the total perforating cost plus pump truck cost is a minimum for the desired pump rate. Unless unusual well conditions exist or a limited entry technique is used, the largest competitively priced perforations should be used to achieve the optimum cost condition in a fracture-treated well.

The cost to overcome friction losses through the perforations increases rapidly if other than the optimum number of perforations is used. The foregoing emphasizes that the perforating program should be correlated with the fracturing treatment design if the total well completion cost is to be a minimum. In old wells, it may be economically desirable to make additional perforations before conducting the fracturing treatment. In all cases, the most efficient perforating program for a specific pump rate can be determined. The optimum pump rate per perforation when fracturing is 1 to 1¼ bbl/min for ⅜-in. perforations, 1.5 to 2 bbl/min for ½-in. perforations, and 2 to 3 bbl/min for ¾-in. perforations; however, these figures will vary slightly, depending upon well depth.

Vertical Spacing. The practice of spacing perforations throughout the entire pay zone is probably based on the premise that if enough openings are made in the pipe, drainage of the entire pay zone will be assured, and that if a zone is overlapped the effect of any errors in depth measurement will be minimized. Major improvements in well logs and their interpretation, plus the correlation of depth measurements to casing collar references have outmoded the need to space the perforations throughout the producing zone.

In planning a perforating program it must be remembered that hydraulic fracturing radically alters the flow of reservoir fluids from the formation to the wellbore. It is therefore essential that the perforations be sized and located in such a way as to assist in the optimum placement of the fracture in the reservoir.

By using the data obtained during drilling of the well, the desired elevation of an induced fracture within a reservoir can usually be determined. If the perforations are concentrated at this depth, within a 1- or 2-ft vertical interval, the probability of placing a fracture at the desired elevation is far greater than if the casing is perforated extensively throughout the entire thickness of the zone.

Communication between zones through the cement sheath behind the casing can be minimized if each zone is selectively perforated. Concentrating the perforations in a narrow vertical interval will result in greater lengths of unperforated casing between the zones, thereby reducing the probability of communication.

Should repair or workover operations or fracture treatments be necessary at a later time, the probability of a successful job is greatly increased if casing perforations are concentrated in a short, vertical interval at each elevation where a fracture is desired.

Hydraulic Horsepower Requirements

A factor that has probably discouraged more realistic planning of the hydraulic phase of stimulation treatments is the lack of data regarding the probable surface pressure for a given set of conditions. The usual basis for this attitude is that a cursory examination of several treatment reports for different wells completed in the same reservoir may reveal that the surface injection pressures vary over a wide range. It is then assumed that a planned treatment would be subject to similar wide variations in basic data and any calculated savings would also be unpredictable, hence not a worthwhile project for an engineering analysis.

However, careful study of the different treatment reports usually will reveal that the wide variation in surface injection pressures is due to such factors as the tubular goods arrangement, the size and number of perforations receiving the injected fluids, the pump rate, and the physical characteristics of the stimulating fluid (viscosity, concentration of friction loss additive, and type and concentration of proppant).

If instantaneous shut-in pressure data have been obtained, it will normally be found that the bottom-hole fracture pressure p_{bf} will be essentially constant for dif-

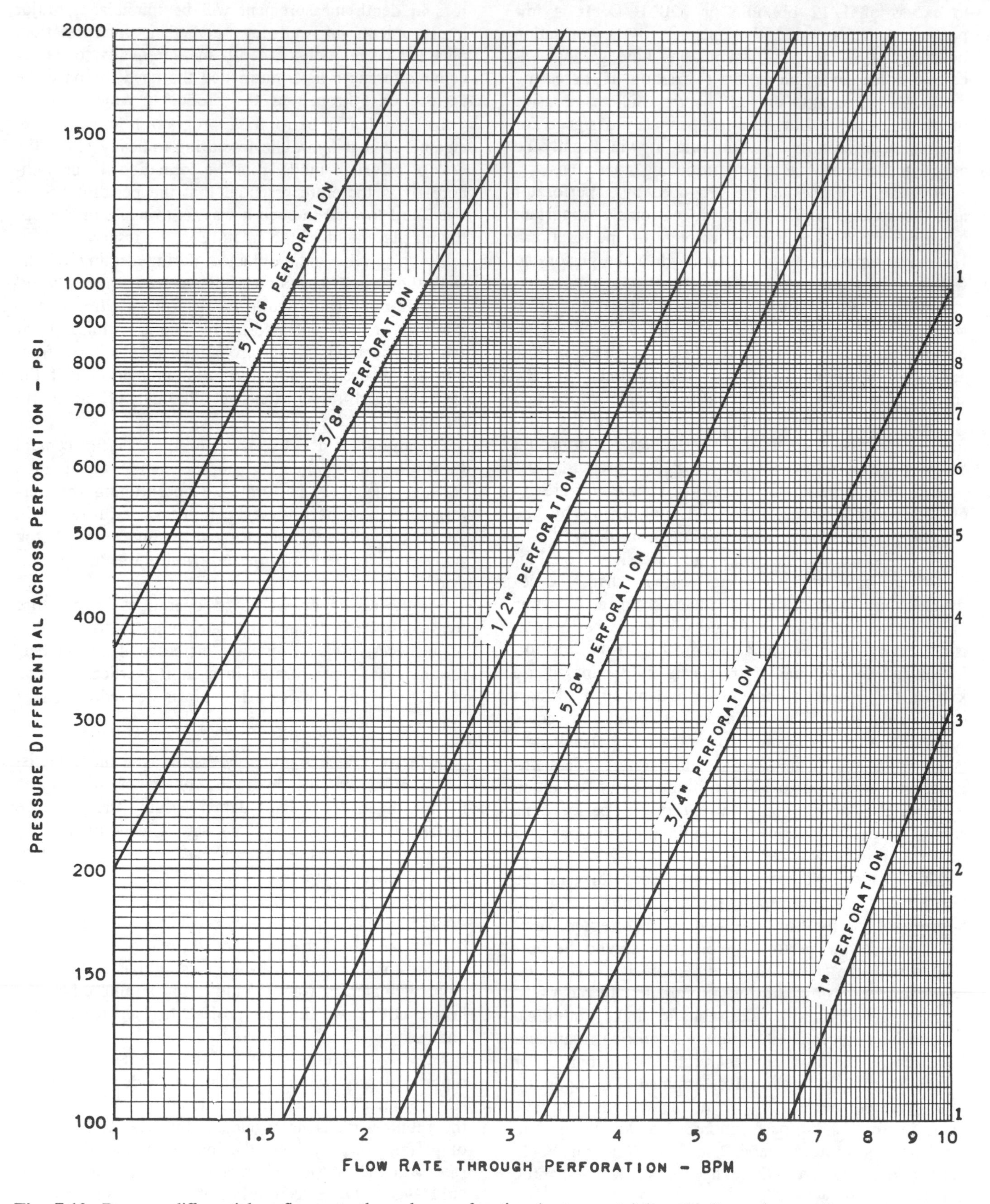

Fig. 7.10 Pressure differential vs flow rate through a perforation (water containing 1½ lb sand per gal).

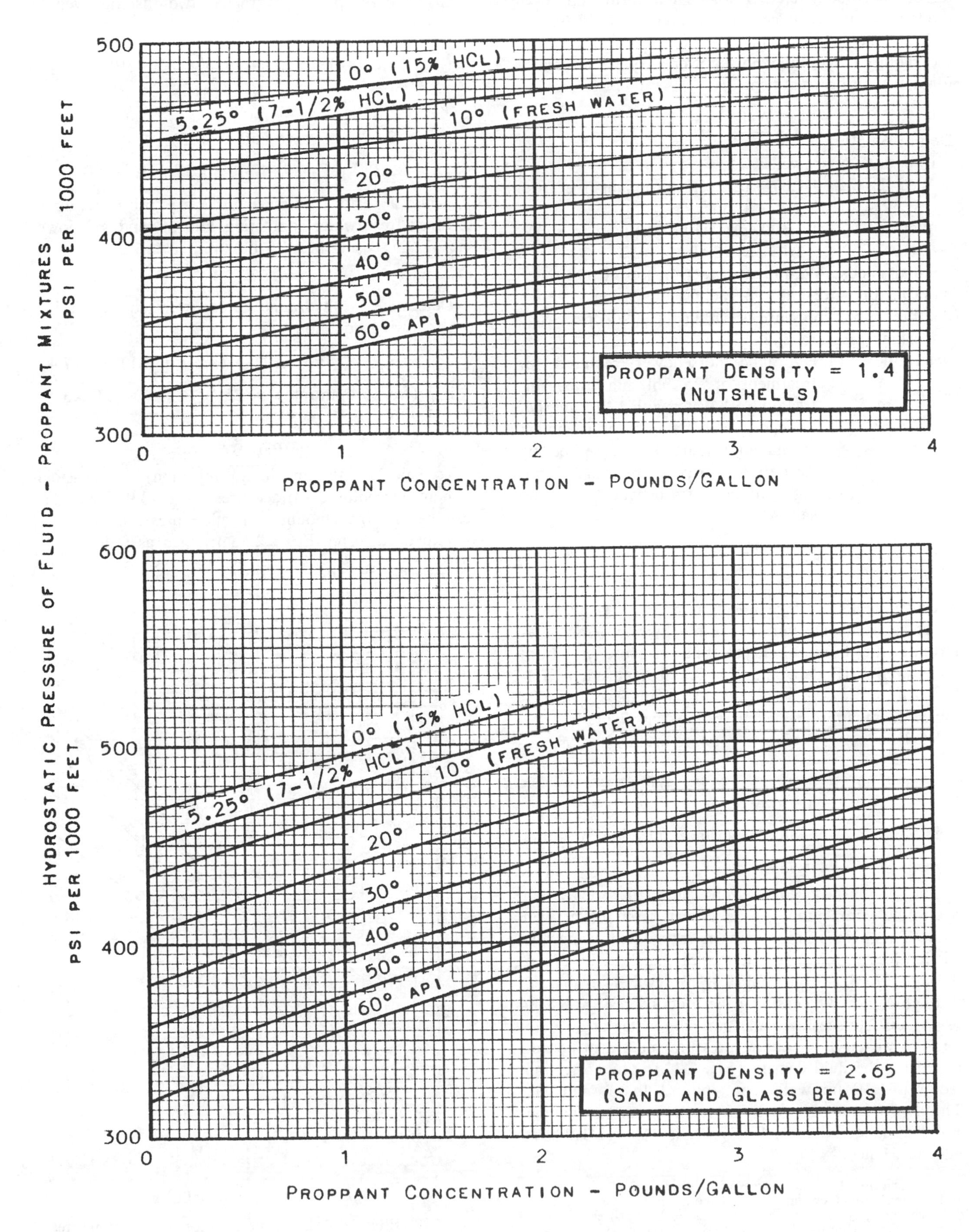

Fig. 7.11 Hydrostatic pressure gradient vs proppant concentration.

ferent wells completed in the same formation, even though the surface injection pressures vary widely.

The following example is presented to show how the required hydraulic horsepower may be calculated for a fracturing treatment.

Well Data

Depth: 6,650 ft

ISIP data on offset wells:

Well 1: p_{ISI} = 1,870 psi with proppant-free fresh water in hole

Well 2: p_{ISI} = 2,300 psi with proppant-free 36° API oil in hole

Tubular goods: 4½-in. casing set on top of pay zone

Working pressure limit: 3,900 psi on casing

The breakdown pressure is within the 3,900 psi casing working pressure limit.

Determine: (1) Hydraulic horsepower required for treatment of 25 bbl/min using untreated fresh water with 1 lb/gal sand.

(2) Injection rate and hhp for treatment at maximum rate without exceeding casing working pressure while pumping untreated fresh water with no sand.

Part A

Step 1

Calculate the surface injection pressure required to inject sand-laden water at a rate of 25 bbl/min. Determine the bottom-hole treating pressure p_{bf} from offset well ISIP data and Eq. 7.2 rearranged, as follows: $p_{bf} = p_{ISI} + p_h$; data to calculate p_h are obtained from appropriate curves on Fig. 7.11.

Well 1: $p_{bf} = 1{,}870 + (433 \times 6.650)$
$= 4{,}750$ psi

Well 2: $p_{bf} = 2{,}300 + (365 \times 6.650)$
$= 4{,}730$ psi

Using Eq. 7.2, obtain the surface injection pressure.

$$p_{ti} = p_{bf} + p_F + p_{pF} - p_h.$$

The p_{pF} term becomes zero since the well is an open-hole completion; p_{bf} is known from offset well data. The pipe friction can be calculated using the friction-loss vs injection-rate curves*, such as Figs 7.14 through 7.18. From Fig. 7.15, (water displaced through 4½-in. OD casing), the friction pressure is 200 psi/1,000 ft of casing while pumping water at 25 bbl/min. However, this friction pressure must also be adjusted for the fluid density (Fig. 7.12). The correction factor is 1.10 for water containing 1 lb of sand/gal. The hydrostatic pressure of fresh water containing 1 lb/gal sand, p_h, from Fig. 7.11, is 465 psi/1,000 ft of hole. Hence:

p_{bf} = 4,740 psi = (average of two offsetting wells)

p_F = 1,463 psi

$$= 6{,}650\ (1{,}000\ \text{ft}) \times \frac{200\ \text{psi}}{1000\ \text{ft}} \times 1.10$$

$p_{pF} = 0$

p_h = 3,090 psi = (6.650 × 465)

p_{ti} = 3,113 psi = 4,740 + 1,463 + 0 − 3,090

Thus, a surface pressure of 3,113 psi is required to inject fresh water containing 1 lb of sand/gal down the 4½-in. casing at a 25 bbl/min rate.

Step 2

Determine the hydraulic horsepower required, using the following equation.

$$\text{hhp} = i_{\text{bpm}} \times p_s \times 0.0245 \text{ (see Eq. 8.1)} \quad . \quad (7.4)$$

hhp (for 25 bpm rate) = 25 × 3,113 × 0.0245 = 1,906.

Part B

To calculate the maximum injection rate possible without exceeding a surface pressure of 3,900 psi, determine first the amount of friction pressure that may be expended, using Eq. 7.2 rearranged as follows:

$$p_F = p_{ti} + p_h - p_{bf} - p_{pF} \quad . \quad . \quad . \quad . \quad . \quad (7.2)$$

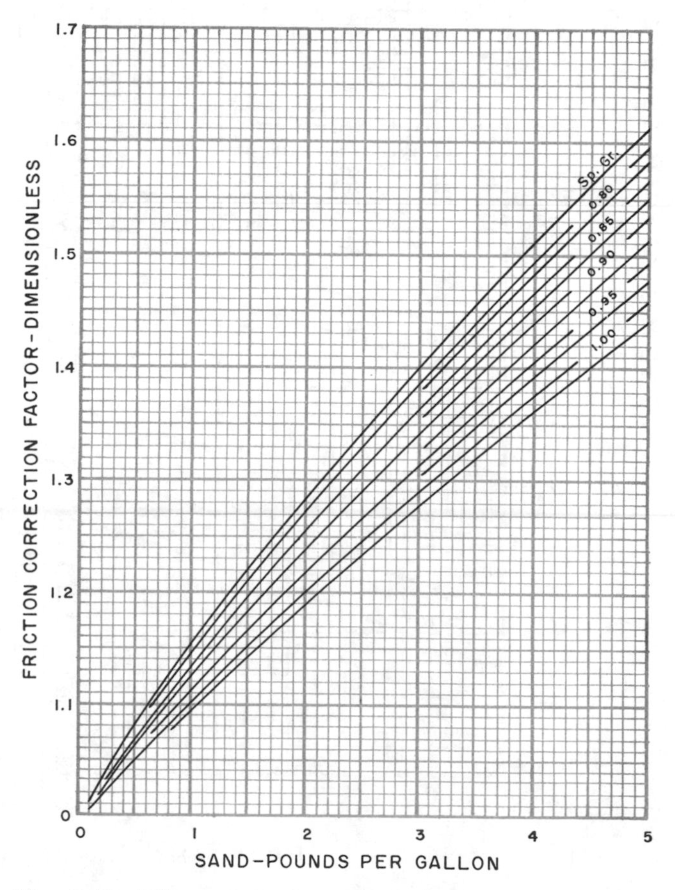

Fig. 7.12 Effect of sand concentration on friction pressure.[25]

*The friction loss curves are for both typical Newtonian and typical non-Newtonian fracturing fluids. Friction loss data for both types have been published by both the Dowell Div. of Dow Chemical Co. and Halliburton Co. and are available upon request.

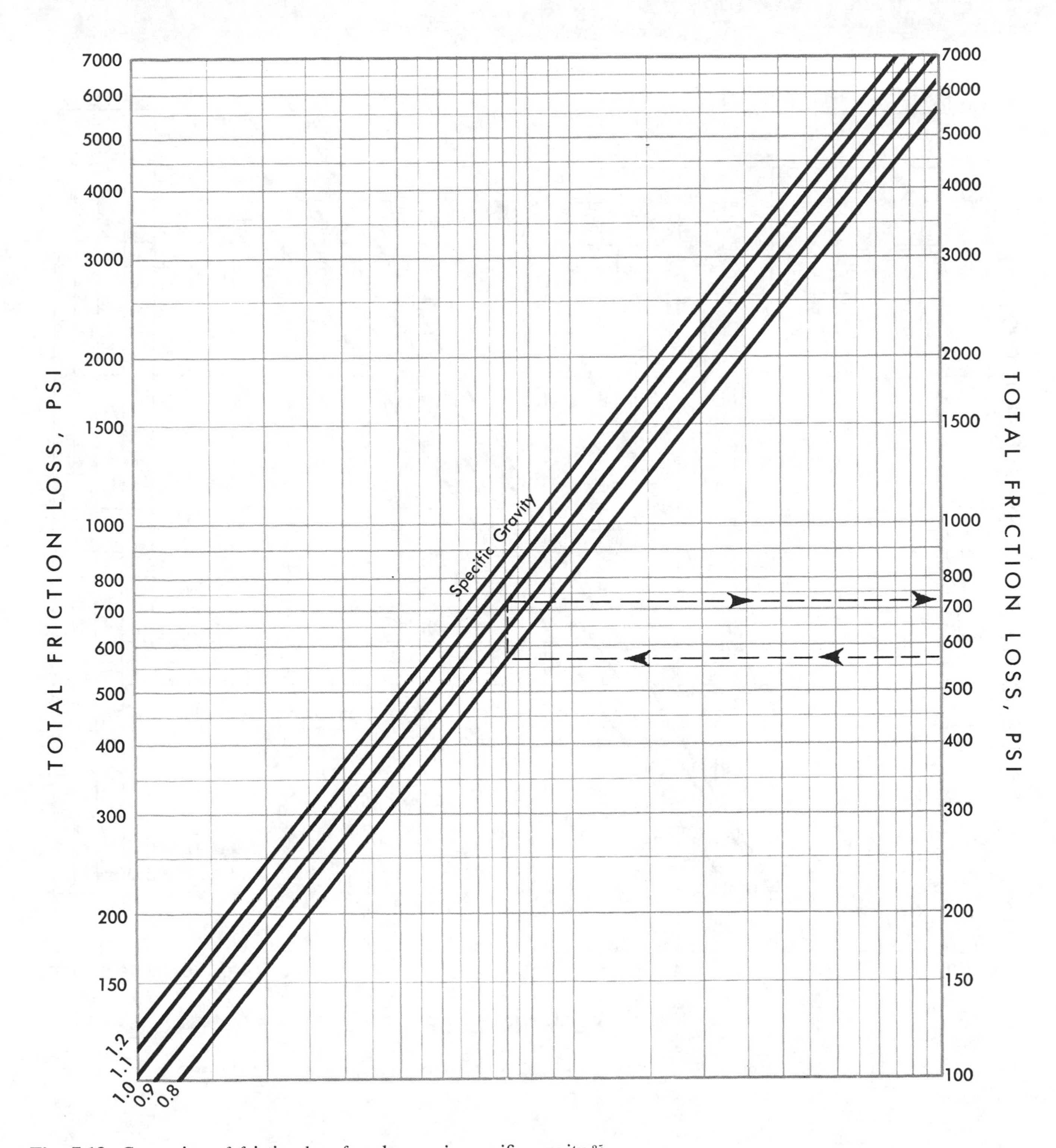

Fig. 7.13 Correction of friction loss for changes in specific gravity.[25]

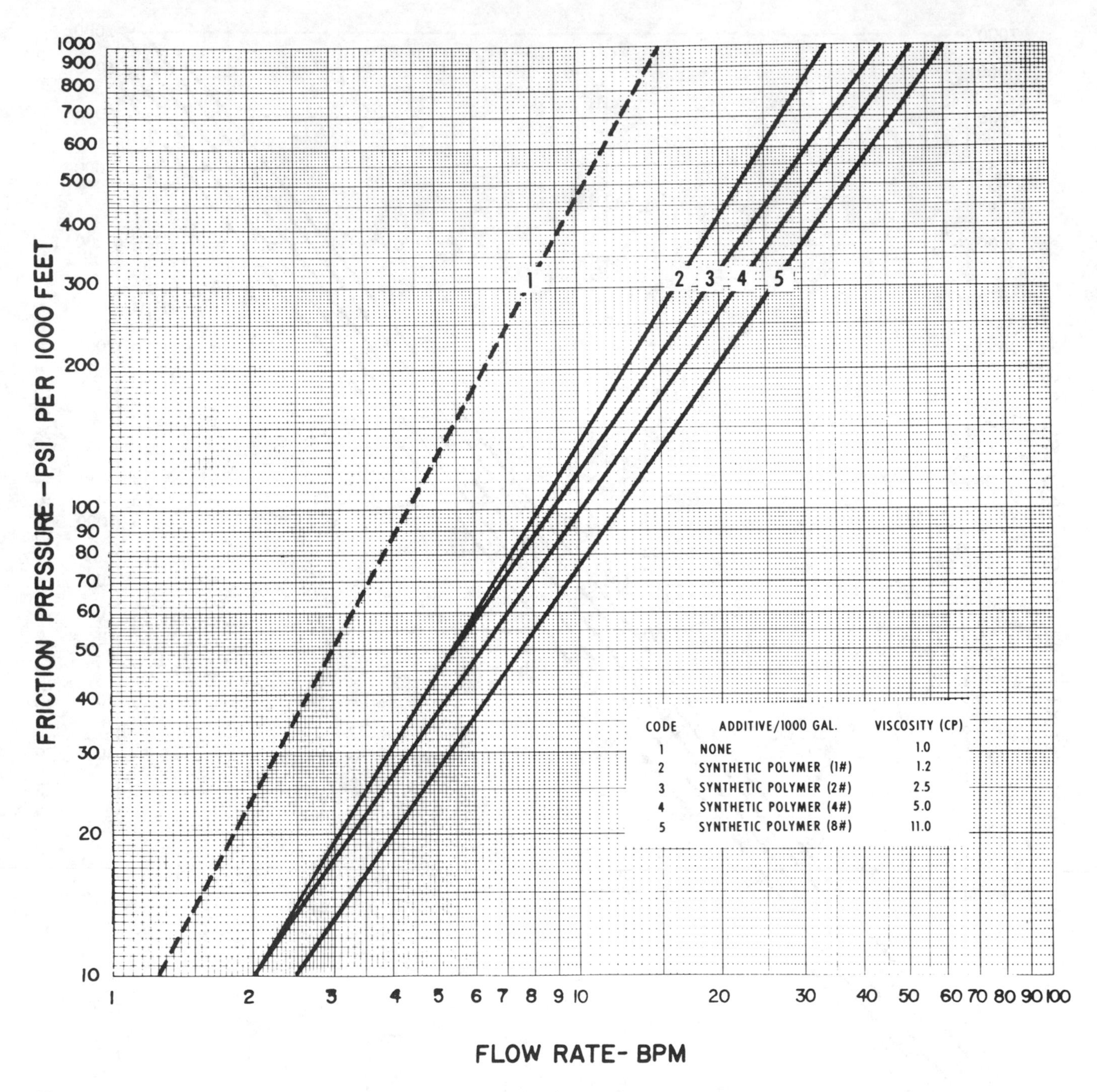

Fig. 7.14 Friction pressure of water-base fluids. (Pipe data: 2⅞-in. OD tubing, 6.5 lb/ft.)[25]

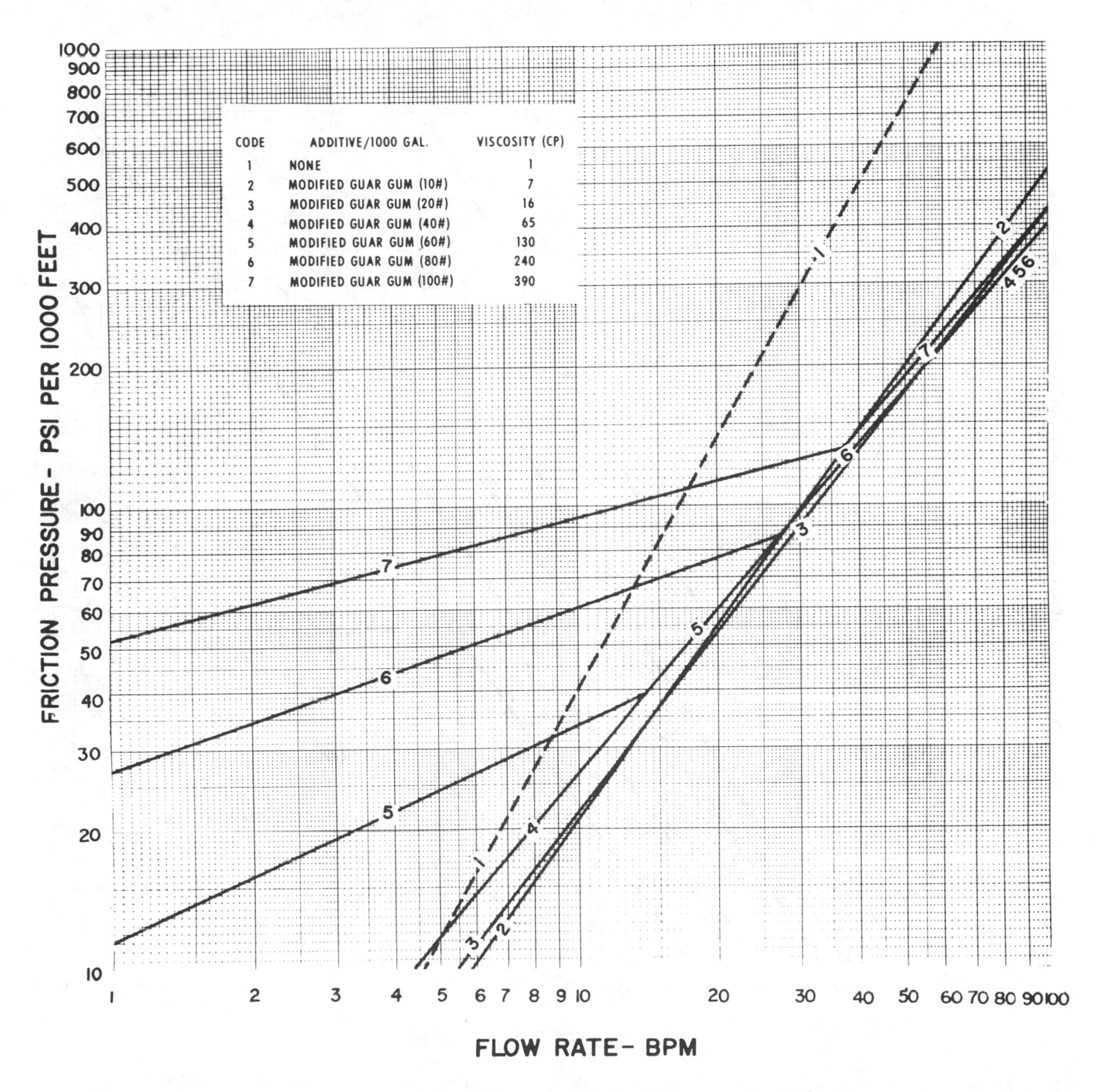

CODE ADDITIVE/1000 GAL. VISCOSITY (CP)
1 NONE 1
2 MODIFIED GUAR GUM (10#) 7
3 MODIFIED GUAR GUM (20#) 16
4 MODIFIED GUAR GUM (40#) 65
5 MODIFIED GUAR GUM (60#) 130
6 MODIFIED GUAR GUM (80#) 240
7 MODIFIED GUAR GUM (100#) 390
FRICTION PRESSURE - PSI PER 1000 FEET
FLOW RATE- BPM

Fig. 7.15 Friction pressure of water-base fluids. (Pipe data: 4½-in. OD casing, 11.6 lb/ft.)[25]

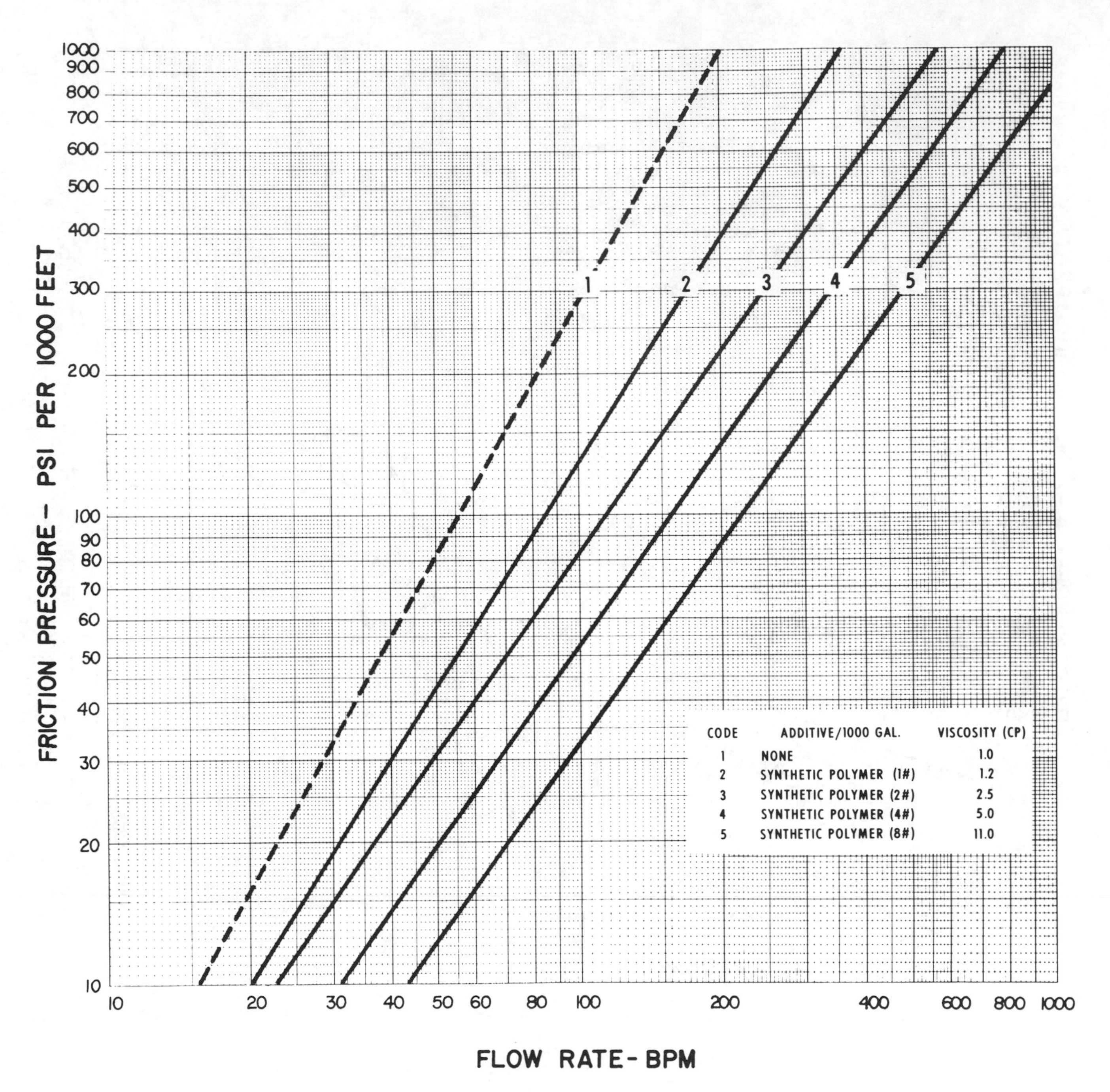

Fig. 7.16 Friction pressure of water-base fluids. (Pipe data: 7-in. OD casing, 23 lb/ft.)[25]

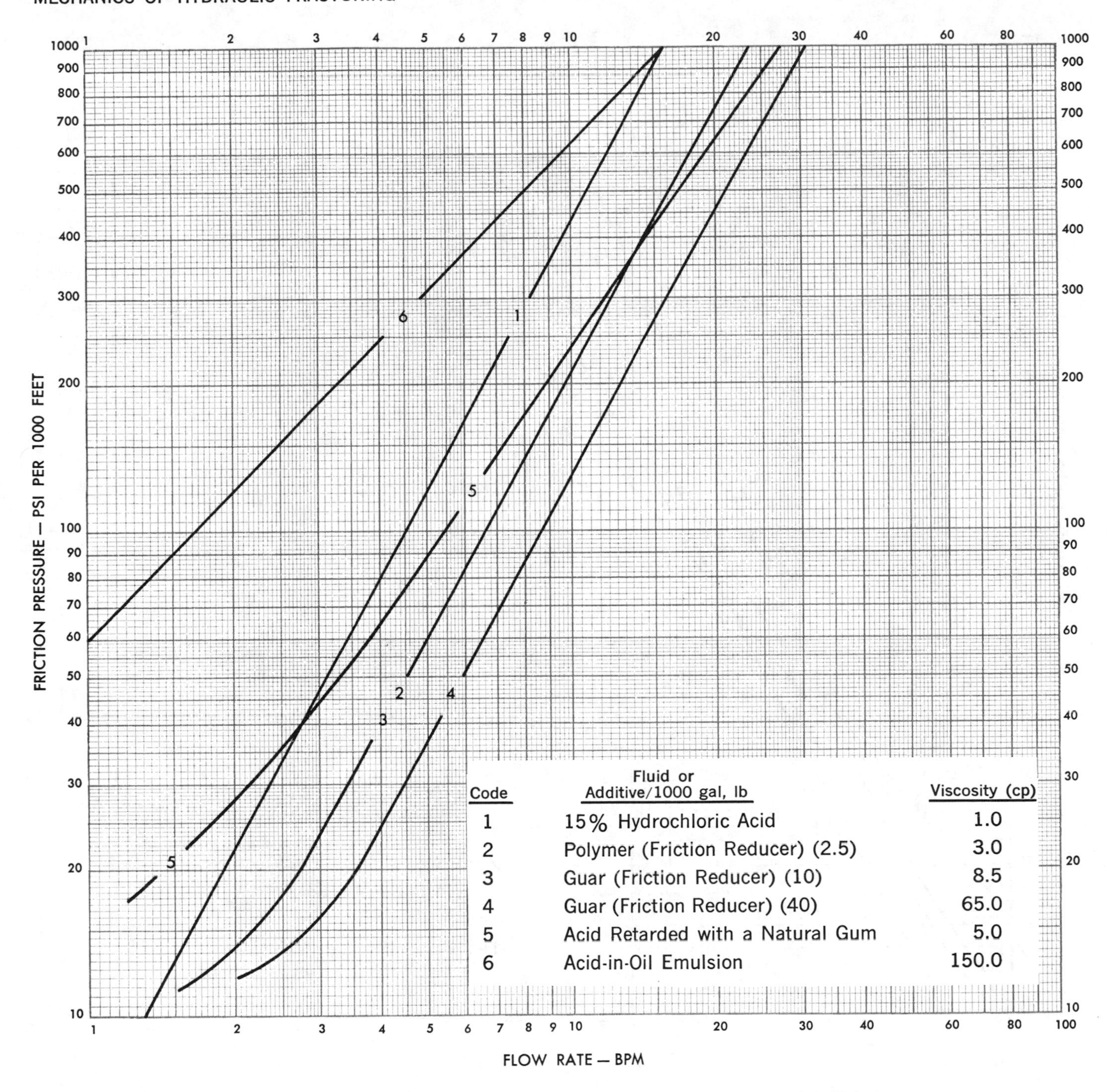

Fig. 7.17 Friction pressure of acid-base fluids. (Pipe data: 2⅞-in. OD EUE tubing, 6.5 lb/ft.)[25]

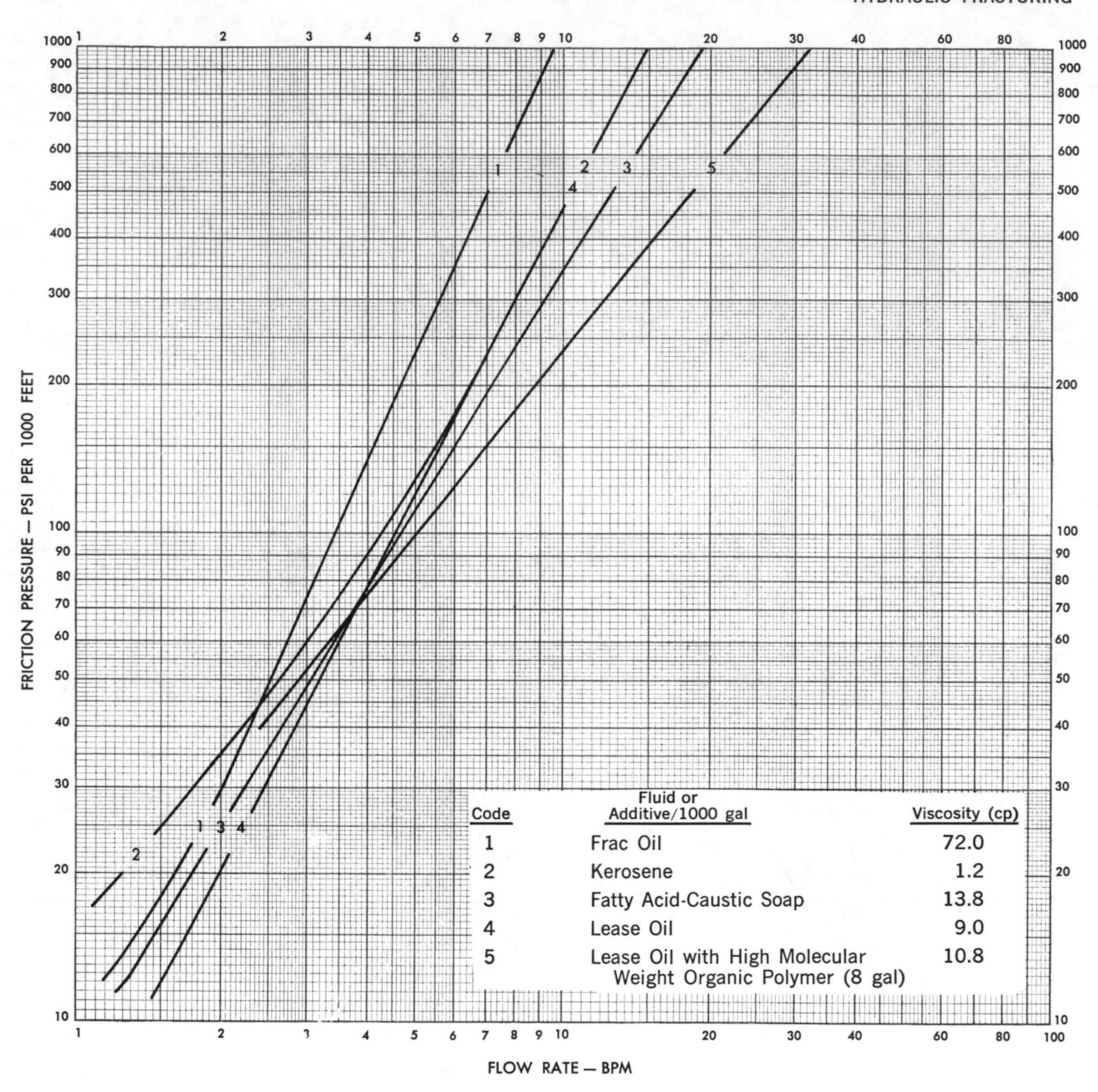

Fig. 7.18 Friction pressure of oil-base fluids. (Pipe data: 2⅞-in. OD EUE tubing, 6.5 lb/ft.)[25]

$$p_F = 3{,}900 + 3{,}090 - 4{,}740 - p_{pF}\ (0)$$
$$= 2{,}250 \text{ psi.}$$

Convert the allowable tubular goods friction to friction loss per 1,000 ft of pipe as follows:

$$p_F = \frac{2{,}250 \text{ psi}}{6.650\ (1000 \text{ ft})} = 353 \text{ psi}/1{,}000 \text{ ft}$$

Since the density of the base fluid (water) is the same as the density used in calculating the friction-loss curves on Fig. 7.15, no density correction factor is necessary, and the maximum pump rate may then be found using the 4½-in.-casing curve on Fig. 7.15. The injection rate corresponding to a friction loss of 353 psi/1,000 ft is 35 bbl/min.

The hydraulic horsepower necessary to inject the fracturing fluid at a rate of 35 bbl/min and 3,900 psi surface pressure is:

$$\text{hhp} = 35 \times 3{,}900 \times 0.0245 = 3{,}344.$$

7.6 Fluid and Proppant Pumping and Proportioning

Pumping units currently available to the industry are capable of delivering over 1,000 hhp. This equipment combines heavy duty engines and highly efficient pumps operating through power trains especially selected to perform under conditions presented by the wide range of pressures and rates demanded. Chapter 8 presents complete information on the various pumping and fluid handling equipment available for hydraulic fracturing.

The average injection rate for fracturing treatments in the U.S. during 1950 was 3.5 bbl/min, but by 1966 this had increased to 25 bpm.[21] Although many high rate jobs are performed today in various fields, there are specific areas, such as the Hugoton, Kans., gas field, where rates in excess of 500 bpm requiring more than 10,000 hhp have been used. Future hydraulic horsepower usage will depend upon the economical analysis of cost vs results.

Gels used initially in fracturing had sufficient viscosity or gel strength to hold sand in suspension without any appreciable settling. Because of this and since volumes were small, the sand was usually premixed with the gel in tanks prior to pumping into the well. Since this was not feasible with the less viscous refined and crude oils, some form of proportioning and mixing was needed. Fig. 8.14 presents a schematic diagram of a mixing unit. Fluid and sand are added continuously in the desired proportion to an agitator tank where they are intimately mixed, then picked up by the pump trucks and pumped into the well. The first units were small, and capable of handling only about 12 bbl/min of sand and fluid mixtures. Equipment is now available to mix and proportion, in a continuous manner, the proppants, fluid and various other additives (dry or liquid), at rates of 175 bpm or more. Chapter 8 presents details of the various types of proportional additive blenders available and describes their operation.

7.7 Deep Well Fracturing

The probability of success in fracture-treating deep wells is dependent to a large degree upon the planning devoted to this technique at the time the drilling and completion program is prepared. In some cases, the possibility of a fracture treatment is not considered until other completion techniques have failed. It may then be too late; the zone may have been extensively perforated and communication established with other nonproductive zones so that selective initiation of a fracture in the desired zone is impossible. In other cases, installation of permanent packers or the tubular goods program selected may seriously limit the injection rate and thereby decrease the probability of success of the fracture treatment. The following factors influence the actual design and execution of the treatment.

Factors Affecting the Design of Hydraulic Fracture Stimulation

The type of formation is an important factor in selecting the type of fracturing fluid and propping material to be used. If the formation is acid-soluble, an acid-base fracturing fluid will enlarge existing or created fractures in addition to transporting the propping material.

The hardness of the formation or the ability of the rock to resist proppant embedment, also influences the type of propping material selected. Sand, unless placed in multilayers, is generally unsatisfactory as a deep-well proppant, as it will pulverize when subjected to high overburden loads. High strength glass is the most widely marketed propping material for large overburden loads.

The fracture flow capacity required to achieve a predetermined post-fracturing productivity is influenced both by the strength of the propping agent and by its capacity to be moved outward from the wellbore to the maximum areal extent of the fracture system.

Rounded nutshell, apricot pit, or plastic proppant deforms under load; however, its density is only about 50 percent that of glass or sand. These lightweight materials are therefore transported more readily throughout the fracture system by the fracturing fluid. This results in more uniform distribution of the propping agent over the created fracture surfaces. However, if glass, which is stronger than walnut shells, is used, the fracture capacity and post-fracturing productivity are affected less by overburden weight, should an optimum concentration of propping agent not be achieved.

Establishing an adequate injection rate is a major problem in deep-well fracturing operations. If the injection rate is less than a critical value, the propping agent may fall out of the carrying fluid and accumulate in the wellbore, covering the perforations and thereby causing a screen-out. In this case the treatment may have to be terminated. In other cases, this injection rate may not exceed the rate at which fluid is lost to the formation, so that the fracture ceases to grow and any additional propping material injected accumulates within the wellbore.

Injection Rates

Restricted injection rates in deep wells are due basically to the high bottom-hole pressure required to initiate and extend deep well fractures. This pressure is the result of overburden loads and friction pressure encountered in pumping through tubular goods at high rates.

Although the column of fracturing fluid will tend to offset this increase in breakdown pressure, there still will be a net increase in surface breakdown and final injection pressures. It is usually necessary to initiate the fracture by injecting the fracturing fluid down tubing isolated from the casing by a packer. This is done so that the casing will not be exposed to hydraulic pressure in excess of its working pressure. However, installation of the relatively small diameter tubing severely restricts the rate at which the fracturing fluid may be pumped. The pressure required to pump fracturing fluids through tubing and/or casing increases directly with well depth. Therefore, in deep wells, a large percentage of the available horsepower is required to overcome friction, and low pump rates result.

There are several solutions to the foregoing problems. One technique is the use of an additive that reduces by 50 to 70 percent the friction loss required to pump the fluid. A second approach is to incorporate a cross-over valve in the tubing string so that the fracturing fluid may then be injected simultaneously down the tubing and tubing-casing annulus after the fracture is initiated. This procedure permits a substantial increase in the injection rate without exceeding the casing's working pressure. The cross-over valve has been successfully used in treating wells as deep as 14,000 ft, and the results of these treatments clearly show that the pump rate can be increased twofold to fourfold over that obtained when treating the well down tubing.

Another technique that has been used successfully in deep well stimulation is to inject in the fluid pumped down the tubing all of the proppant that will be required for the fracture treatment. Since all the propping material is entrained in the tubing fluid, there is no hazard involved to the packer or cross-over tool if, for some unanticipated reason, pumping operations are halted.

Should a screen-out appear imminent, large volumes of proppant-free fluid may be immediately pumped into the formation fracture to clear up the difficulty. This is accomplished by decreasing or discontinuing pumping into the tubing and continuing the injection of proppant-free fluid into the annulus. If a screen-out occurs, the relatively small volume of fluid containing the proppant in the tubing can be reverse circulated from the well. Also, backflowing fluid from the formation, while continuing reversing operations, will often clear the formation fracture and allow the treatment to be resumed.

This combination of tubing, cross-over tool and packer provides treating procedure flexibility, whereas if the well is being treated down casing (no tubing in the well and the breakdown pressure within the casing working pressure) and a screen-out occurs, the job would have to be terminated immediately. This would result in a costly clean-out and the expense of a second treatment.

Treatment Hydraulics

When pumping liquids at high rates and pressures through tubular goods, it is important that the hydraulics be preplanned. This is done to determine if a high enough rate can be attained down the tubing to entrain the necessary concentration of propping material so that the desired bottom-hole proppant:fluid ratio is achieved when the tubing and the annulus fluids are mixed.

The perforating program greatly influences the hydraulic horsepower required for a deep-well fracturing operation. This phase of the operation should be carefully engineered, as described in Sec. 7.5.

7.8 Nitrogen and Carbon Dioxide as an Aid in Fracturing

Gases as an aid to stimulation treatment have been used for many years.[22] In its original form, this technique consisted of injecting, into a subsurface formation, natural gas (compressed by mobile field compressors), together with liquid well stimulation chemicals, to improve well productivity. The object was to lighten the fluid column sufficiently so that, during the "clean-up" of stimulation fluids following the treatment, the available energy would be great enough to flow the fluids from the wellbore without the use of swabs or other mechanical clean-up tools.

A more recent innovation has been the use of liquefied nitrogen or carbon dioxide. This has proved to be particularly applicable in conjunction with fracturing treatments, especially in reservoirs having low bottom-hole pressure. Results utilizing these techniques have been extremely encouraging, and in more than 75 percent of the treatments the stimulation fluids have been returned to the surface without the use of mechanical aids.

Unfortunately, in many cases the added advantage of gasifying the stimulation liquids cannot be economically justified merely on the basis of improved fluid returns following treatment. However, laboratory and field studies have revealed that additional benefits are derived from using these gaseous materials as well stimulation aids, and this may justify their use.

For example, laboratory tests[23] have shown that a multiphase oil-water-gas mixture exhibits definite fluid-loss-control characteristics without the use of conventional solid fluid-loss additives. Furthermore, the degree of fluid-loss control obtained can be regulated by the proportions of gas and liquids used and by the concentration of a nonemulsifying surfactant included to improve phase dispersion. Apparently, fluid-loss control is achieved using this technique by setting up a temporary gas block and/or water block in the formation flow channels. Other investigators[24] have reported that, to a minor extent, carbon dioxide reduces the effectiveness of fluid-loss additives. They pointed out, however, that increasing the concentration of the fluid-loss additive readily overcomes this deficiency. Because of the differences in physical and chemical properties between nitrogen and carbon dioxide, one gas is usually better suited for a specific application than the other. Generally speaking, nitrogen is superior when injection rates are low and when precise volume control is critical. Carbon dioxide, on the other hand, is better adaptable to high rate fracturing treatments.

The physical properties of liquid carbon dioxide make it easy to handle, and thus ideal for use in conjunction with stimulation treatments. With the general trend toward higher fracturing pump rates, the injection rates of the gases must keep pace. Pumped as a liquid with standard fracturing equipment, carbon dioxide places no limits on flow rates. Injection in excess of 45,000 scf/min is not uncommon. However, because carbon dioxide is highly soluble in oil, it is less desirable for use in oil-base treatments than nitrogen, since larger quantities are required to be effective.

The mobile units for CO_2 service consist of a charging trailer and a high-pressure pumper. Mounted on the charging trailer are a power unit and a centrifugal pump. The centrifugal pump is designed to boost the pressure of liquid CO_2 (brought to the well site in a transport truck) a minimum of 130 psi at 12 bbl/min, and to feed the CO_2 to the suction of the high-pressure triplex pump on the pumper unit. The high-pressure pumper is a standard unit on which the low-pressure suction header has been replaced with a high-pressure suction header.

The mobile units used for nitrogen service are especially built and engineered for this purpose. These units are used to transport liquid nitrogen and to convert it to high pressure gas. The cryogenic storage tank has a capacity of 2,000 gal of liquid nitrogen (186,000 scf of gas) plus 5 percent vapor space. Each of the two high-pressure, direct-fired diesel vaporizers and each of the two high-pressure cryogenic pumps is designed to discharge 1,800 scf of nitrogen per minute. They have a maximum working pressure of 10,000 psi. To insure the safest possible operation, all lines and vessels are equipped with burst disks and pressure relief valves.

In order to obtain the maximum benefit from using either nitrogen or carbon dioxide commingled with fracturing fluids injected into a well, it is very important that each job be engineer-designed. By properly considering the many well variables and combining this information with thermodynamic data on the gases, the proper ratio of gas to liquid for a particular well may be quickly determined.

Both nitrogen and carbon dioxide have proved to be useful aids in oilwell stimulation. Carbon dioxide and foaming agents remove fluids from the formation better when they are combined in water than when they are used separately. Carbon dioxide can reduce clay swelling, and can reduce the friction loss of oil by an estimated 29 to 60 percent.

Nitrogen and carbon dioxide are both effective in returning stimulation fluids to the surface. Both have been used as a low viscosity breakdown fluid, and carbon dioxide has proved effective in removing water or emulsion blocks.

7.9 Summary

Field data have shown that hydraulically induced fractures in shallow wells can be either horizontal or vertical. Indications are that geologic conditions and well depth will control the fracture plane inclination. Additional correlation of theoretical and actual field results is needed before an accurate method of predicting fracture plane inclination in new areas can be developed.

Many methods of fracture placement have been proposed and successfully field tested. However, a technique can be successful in the placement of fractures only if the in-situ formation stresses are favorable to the type of fracture desired.

Multiple-fracturing or selective-fracturing techniques using diverting agents, ball sealers, limited entry, bridge plugs and sand plugback have been successful. All have advantages and disadvantages that must be analyzed in light of specific well conditions before the proper method can be selected.

Appreciable money can be saved and efficiency can be increased if the hydraulics and the mechanics of hydraulic fracturing treatments are carefully planned. Correlating the tubular goods arrangement, perforating program, pump rate and additives used is necessary if the treatment is to be properly engineered.

Successful fracture stimulation of wells deeper than 10,000 ft requires a thorough understanding of treatment hydraulics and of fracturing fluid behavior. The tubular goods friction losses increase directly with well depth, and both breakdown and fracture extension pressures also increase. These factors combine to reduce injection rates and to increase the amount of fluid lost from the fracture system. In turn, the probability of a proppant screen-out in the fracture system and wellbore is increased. These difficulties can be overcome through the use of the cross-over tool, friction reducing additives, and special proppant handling methods.

Gaseous phase fracturing using nitrogen and carbon dioxide can result in much faster fracturing fluid removal and more efficient fracturing.

In general, the key to more efficient and more profitable fracturing is proper engineering.

References

1. Howard, G. C.: "Special Fracturing Test, Pine Island Field, La.", unpublished report, Pan American Petroleum Corp. (1954).
2. Reynolds, J. J., Popham, J. L., Scott, J. B. and Coffer, H. F.: "Hydraulic Fracture—Field Test to Determine Areal Extent and Orientation", *J. Pet. Tech.* (April, 1961) 371-376.
3. Howard, G. C. and Fast, C. R.: "Factors Controlling Fracture Extension", paper presented at the Spring Meeting of the Petroleum and Natural Gas Div. of CIM, Edmonton, Alta., May, 1957.
4. Fraser, C. D. and Pettitt, B. E.: "Results of a Field Test to Determine the Type and Orientation of a Hydraulically Induced Formation Fracture", *J. Pet. Tech.* (May, 1962) 463-466.
5. Anderson, Terry O. and Stahl, Edwin J.: "A Study of Induced Fracturing Using an Instrumental Approach", *J. Pet. Tech.* (Feb., 1967) 261-267.
6. Jennings, E. R. and Vincent R. P.: "A Glass Fabric-Plastic Liner Casing Repair Method", *Drill. and Prod. Prac.*, API (1959) 67.
7. Howard, G. C. and Fast, C. R.: "Squeeze Cementing Operations", *Trans.*, AIME (1950) **189**, 53-64.
8. Gilbert, Bruce: "The F I Process . . . Theory and Practice", paper 1021-G presented at Fourth Annual Joint Meeting, Rocky Mountain Petroleum Sections of AIME, Denver, Colo., March 3-4, 1958.
9. Huitt, J. L., Pekarek, J. L., Swift, V. N. and Strider, H. L.: "Mechanical Tool for Preparing a Well Bore for Hydraulic Fracturing", *Drill. and Prod. Prac.*, API (1960) 129.
10. Scott, P. P., Jr., Bearden, W. G. and Howard, G. C.: "Rock Rupture as Affected by Fluid Properties", *Trans.*, AIME (1953) **198**, 111-124.
11. Clark, J. B., Fast, C. R. and Howard G. C.: "A Multiple-Fracturing Process for Increasing the Productivity of Wells", *Drill. and Prod. Prac.*, API (1952) 104.
12. Howard, G. C. and Scott, P. P., Jr.: "An Analysis and the Control of Lost Circulation", *Trans.*, AIME (1951) **192**, 171-182.
13. Neill, G. H., Brown, R. W. and Simmons, C. M.: "An Inexpensive Method of Multiple Fracturing", *Drill. and Prod. Prac.*, API (1957) 27.
14. Kastrop, J. E.: "Newest Aid to Multi-Stage Fracturing", *Pet. Eng.* (Dec., 1956) B-40.
15. Brown, R. W., Neill, G. H. and Loper, R. G.: "Factors Influencing Optimum Ball Sealer Performance", *J. Pet. Tech.* (April, 1963) 450-454.
16. Stekoll, Marion H.: "New Light on Fracturing Through Perforations", *Oil and Gas J.* (Oct. 29, 1956) 95.
17. Coburn, R. W.: "Custom Designed Well Stimulation", *Oil and Gas J.* (March 4, 1957) 102.
18. Murphy, W. B. and Juch, A. H.: "Pin-Point Sandfracturing — A Method of Simultaneous Injection Into Selected Sands", *J. Pet. Tech.* (Nov., 1960) 21-24.
19. Lagrone, K. W. and Rasmussen, J. W.: "A New Development in Completion Methods — The Limited Entry Technique", *J. Pet. Tech.* (July, 1963) 695-702.
20. Flickinger, D. H. and Fast, C. R.: "The Engineering Design of Well Stimulation Treatments", paper presented at Annual Meeting of Petroleum Branch of the CIM, Edmonton, Alta., April 1-3, 1963.
21. Hassebroek, W. E. and Waters, A. B.: "Advancements Through 15 Years of Fracturing", *J. Pet. Tech.* (July, 1964) 760-764.
22. Foshee, W. C. and Hurst, R. E.: "Improvement of Well Stimulation Fluids by Including a Gas Phase", *J. Pet. Tech.* (July, 1965) 768-772.
23. Hall, C. D., Jr., and Dollarhide, F. E.: "Effects of Fracturing Fluid Velocity on Fluid-Loss Agent Performance", *J. Pet. Tech.* (May, 1964) 555-560.
24. Neill, G. H., Dobbs, J. B., Pruitt, G. T. and Crawford H. R.: "Field and Laboratory Results of Carbon Dioxide and Nitrogen in Well Stimulation", *J. Pet. Tech.* (March, 1964) 243-248.
25. *Frac Guide Data Book*, Dowell Div. of Dow Chemical Co., Tulsa, Okla. (1965).

Chapter 8

Mechanical Equipment for Hydraulic Fracturing

8.1 Introduction

After the specifications of a fracturing treatment have been established, the method of placing the treating fluid and the propping agent in the well must be considered. Rate of placement, addition of chemicals, proper mixing, and control of the concentration of proppants in the fracturing fluid are important in planning the fracturing job.

In the early treatments, the equipment used had been designed primarily for cementing or acidizing. Continued demand for increased pump rates at high operating pressure has stimulated a continuing development program to meet job requirements. Each new piece of equipment was designed to fulfill a specific need.

This chapter describes each category of equipment, its purpose, design features and capabilities. This information is required if a fracturing operation is to be planned adequately. In studying mechanical equipment, it must be remembered that all components (electrical and mechanical) must perform as a system.

8.2 Historical Development

Service company equipment available for pumping in 1949 had been designed for cementing and acidizing. It was largely confined to a single truck-mounted pump of low-horsepower, low-pump-rate design. Engines available at that time ranged from 75 to 125 hhp and were usually used to power the truck itself.

Inasmuch as the first hydraulic fracturing jobs employed small volumes of fracturing fluid, and the calculated rates and pressures were similar to those encountered on cementing and acidizing jobs, this pumping equipment was adequate with only slight modifications.

The fracturing fluid was prepared by circulating either gasoline or kerosene in the open displacement tank mounted on the pump truck while the gelling agent was added through the cement hopper or directly into the mixing tank. After the desired viscosity and fluid-loss properties were imparted to the fracturing fluid, the sand propping agent was poured from sacks into the tank above the pump suction while the mixture was pumped into the well. Fig. 8.1 shows a cementing truck being used for this type of job in 1950.

Process Requirements

In 1950 the Hydrafrac process consisted of two basic steps: (1) injecting a viscous liquid containing a granular propping agent such as sand into a well under high hydraulic pressure and high injection rates to fracture the formation; and (2) changing the viscous fracturing liquid from a high to a low viscosity so that it could be displaced easily from the formation.[2] Both steps depended on surface equipment to pump the fluids into the formed fractures. The high formation breakdown pressures and fast displacement rates encountered in the first step required that the greatest emphasis be placed on the performance of the pump truck. The pump unit was required to mix the fluid and propping agent, apply the pressure, and provide the injection rate to fracture the formation. Fig. 8.2 is a schematic of equipment hook-up for this early work.

An empirical method was used to calculate fluid viscosity and pump rates required to break down a formation and to extend the fracture. Table 8.1 is a typical example of the results of some of this work. The use of very high viscosity fracturing fluids allowed the displacement rates to be within the range of available pumping equipment.

Available Equipment

A typical pump in 1949 was a vertical duplex or a horizontal triplex pump. Either of these pumps was driven by a four cylinder diesel or gasoline engine, usually through a 10-speed transmission. (Fig. 8.3 is an example of a pressure-volume curve for this setup. For example, an average Hydrafrac operation in late 1949 required a surface pressure of 1,000 to 1,300 psi; and an injection rate of 2.5 to 3 bbl/min.) The engine also served as the prime mover for the truck. Normally, there

Editors note: This chapter was prepared with the assistance of E. H. Gras of Halliburton Services.[1]

was one displacement pump per truck. The pump was mounted on the truck in conjunction with a small mixing pump driven by a 40 hp auxiliary engine. As the fracture job requirements increased, multiple trucks were used to achieve an increase in pump rates and mixing-tank volume.

Developments of Pumping Rate, Fluid Volume and Fluids

The first fracturing jobs were performed at rates of 1 to 4 bbl/min with 10 to 20 bbl of fluid. The fluids were refined petroleum products, gelled to attain viscosity and fluid-loss properties.

In 1952 the use of crude oil and refined Nos. 5 and 6 API fuel oils became more popular. By early 1953 these fluids were used in most of the hydraulic fracturing operations. The relatively low cost of the crude oils and fuel oils allowed several trends to develop that changed equipment requirements. By 1956 the average pumping rate had increased to 13 bbl/min with a maximum of 95 bbl/min obtained on some jobs.

In 1956 the use of water-base fracturing fluids began to increase. This, in turn, caused a marked increase in the volume of fracturing fluid employed per job. Fig. 8.4 illustrates the change in base fluids used, and Fig. 8.5 shows the corresponding change in pumping rates.

The demand for high injection rates was accompanied

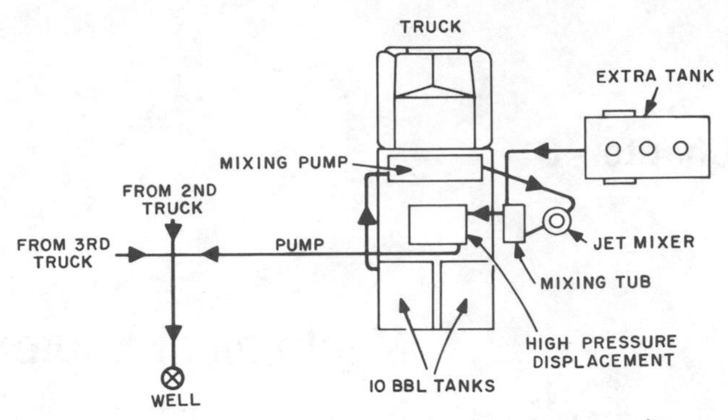

Fig. 8.2 Well hook-up using service company equipment.

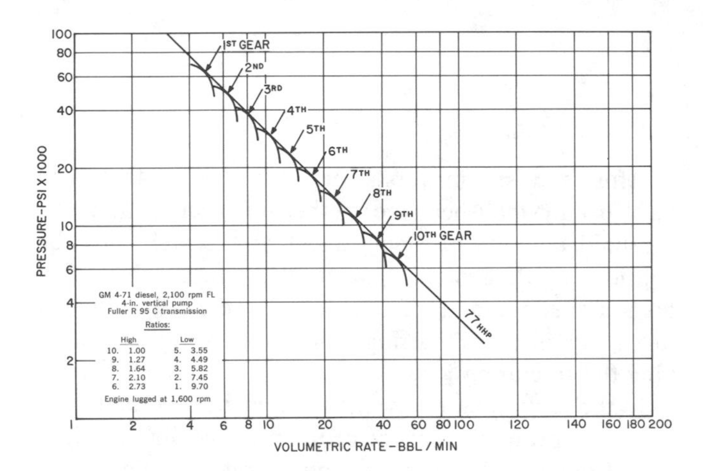

Fig. 8.3 Pressure-volume curve.

Fig. 8.1 Cementing truck used for hydraulic fracturing operations, 1950.

TABLE 8.1 — PUMP RATE FOR HYDRAULIC FRACTURING

	East Sasakwa Field, Booch Sand	Hugoton Field, Fort Riley Lime	East Texas Field, Woodbine Sand
Amount of open hole, ft	35	35	20
Average formation permeability, md	150	13	2,000
Bbl/min to fracture			
Crude oil — 5 cp	2.0	0.13	21.0
Gel — 100 cp	0.1	0.007	1.0
Bbl/min to extend fracture to 50-ft radius			
Crude oil — 5 cp	11.0	1.0	213.0
Gel — 100 cp	0.6	0.05	10.7
Pressure to fracture, psi	2,100* 1,250**	1,700* 900**	2,900* 1,700*
Pressure to extend fracture	1,700* 850**	1,350* 550**	2,600* 1,400**

*Bottom-hole pressure
**Surface pressure

by a similar demand for storage facilities to accommodate the large volumes of fluid. It became necessary to devise methods of adding the propping agent to the fluid as it was being pumped and of hauling and dispersing sand in bulk quantities. Changing the base fracturing fluid from refined and lease crudes to gelled crudes, emulsified crudes and thickened water forced the redesign of mixing equipment. The number of supplementary additives, both liquid and solid, also increased. These additives included fluid-loss agents, non-emulsifying chemicals, and friction reducers.

The pumping equipment in general use in 1954 was capable of providing 300 hhp/pump, and by 1967 pumping units could deliver as much as 850 hhp/pump if powered with a gas turbine engine, or 550 hhp if powered with a conventional diesel engine. Liquid-propping-agent proportioners were also developed. The deliverabilities of these proportioners, designed to continuously mix liquid and propping agents in the correct ratio, increased from 12 to 300 bbl/min. Propping agents were added in ratios varying from 0.1 to 8.0 lb/gal.

8.3 Pumping Units

Purpose

The demand for high injection rates and pressures for oilfield fracturing dictated the development of special pumping equipment for injecting a prepared propping agent slurry. So that they would be economical and portable, it was necessary that the pumping components be lightweight and capable of developing, transmitting and absorbing the required horsepower.

Configuration

Fig. 8.6 shows a manually operated, diesel powered truck pumping unit, typical of the 1969 equipment specifically designed for hydraulic fracturing operations. Lightweight components permit the use of two pumping units for full utilization of truck capacity. Fig. 8.7

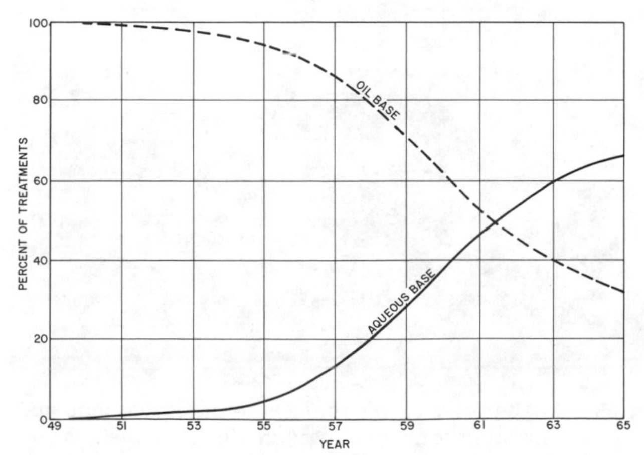

Fig. 8.4 Usage trend, base fluids for fracturing.

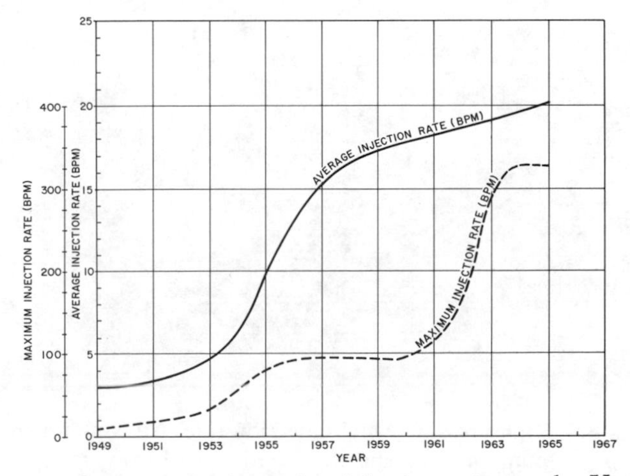

Fig. 8.5 Hydraulic fracturing injection rate trends, U. S.

shows a diesel powered pump that is typical of the type operated by remote control. Trailer mounted pump units have been used in some areas; however, the truck unit is generally preferred where vehicle maneuverability is a factor. The lower weight per axle on the trailer units, however, allows the additional installation of discharge lines, fittings and hoses that are required for tying pumps into wellheads and blenders. The displacement pumps on these fracturing units are equipped with a variety of plunger sizes, making it possible to attain injection rates of 42 bbl/min and pumping pressures as high as 20,000 psi.

Engines

The basic considerations in selecting prime movers for installation on fracturing units are reliability, compactness and cost. Reliability is the most important factor because engine failure during a fracturing job could result in an inability to complete the job in an expensive well workover. To allow mobile pumping equipment to stay within legal highway limits, prime movers must be lightweight and compact. Initial price, cost of maintenance, overhaul expense, and fuel consumption are important factors that control the economics of the fracturing job.

The four basic types of prime movers used on oilfield fracturing units have the following general brake horsepower characteristics:

1. Heavy-duty gasoline engine — 400 bhp
2. Lightweight, high-speed diesel engine — 700 bhp
3. Liquid-coolant gasoline engines (aircraft) — 1,100 bhp
4. Two-shaft gas turbine engines — 1,200 bhp

The number of gasoline engines installed on new pumping units has decreased greatly in the last few years. This downward trend resulted from the general inability of the gasoline engine to match the high horsepower output of the diesel engine.

The lightweight diesel of the 1955 to 1965 era was more rugged than the gasoline engine and was thus capable of being operated at or near maximum rated horsepower for the intermittent service required during hydraulic fracturing operations. The high compression ratio of the diesel engine makes it more efficient than the gasoline engine at full power, and gives it an even greater advantage at partial load. The high operating efficiency has resulted in low fuel consumption. For intermittent service, fuel economy has not been a major consideration, but for portable units the weight of fuel that must be carried in order to sustain a pumping unit is an important design factor.

From 1950 to 1966 the total diesel engine horsepower used for driving pumps increased from about 250 to more than 1,200 per vehicle. With diesel-weight; horsepower ratios of approximately 7 to 10 lb/hp, the total horsepower that can be truck-mounted using diesel engines has been limited.

The weight-horsepower limitation and the desire for self-sufficient, high-horsepower pumping units under the control of one operator have dictated the use of gas turbine engines. The first such pumping units were placed in field service during 1960. They were primarily built and field tested to determine if a turbine engine could withstand oilfield service work. To date, the lightweight turbines have performed well.

Initial cost of the gas turbines in 1966 was high in comparison with that of the lightweight reciprocating engines. Diesel engines with horsepowers ranging from 600 to 1,000 cost approximately $25/bhp/engine, when equipped with required accessories. The gas turbines of equivalent horsepower cost approximately $40/bhp/engine with the required accessories. The major service companies are using more turbine engines each year.

Fig. 8.8 is a schematic comparison of the torque curves of a typical diesel engine and a two-shaft gas turbine engine. Torque of the gas turbine engine increases approximately 200 percent under lugged-back or stalled conditions as compared with approximately 10 percent for the diesel engine. (The gas turbine torque curve is extremely helpful when a pump is being

Fig. 8.6 High-pressure, high-injection-rate truck-mounted pumping equipment used in 1967; manual control.

Fig. 8.7 High-pressure, high-injection-rate truck-mounted pumping equipment used in 1967; remote control.

placed in operation against the pressure of other pumps on a common discharge line.) Also, the wider utility range of the gas turbine allows reduction in the number of transmission ratios required to meet the varying pressure-volume demands of fracturing.

Fig. 8.9 shows a two-shaft gas turbine powered pump unit that was placed in field service during 1966. This unit provides 850 hhp. Experimental pump trucks have been built that carry four gas turbine driven pumps capable of delivering approximately 1,700 hhp.

As the demand for higher horsepower pumping units continues, the use of gas turbine engines will increase. This increased demand will allow the gas turbine to be competitive with the diesel engine.

Power Trains

In 1969 the power train on a fracturing unit normally consisted of a clutch, multiple speed mechanical transmission or torque converter, and a flexible universal joint or spherically coupled drive shaft. These components had to be lightweight and capable of transmitting the engine horsepower.

Most of the clutches used on fracturing equipment are of the dry, heavy-duty automotive type. Field operation has shown that this type of clutch performs satisfactorily and has sufficient life for intermittent oilfield service work.

The transmission of power from engine to pump must be designed so that the pump satisfies its complete pressure-volume requirement with a given input engine horsepower. Mechanical transmissions with up to 15 torque multiplying ratios are normally used on fracturing units. Mechanical-type transmissions are frequently used, since they have a weight and mechanical efficiency advantage over the torque-converter type of transmission. The full torque shaft mechanical transmission is particularly advantageous. The torque converter is generally considered only 80 percent as efficient as a mechanical transmission. However, the need to perform fracturing treatments without shut-down or abrupt changes in injection rates and pressures has resulted in the continued use of the multistage torque converter. The inherent torque multiplication characteristic of a two-shaft gas turbine makes this engine particularly applicable to hydraulic fracturing operations. To achieve the smoother operation, the torque converter sacrifices some horsepower; therefore, excess horsepower must be supplied to deliver a prescribed pressure and volume.

Drive couplings such as those used on stationary engines are not suitable on mobile equipment because of large chassis frame deflections. The weight restrictions make it impractical or uneconomical to install a subframe that would be rigid enough to maintain the alignment of pumping components during moving and pumping operations. For this reason, heavy-duty flexible couplings are used between transmission and pump because they can accept considerable misalignment and still deliver the required horsepower.

Pumps

Well designed, rugged, portable pumps are essential to the fracturing process. Hydraulic fracturing depends upon the ability of pumps to handle various fluids containing propping agents at the high displacement rates and pressures required to meet the well treating conditions encountered.

The characteristics of the fracturing jobs vary, depending upon locality, depth of well, formation to be fractured, bottom-hole conditions, and formation pressures. Each well should be carefully analyzed and a fracturing program should be developed to meet the conditions present. This requires flexibility in pump performance; i.e., a pump must be capable of covering a wide range of pressure and volume requirements. In the mid-1960's, fracturing job pumping rates ranged from 10 to 100 bbl/min, with injection pressures varying from 1,500 to 12,000 psi. To do this work, pumps were designed to operate with an engine of a given

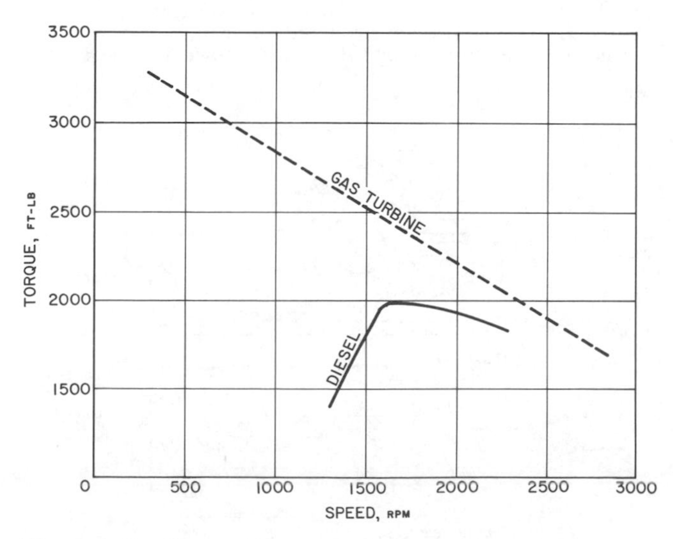

Fig. 8.8 Engine torque curves, gas turbine vs diesel.

Fig. 8.9 Two-shaft gas turbine engine driving a positive displacement pump.

horsepower. By changing pump plunger sizes, flexibility in the choice of pressures and volumes also was achieved.

Pumps generally used for fracturing services are horizontal, single-acting plunger units. For example, the pumps shown in Figs. 8.6 and 8.7 are horizontal triplex plunger pumps designed to absorb 600 to 750 bhp. These pumps typically have an 8-in. stroke and can be fitted with fluid ends having 3- to 7¾-in.-diameter plungers. By 1967, the various sizes of fluid ends gave the fracturing pump a pressure-volume range as high as 15,000 psi and 2 bbl/min and a maximum volume output of 27 bbl/min at 1,500 psi. Pumps capable of 20,000 psi and 1½ bbl/min were available on special order. Pumps are produced in left-hand and right-hand power end models to facilitate the mounting of pumping equipment side by side on truck or trailer chassis. Pump-engine combinations are frequently mounted as independent units; this provides an added service safety factor. A typical pump for fracturing service weighs approximately 5,250 lb for a weight ratio of 8.75 lb/hp. By comparison, the weight ratio of a typical oilfield slush pump is four to six times greater than this design.

The pump power-end case is fabricated from heat treated steel plate having a tensile strength of 115,000 psi. The case was designed to have a maximum fatigue life of 180,000 when operating under maximum plunger-end loading (see Sec. 8.3).

The typical horizontal pump designed for hydraulic fracturing operations utilizes a lightweight worm gear drive for power-end torque multiplication. The right-angle drive feature permits compact side-by-side mounting of two pumps. A standard pump worm-gear set has a pinnon:gear ratio of 8.6:1. However, other ratios are available to meet specific applications.

Pump fluid-ends used for oilfield service must be capable of handling acids, sand and oil mixtures, cement, and mud at high pressures. Pumping of such materials dictates that fluid-ends be designed for easy maintenance.

Fig. 8.10 shows a section through a valve chamber. Discharge valve and cylinder head covers are typically threaded and can be hammered on and off rapidly for servicing of the fluid-end. Valves, seals, seats and springs are designed to withstand the abrasive action of propping agents pumped at high speeds and high pressures.

Pressure-Volume and Horsepower Calculations

Pressure and volume requirements for oilfield hydraulic fracturing vary for various formations and locations. Based on volume of fluid and propping agent being pumped at the required fracturing pressure, hydraulic horsepower can be calculated.

$$hhp = \frac{pq\gamma}{C} \qquad (8.1)$$

where

C = conversion constant depending on units of q
= 1,714, if q = gal/min and $\gamma = 1$
= 40.8 if q = bbl/min and $\gamma = 1$

Volumetric rate q is the actual volume being pumped, and this value is affected by the volumetric efficiency of the pump. Theoretical volumetric rate is reduced by volumetric efficiency, which, in turn, is dependent upon pump speed, pressure, internal characteristics of fluid-end, and fluid being pumped.

$$q = (A_p)(L_s)(N_p)(\text{Prpm})(\text{VE}) \qquad (8.2)$$

where

q = volumetric rate (units of q depend on units of A_p and L_s and resulting conversion factor)
Prpm = pump revolutions per minute*

The following example illustrates the use of these formulas for calculating volumetric rate and hydraulic horsepower.

Example 1

Determine the hhp of an 8-in. stroke, 5-in. diameter plunger triplex pump on a hydraulic fracturing pump unit operating at 1,500 psi at 220 Prpm. Assume volumetric efficiency of the pump at this speed to be 97 percent. Specific gravity of fluid being pumped is 1.

$$q = (A_p)(L_s)(N_p)(\text{Prpm})(\text{VE}) \qquad (8.2)$$

$$= (19.63 \text{ in.}^2)(8 \text{ in.})(3 \text{ plungers})(220 \text{ Prpm})(.97)$$
$$= 100{,}537 \text{ cu in./min}$$
$$= (100{,}537 \text{ cu in./min}) \frac{1 \text{ gal}}{231 \text{ cu in.}} = 435 \text{ gal/min}$$
$$= (435 \text{ gal/min}) \frac{1 \text{ bbl}}{42 \text{ gal}} = 10.36 \text{ bbl/min}$$

*This value represents actual revolutions of the pump crankshaft and may also be referred to as crank rpm, or pump strokes per minute. Calculation of this speed from engine speed and transmission reduction is discussed in the section on pressure-volume data and shown in Eq. 8.6.

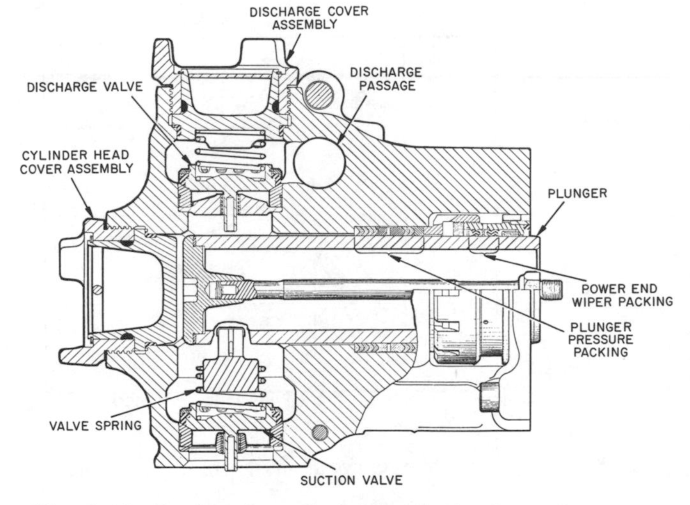

Fig. 8.10 Section through fluid end of a fracturing pump.

$$hhp = \frac{pq^2r}{40.8} = \frac{(1500 \text{ psi})\,(10.36 \text{ bbl/min})\,(1)}{40.8}$$

$$= 396 \text{ hp}$$

A variation of Eq. 8.2 is shown below. This is useful in pressure-volume calculations also, which will be discussed later.

$$\text{VPR} = (A)(L)(N)(\text{VE}) \quad . \quad . \quad . \quad . \quad . \quad . \quad (8.3)$$

$$q = (\text{VPR})\ (\text{Prpm}) \quad . \quad . \quad . \quad . \quad . \quad . \quad (8.4)$$

With the hydraulic horsepower output of the pump known, the necessary engine horsepower or brake horsepower can be calculated. The actual brake horsepower required is affected by mechanical losses through the pump drive, transmission and engine. These losses may be described as a percentage reduction of input engine horsepower, and may be termed as over-all mechanical efficiency.

$$\text{ehp} = \frac{\text{hhp}}{\text{ME}} \quad . \quad . \quad . \quad . \quad . \quad . \quad . \quad . \quad . \quad . \quad (8.5)$$

The actual engine horsepower required to deliver the 396 hhp calculated in Example 1, assuming an over-all mechanical efficiency of 85 percent, may be determined as follows:

$$\text{ehp} = \frac{\text{hhp}}{\text{ME}},$$

$$\text{ehp} = \frac{396}{0.85} = 467 \text{ hp delivered} \quad . \quad . \quad . \quad . \quad . \quad . \quad (8.5)$$

Thus, because of the demands of the cooling fan and accessories, this service would require a fracturing unit with an engine capable of 600 ehp.

For performance of the job, the operator must establish the speed-torque-horsepower range that the engine will transmit through the power train. For this information, he consults the P-V* curve for his pump-transmission engine combination. Fig. 8.11 is a typical pressure-volume* curve, plotted on log-log paper for a 4-in. plunger, 8-in. stroke, triplex pump driven through a 10-speed transmission by a 600-ehp engine.

This curve shows pump pressure-volumetric capabilities through the various transmission ranges, at rated engine speed and power. To calculate these data, the engine horsepower-torque vs engine-speed curves (similar to Fig. 8.12) must be available. These curves, for a 600-hp diesel engine, show the horsepower and torque developed vs engine speed. The torque point is usually determined at the maximum lug-back engine speed. This information permits calculation of the maximum volumetric rate and maximum pressure capabilities for a given transmission speed. Fig. 8.11 shows the points at maximum engine and lug-back speeds at which the operator must shift to a lower gear in order to meet the pressure requirements. These points occur when pressure output is directly proportional to pump input torque. The following example illustrates calculation of a P-V curve. Previous formulas and terminology will apply.

New terms introduced are:

PR = pump reduction ratio (internal gear or chain reduction)
TR = transmission reduction ratio
erpm = engine revolutions per minute

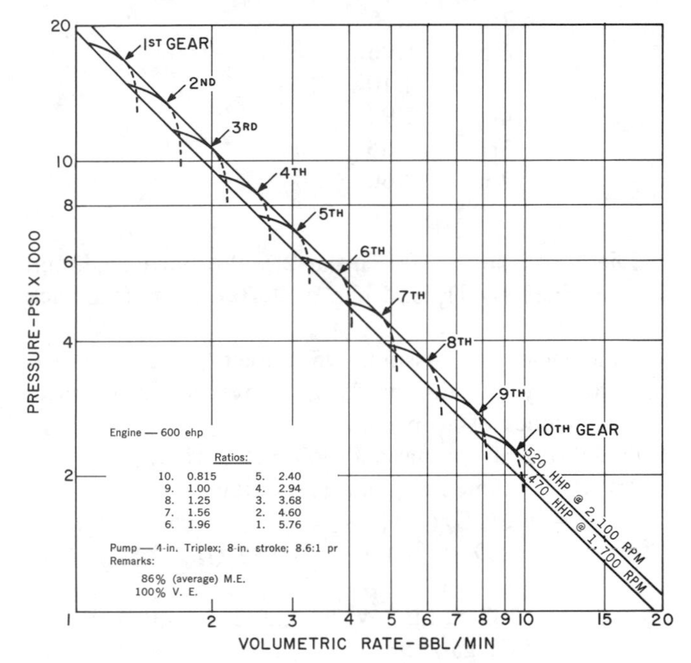

Fig. 8.11 Pressure-volumetric rate curve.

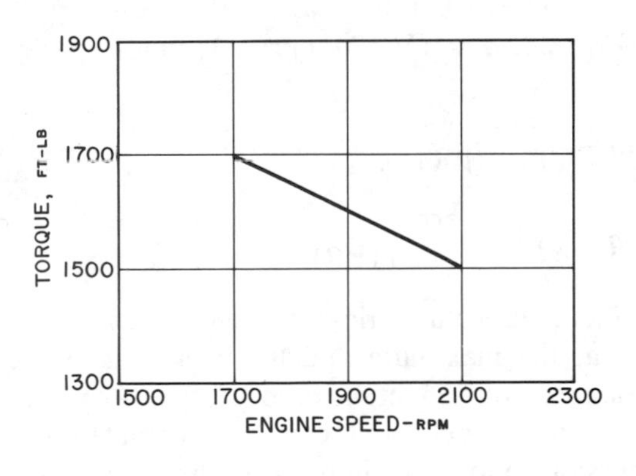

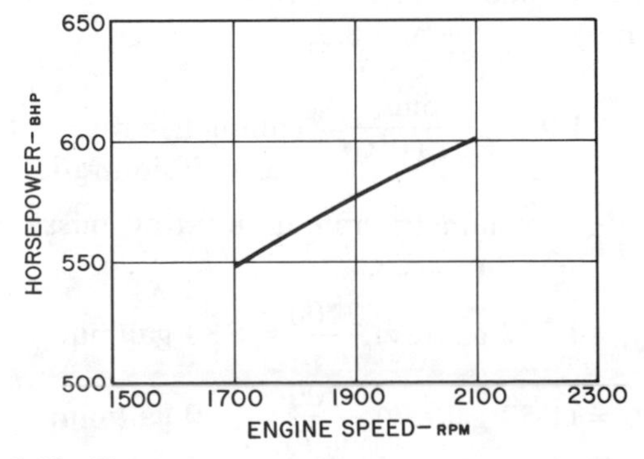

Fig. 8.12 Torque and power curves for 600-hp diesel engine.

*P-V refers to pressure-volumetric rate data. However, in field use this is usually shortened to pressure-volume.

Example 2

Pump — Triplex, 4½-in. diameter plunger, 8-in. stroke, 8.6:1 pump gear reduction ratio.
Transmission — 10 speed as listed below.

Conditions:

Gear	TR
10th	0.815
9th	1.0
8th	1.25
7th	1.56
6th	1.96
5th	2.40
4th	2.94
3rd	3.68
2nd	4.60
1st	5.76

Diesel engine — 600 ehp at 2,100 erpm (maximum engine speed), 554 ehp at 1,700 erpm (lug-back speed)
Efficiencies — Assume 95 percent volumetric efficiency at all speeds and over-all mechanical efficiency of 86 percent.

Calculations of volume and volumetric rate:

VPR = volume per pump revolution
= $(A)(L)(N)(VE)$
= (15.9 sq in.) (8 in.) (3) (0.95)
= 363 cu in./rev
= (363 cu in./rev) (1 gal/231 cu in.)
= 1.57 gal/rev
= (1.57 gal/rev) (1 bbl/42 gal)
= 0.0374 bbl/rev

From Eq. 8.4, q = (VRP) (Prpm), but

$$\text{Prpm} = \frac{\text{erpm}}{(\text{TR})\,(\text{PR})}, \quad \ldots \quad (8.6)$$

$$q = (\text{VPR})\frac{\text{erpm}}{(\text{TR})\,(\text{PR})} \quad \ldots \quad (8.7)$$

To facilitate volumetric rate calculations, q is calculated at the maximum and lug-back engine speeds* at a transmission reduction ratio of 1:1. Then the volume q actually delivered can be calculated for any transmission ratio by dividing q by the actual TR. This is shown by Eqs. 8.8 and 8.9.

$$q^1 = (\text{VPR})\frac{\text{erpm}}{(1)\;(\text{PR})} = \text{volumetric rate at 1:1 transmission ratio.} \quad (8.8)$$

$$q = \frac{q^1}{\text{TR}} = \text{volumetric rate at any transmission ratio.} \quad \ldots \quad (8.9)$$

$$q^1_{2100} = (1.57\ \text{gal/rev})\frac{2100}{8.6} = 384\ \text{gal/min},$$

$$q^1_{1700} = (1.57\ \text{gal/rev})\frac{1700}{8.6} = 310\ \text{gal/min};$$

*These are shown as subscripts in calculations.

for example,

$$q = \frac{q^1_{2100}}{\text{TR}} = \frac{384}{1.56} = 246\ \text{gal/min in 7th gear}, \quad (8.9)$$

$$q = \frac{q^1_{1700}}{\text{TR}} = \frac{310}{3.68} = 84.2\ \text{gal/min in 3rd gear.}$$

Horsepower and pressure calculations:

$$\text{hhp} = (\text{ehp})\,(\text{ME}), \quad \ldots \quad (8.5)$$
$$\text{hhp}_{2100} = (600)\,(0.86) = 516\ \text{hp},$$
$$\text{hhp}_{1700} = (554)\,(0.86) = 476\ \text{hp};$$

but

$$p = \text{hhp}\,\frac{(1714)}{q}, \quad \ldots \quad (8.1)$$

$$p_{2100} = \frac{(516)\,(1714)}{q_{2100}} = \frac{88.44 \times 10^4}{q_{2100}},$$

and

$$p_{1700} = \frac{(476)\,(1714)}{q_{1700}} = \frac{81.58 \times 10^4}{q_{1700}};$$

for example, $p_{2100} = \dfrac{88.44 \times 10^4}{246\ \text{gpm}} = 3{,}595$ psi in 7th gear,

$p_{1700} = \dfrac{81.58 \times 10^4}{842\ \text{gpm}} = 9{,}689$ psi in 3rd gear.

From the above calculations, Table 8.2 is made and a P-V curve similar to Fig. 8.11 is plotted. Engine speeds between maximum and lug-back may also be used to plot the actual curve. Thus, the capabilities of any pumping unit combination can be determined.

While the calculations in Table 8.2 were based on a specific engine-transmission-pump combination, the method is similar for gasoline engines, gas turbines, electric motors, torque converters and pumps other than triplex.

Flexibility in P-V requirements in a pump is achieved by changing the plunger size. On units requiring increased volume output at lower pressures, the larger plungers are used, and on high-pressure pumps, the smaller plungers are used. Horsepower capability is the same within P-V requirements.

Fig. 8.13 is a theoretical P-V curve showing the maximum pressure and volume capability of an 8-in. stroke triplex with various-sized plungers. Maximum pressure is determined by plunger load furnished by the power end of the pump.

$$\text{PL} = (p)\,(\text{PA}) \quad \ldots \quad (8.10)$$

Hydraulic horsepower and volumetric rate are calculated using Eqs. 8.1 and 8.2, respectively.

It should be noted that by using, within the limitations of the pump, a hydraulic horsepower governor and a suitable torque converter, a pumping unit can be designed to deliver a constant hydraulic horsepower. This configuration is finding increased acceptance, particularly with the gas-turbine powered pump.

8.4 Liquid-Propping-Agent Proportioners

Purpose

Once the advantages and the need for hydraulic fracturing were established by the oil industry, a method of proportioning the fracturing fluids and propping agents had to be developed. The importance of controlling the ratio of these materials within close limits and the ability to vary this ratio throughout the treatment were established in the early stages of commercial development. Treatment size and rate of displacement increased, necessitating the development of a procedure to continuously mix and proportion the proppant in the fracturing fluid.

As fluid types and characteristics changed, the proportioning equipment was modified so that either or both dry and liquid chemical additives could be proportioned and mixed with the base fracturing fluid at varying rates and in varying ratios.

Proportioning equipment more recently has been used to meter accurately the fracturing fluid and propping agent and chemicals from several supply sources, combine and mix them in varying ratios with the base fracturing fluid, and then deliver the mixture to the high-pressure pumps.

TABLE 8.2 — PRESSURE-VOLUME ENGINE SPEED RELATIONSHIP

Engine: 600 ehp
Transmission: 10 speeds as listed
Pump: 4.50-in. plunger, 8-in. stroke triplex
8.6:1 pump reduction (internal)
Efficiencies: 86 percent mechanical, 95 percent volumetric
Hydraulic horsepower: 516 hp at 2100 erpm (maximum engine speed)
476 hp at 1700 erpm (lug-back speed)

Engine rpm (erpm)	Transmission Reduction TR	Transmission Reduction Gear	Volumetric Rate, q (gal/min)	Pressure, p (psi)
2100	0.815	10th	471	1,878
1700			476	2,141
2100	1.00	9th	384	2,303
1700			311	2,623
2100	1.25	8th	307	2,881
1700			249	3,276
2100	1.56	7th	246	3,595
1700			199	4,099
2100	1.96	6th	196	4,512
1700			159	5,131
2100	2.40	5th	160	5,527
1700			129	6,324
2100	2.94	4th	130	6,803
1700			106	7,696
2100	3.68	3rd	104	8,504
1700			84.2	9,689
2100	4.60	2nd	83.5	10,592
1700			67.6	12,068
2100	5.76	1st	66.7	13,259
1700			54.0	15,107

Configuration

Proportioning is accomplished by several methods. In most cases the fluid is moved by positive displacement gear pumps into a mixing tank; however, in some configurations, centrifugal pumps are used. At the same time, the propping agents and chemicals are metered into the same tank, where they are agitated. This mixture is then moved by a centrifugal pump into the suction manifold of the high-pressure positive displacement pumps. Since the speed at which the materials are mixed must be variable, while the speed of the pumps pressurizing the high-pressure-rate fracturing pumps is usually constant, many proportioning units are powered by two engines. Variations of this design, however, are common. For example, one company utilizes one engine to drive two centrifugal pumps, with proportioning being accomplished at the point of sand injection. Fig. 8.14 is a schematic diagram of an early liquid-propping-agent proportioner. Fig. 8.15 shows a 100-bbl/min trailer-mounted, liquid-propping-agent proportioner.

Pumps

Generally, there are two classes of pumps on a proportioner — the metering pump and the transfer pump. The positive displacement gear pump meters the clean fracturing fluid (fluid without propping agent) into the mixing tank. Some equipment designers prefer to use a

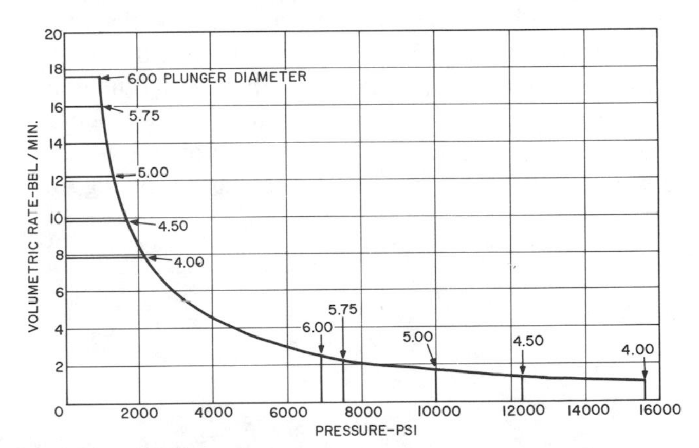

Fig. 8.13 Effect of plunger diameter on maximum pressure and volumetric rate capabilities of a reciprocating pump based on same plunger load and same hhp output.

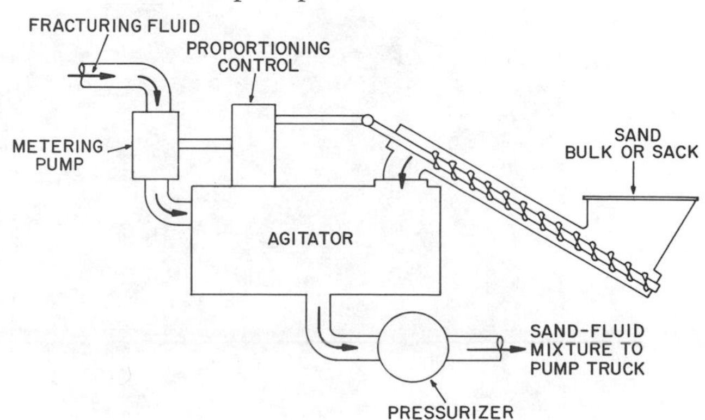

Fig. 8.14 Schematic diagram of sand fluid proportioner.

centrifugal pump for metering. For example, in areas where the fluid is extremely dirty (as water from an earthen pit), a centrifugal pump provides better service. For this type of pump, a priming device must be provided because the fracturing fluid stored in an earthen pit is usually below the level of the pump. Metering with the gear pump is accomplished by running the pump at a known speed or, with a centrifugal pump, by sending the fracturing fluid through a flow meter.

The transfer pump is normally a centrifugal pump. It provides positive pressure on the suction of the high-pressure pump, reduces pump vibration and increases pump efficiency. To provide steady pressure, this pump usually runs at a constant speed and thus is able to furnish the mixed fluid on a demand basis. The mixture may also be passed through a flow meter to indicate total rate and volume being pumped into the well system by the high-pressure pumps.

Propping Agent Systems

Two systems are used to meter propping agents — the conveyor screw, and the gravity feed through a calibrated butterfly-type valve.

The conveyor screw serves two purposes: (1) it proportions or meters the proppant; and (2) it transfers the material into the mixing tank. The rate of delivery of a propping agent by a conveyor screw turning in a tube is approximately proportional to the revolutions per minute. The screw may be driven mechanically or hydraulically. With the mechanical drive train for the screw connected to the pump shaft through an infinitely variable transmission, the ratio of screw speed to pump speed may be controlled easily. The propping agent can be transferred to a higher elevation by inclining the tube. The conveyor screw is frequently driven by a hydraulic pump. The signal from a tachometer generator turned by the screw may be fed to an indicating meter on the control stand. The rate is adjusted by varying the shaft speed of the screw conveyor. Propping agent ratio is normally shown on a dial, and ratio of metering pump speed to screw speed is controlled by matching the calibrated tachometer generator signal from the screw to a signal from the turbine flow meter. The match meter system and a typical control box are shown in Fig. 8.16.

The gravity feed proportioner dumps the additive directly into the tank where it is mixed with the fracturing fluid. By precalibrating the butterfly control valve, the rate of discharge of the additive or proppant is established for each setting of the valve opening. This controlled rate of proppant addition must be added to a metered input of fracturing fluid. Sand can be dumped without being controlled if a densitometer is placed between the mixing tank and the high-pressure pump suction manifold to monitor the proportioner output.

Chemical Additive Systems

Progress in hydraulic fracturing has required the introduction of many chemical additives, both dry and liquid. This has brought about the addition of the dry chemical additive and liquid additive systems on proportioners. For dry chemicals, calibration and addition usually are accomplished by a rotating set of vane feeders housed in a special case located under the hopper feeder. By precalibrating each chemical, the speed of this feeder can be controlled and matched against the reading of a flow meter to get the proper ratio of chemical to fluid. The vane feed frequently is linked me-

Fig. 8.15 Trailer-mounted sand-fluid proportioning unit.

chanically to the drive of a positive displacement gear pump for control. Liquid chemicals are customarily handled with smaller gear pumps, and the rate of feed is controlled in much the same manner as that of the dry chemicals.

Metering and Control Devices

Flow Meters. Flow meters of various designs are used in hydraulic fracturing operations to determine the rate of flow and the total volume of fracturing fluid injected into the well. The rate of flow is important to the operator of the blender or proportioning unit that is used to mix the fluid and propping agent. By knowing the total volume of fluid pumped into the well, the supervising engineer is able to change concentrations of propping agent and other additives at the proper time. Flow meters are used with fluids, with or without propping agents. These meters normally are placed in the suction and/or discharge lines of the unit that blends the fracturing fluid. Fig. 8.17 shows the flow meter installed in the discharge line of a proportioning unit.

The turbine-type flow meter, equipped with special bearings to permit operation in fluids carrying abrasives in suspension, consists of a housing, stationary vanes and a rotor element. Fig. 8.18 shows a cut-away view of a typical flow meter. The rotor element has a number of blades that are pitched at an angle to the direction of the fluid flow through the housing. As fluid flows through the housing, it strikes the rotor blades, thus causing the rotor to spin at a speed that is proportional to the velocity of the fluid stream. The speed of the rotor is sensed by a magnetic pickup and converted to flow rate, and the total fluid passing through the flow meter is shown by the readout instrument.

Fig. 8.17 Flow-meter installation on fluid proportioner unit.

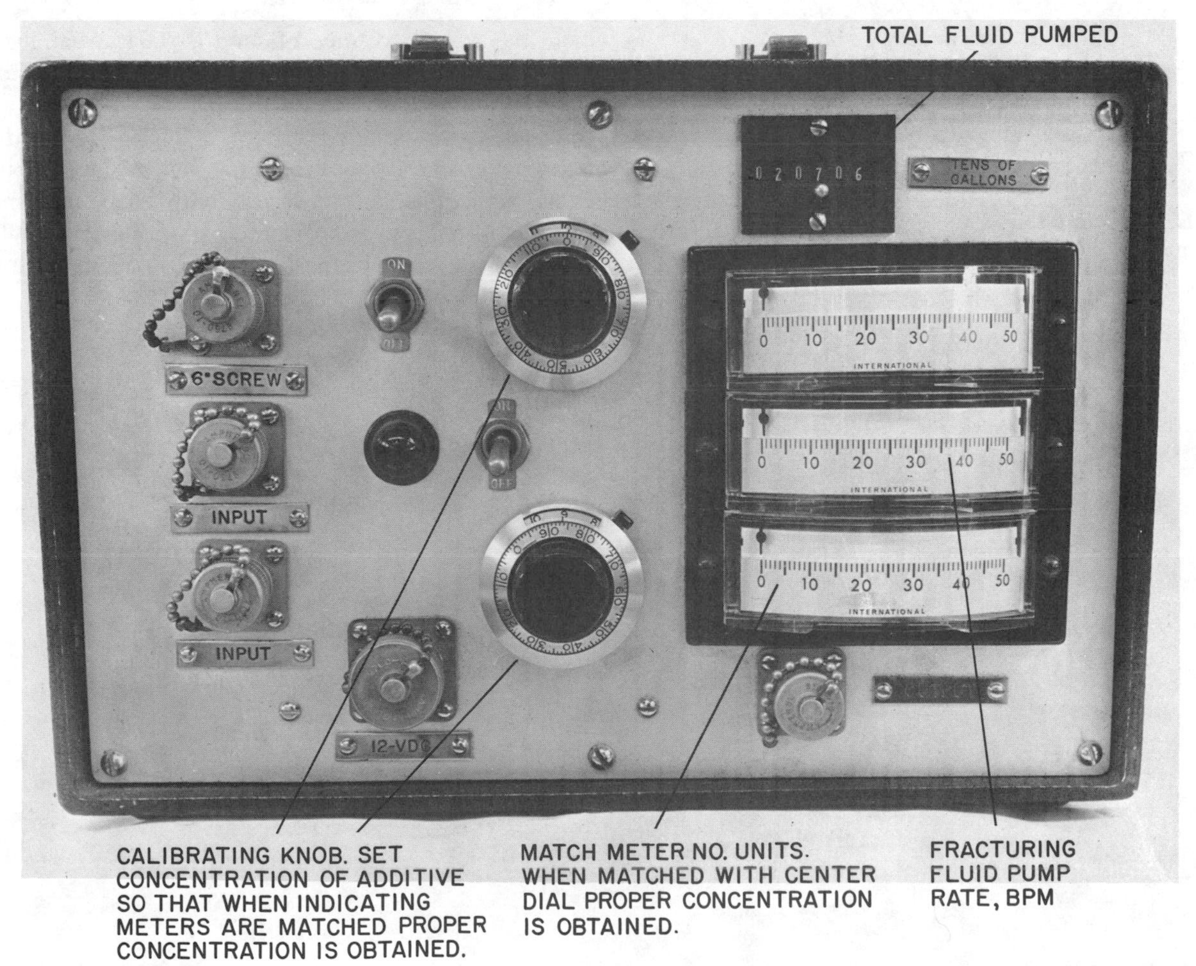

Fig. 8.16 Match-meter-control box.

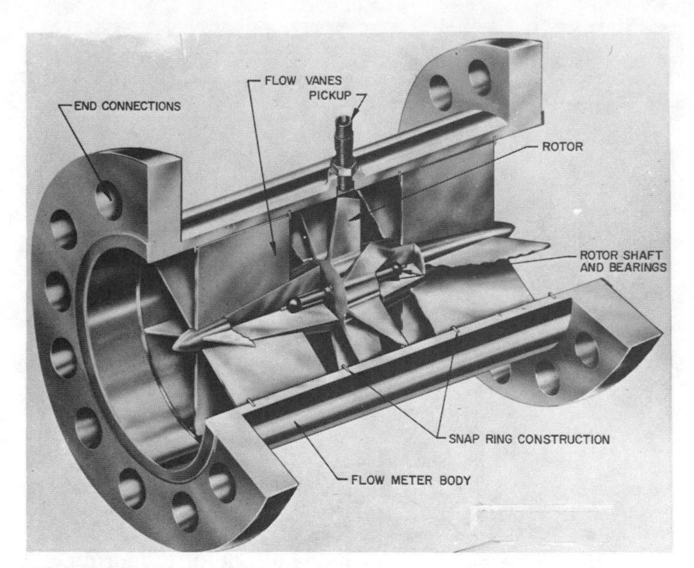

Fig. 8.18 Flow meter cutaway.

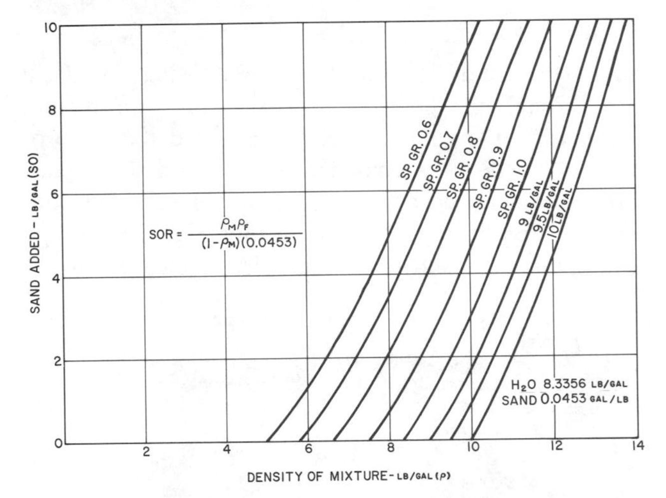

Fig. 8.19 Lb/gal sand added vs lb/gal mixture for various specific gravities.

The fracturing fluid may also be metered by calculating or measuring the volume displaced per revolution of the drive shaft of the pump used to move the fracturing fluid into the well. The number of revolutions (or displaced volume) may be recorded or totaled on strip charts or on meters. Injection pressure frequently is recorded on the same charts to facilitate hydraulic horsepower calculations. This system is applicable to either single or multiple pump unit fracturing operations.

Density Meter. Hydraulic fracturing treatments require that a specific and uniform concentration of the sand, or other propping agent, be injected into a particular fracture zone. Excessive sand concentrations may cause sand to bridge in the well or in the fracture, preventing further injection of proppant. An inadequate concentration of the propping agent allows the fracture to close after treatment.

One method of controlling proppant concentration is to measure the fluid-proppant density. Since most propping agents have a density different from that of the base fracturing fluid, changes in concentration will be detected as changes in density.

The following is the relation between density (ρ) and sand-oil ratio (SOR) in pounds per gallon:

$$\text{SOR} = \frac{\rho_m - \rho_f}{(1 - \rho_m)(0.0453)} \quad . . . . (8.11)$$

where the specific volume of sand is 0.0453 gal/lb.

A plot of this equation for various gravities and fracturing fluids is shown on Fig. 8.19.

Fig. 8.20 is a photograph of a device designed to measure the density of fracturing fluids. Two tubular beams are connected in series, with their outermost ends supported by a special cross-spring pivot. The tubes are supported in the center by a pneumatic force

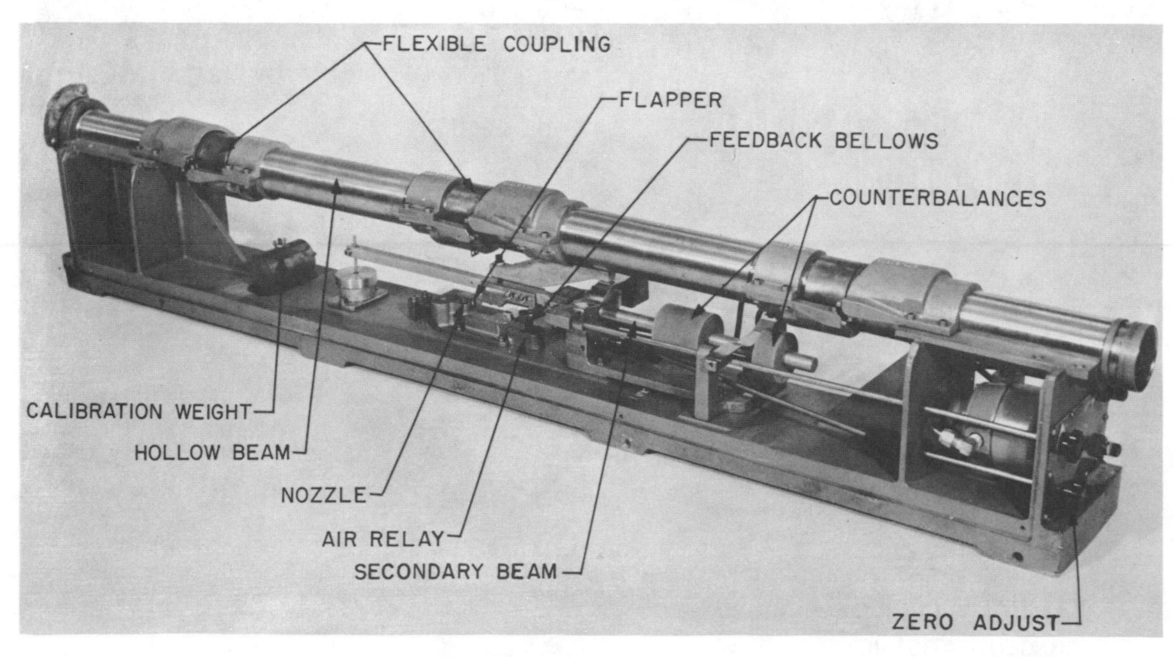

Fig. 8.20 Density meter.

balance sensor. Flexible couplings are connected at the three flex joints to provide a pressure-tight seal. A continuous sampling of the mixture of fracturing fluid and propping agent being fed to the displacement pumps is diverted through the density meter. Within the instrument itself, a constant volume of fluid is confined in the tube between the outer flex pivots. A force equal to the weight of the fixed volume of fluid is exerted downward, and is measured by a 3- to 15-psi pneumatic force balance transmitter. The pneumatic output of the force transmitter is proportional to the density of the fluid circulated throughout the instrument. The friction in the flexible couplings must be calibrated out of the force measurements.

Assuming that the specific gravity of the fracturing fluid is constant and that the sample being measured is representative of the fluid being pumped into the well, changes in density can be interpreted as changes in sand concentration. By using a pneumatic controller and control valve, the addition of sand to the fluid can be set and automatically controlled as desired.

Fig. 8.21 is a photograph of a setup for a typical field fracturing treatment. Note the density meter on the sand proportioner and the control valve on the sand truck. The pneumatic controller is located near the operator's console, which is not shown here.

A widely used gamma-ray absorption technique for measuring fluid density involves placing a gamma-ray source and a gamma-ray counter on opposite sides of pipe through which fracturing fluid is to be pumped. If the emission rate of the gamma-ray source material is known and the transmission through fluid containing no propping agent is known, then the reduction in gamma-ray transmission with propping agent in the fluid can be determined. The device used in this technique is illustrated schematically in Fig. 8.22.

Fracture Parameter Meter. The fracture parameter meter is a portable instrument designed to provide a remote indication and recording of job control variables. The data indicated and recorded consist of injection rate, fluid density, total volume and wellhead pressure. Fracturing fluid injection rate and pressure data are normally recorded on a strip chart; however, as many as five parameters may be recorded. Fig. 8.23 illustrates an instrument of this type.

A turbine flow meter in the proportioner discharge manifold generates an output signal that is proportional to the rate of flow. This signal is transmitted to the recording meter where the flow rate is recorded and totaled.

A pneumatic pressure transducer connected to the wellhead converts hydraulic pressure to an air signal. This signal is transmitted to the meter where it actuates a visual pressure gauge and recorder pen.

The density meter on the proportioner unit samples the fluid in the discharge manifold and then produces an air signal representing the fluid density. This signal is transmitted to the fracturing control meter where it

Fig. 8.21 Density meter mounted on fluid proportioner unit.

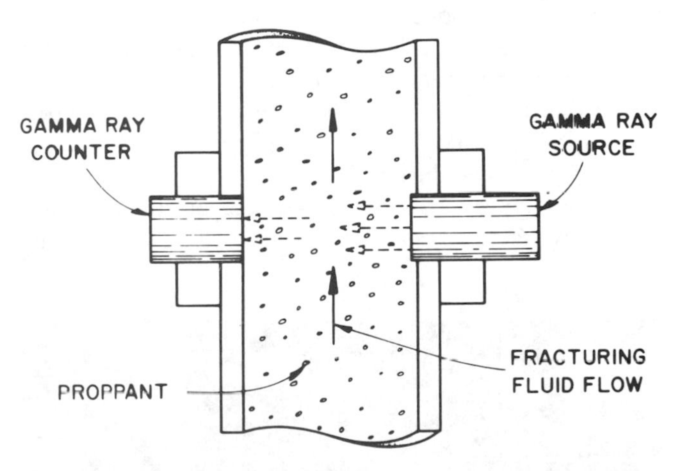

Fig. 8.22 Gamma-ray absorption-type fluid density meter.

Fig. 8.23 Fracture monitor.

operates a density gauge and recorder pen.

Remote Controls. A remote control console allows the fracturing job to be controlled and monitored at a distance from the equipment. Typical units are shown in Fig. 8.24. Use of control consoles alleviates the hazard to personnel working close to very high fluid pressures and allows the equipment operators and supervising engineer to be in a more desirable location for controlling the entire fracturing operation. Controls are conveniently located for running the engines and regulating the pumps, as well as for controlling the addition of propping agent and maintaining the desired fluid density. Visual gauges are used to display engine and pump performance as well as fracturing job parameters. Gear shifting, when required, is performed pneumatically.

Special Devices

Many special devices are used with the proportioners. Special gelled oil has been developed to help prevent propping-agent fallout while pumping into the well. The crude oil is gelled by adding chemicals and shearing the mixture in a homogenizer or stabilizer. An eductor is used with the dry additive system. This permits dry chemicals to be introduced through an eduction chamber so that they may be wetted before they are dumped into the mixer tank. This procedure prevents "lumping" of the dry powder during mixing.

Various types and sizes of portable mechanical mixers are available for premixing and pregelling hydraulic fracturing fluids. The capacity of mechanical mixer tanks ranges from 500 to 4,200 gal. Mechanical mixers are normally installed in vertical tanks with the shaft extending through the top or in horizontal tanks with the shaft extending through each end. The drive shafts are usually carried on outboard bearings. The mixing shafts are driven from one end by either a mechanical or a hydraulic drive. Horizontal tanks generally are preferred because they are easier to haul than vertical tanks.

Various mixing shaft configurations are used in mechanical mixing. Some horizontal mechanical mixers have straight horizontal blades or paddles connected to the shaft by small radial tubes. Mixing is accomplished in the straight paddle mixer by a simple rolling of the fluid. To obtain maximum mixing action in a horizontal tank, the position of the paddle blades forms a continuous helix. These mechanical mixers usually are referred to as "ribbon mixers," most of which have inner and outer "ribbon" blades, or flights, installed in opposite directions to further increase mixing action.

Propellers are used for mixing fluids in small vertical tanks. This arrangement is used for either batch or continuous mixing in sand proportioner tanks. The propeller is typically driven by a small hydraulic or pneumatic motor.

Large, vertical batch tanks equipped with a top-driven, curved-blade turbine are used for batch mixing hydraulic fracturing fluids. These tanks are designed primarily to help eliminate shaft sealing problems that arise from handling both corrosive and abrasive fluids (acid or cement) in horizontal tanks. The capacity of vertical mixing tanks ranges from 2,000 to 2,500 gal. For minimizing fluid shear, a curved-blade turbine works better than a flat-blade turbine.

Fig. 8.24 Remote control console.

8.5 Bulk-Handling Equipment

Bulk handling of propping agents used in hydraulic fracturing provides an efficient means of delivering the material to the well site. By eliminating small containers and sacks, handling costs and waste are reduced. Delivery rates must be great enough to keep pace with high-rate mixing and pumping units.

In 1969, various types of bulk transports were in use including dump trucks and dump trailers with tilting storage bins that allow the material to gravity-feed through a discharge valve. Material discharge rates are controlled by the valve, which permits rates up to 12,000 lb/min.

Fig. 8.25 shows a 20,000-lb capacity dump truck serving as a propping-agent transporter.

A second type of bulk transport is unloaded by two 9-in.-diameter conveyor screws running the full length of the bin bottom. The screws are driven by a power takeoff. The rate of discharge is controlled by varying the truck's engine speed. The propping agent can be discharged at the rear of the truck at rates up to 6,000 lb/min.

Fig. 8.26 shows another type of bulk transport; this one uses compressed air to discharge material from pressurized tanks. The propping agent is carried in the two front tanks and is blown to the rear surge tank where a valve controls the rate of discharge. The material can be transferred from this unit into storage bins or other bulk material units. Discharge rates of 2,000 to 4,000 lb/min are obtainable, depending upon the distance the material is transferred.

8.6 Surface Piping and Wellhead Equipment

Suction Systems

Suction systems for the delivery of fracturing fluid to the proportioners vary with the type of fluid used and with the particular area of operation. The three basic methods of storing fracturing fluid are: (1) truck transports, (2) fixed storage tanks of various designs and capacities, and (3) earthen pits. Standard suction manifolding consists of flexible hose between the fluid source and the proportioner. Several systems are employed as gathering manifolds when more than one tank or one transport is used. A major operating problem in maintaining injection rate has been the lack of adequate suction line capacity at the fluid source. To overcome this, service company horizontal tanks have been equipped across the front with large ID manifolds containing several outlets for connecting the flexible hoses. These large sections can be interconnected so that the final hookup will allow a balanced suction head in all tanks during the course of the job. Ground headers of various designs are used as a gathering manifold for connecting transports and when using earthen pits. To provide the specialized equipment needed in many areas, trailer-mounted headers are employed. Fig. 8.27 is an example of one of these headers.

Discharge Systems

Pressures and injection rates of the individual pump control the size, strength, and number of discharge lines between pump and wellhead. Basically, two methods have been used for laying discharge lines for a fracturing treatment. One way is to run a single line from each fracturing unit injection pump to a ground header, then a larger single line from there to the wellhead. The other is to run single lines from each pump all the way to the wellhead, where they connect to a multiconnection header. Pressures and injection rates, sand-fluid ratios, type of fluid, and number of pumps govern the configuration and size of the setup used.

To minimize erosion caused by sand-laden slurries, piping should be sized to hold pumping rates below 40 ft/sec. It should be designed so that the yield pressure is at least one and one half times the discharge-line working pressure for service at greater than 10,000 psi, and two times working pressure (minimum) for pumping operations at less than 10,000 psi. Steel with high resistance to impact must be used to combat the high fatigue forces encountered in multiple pump hookups.

Fig. 8.25 Propping agent transport unit: 20,000-lb capacity.

Fig. 8.26 Propping agent transport unit, 35,000-lb capacity, pneumatic.

Fig. 8.27 Ground header (suction manifold).

Check valves are installed in fracturing discharge lines as a safety device in case of line or connection rupture. The valves are placed as close to the fracturing wellhead as possible to provide maximum protection. Most fracturing wellheads have check valves built into them. Two types of check valves have been used. The first type, a plunger valve, uses a spring-loaded plunger that is actuated by the movement of fluid — the valve opens against a spring and closes with a spring. The second type, or swing check valve, is required when "ball sealing" devices are injected into the flow line to seal perforations. The swing valves provide a larger opening to allow passage of the balls. These valves are closed by gravity and must be upright.

Both valves employ a metal-to-metal seat with a resilient lip seal on the pressure side for 100 percent shutoff.

Wellhead Manifolds

Special wellhead manifolds and connections have been developed to meet the requirements of hydraulic fracturing. These form an intermediate manifold between the tubing or casing in the well and the discharge lines carrying fracturing materials from the pumps. Construction and design requirements of wellhead manifolds include safety, pressure capacity, mechanical strength and adaptability. Manifolds have ranged in size from lightweight, 2-connection tubing manifolds made from steel forgings to heavy, 12-connection manifolds made of steel weldments.

One example of a wellhead manifold is shown on Fig. 8.28. This is a cast steel manifold rated at 10,000 psi test pressure. Each manifold body has four inlet connections, and multiple bodies can be stacked vertically to provide more inlet connections from the pumps. Each inlet connection is provided with a check valve (shown on the cut-away view of Fig. 8.29) to help prevent backflow of fracturing materials from the manifold. The check valve in every flow line is an important safety feature to minimize hazards in the event a supply line from one of the pumps develops a break. Special forged steel components for wellhead manifolds are being used for fracturing operations when pressures greater than 10,000 psi are encountered.

Adapters used to connect the wellhead manifold body to the casing thread contain a seat in the upper end for a pressure test plug. This arrangement permits

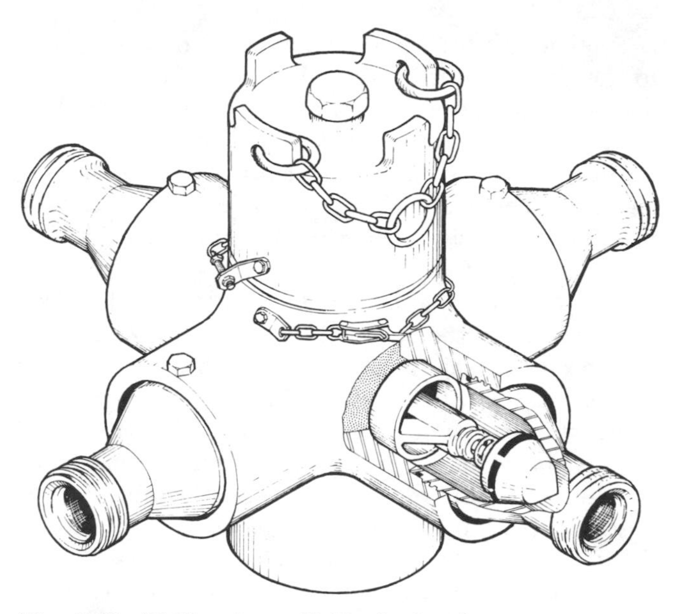

Fig. 8.29 Wellhead manifold check-valve.

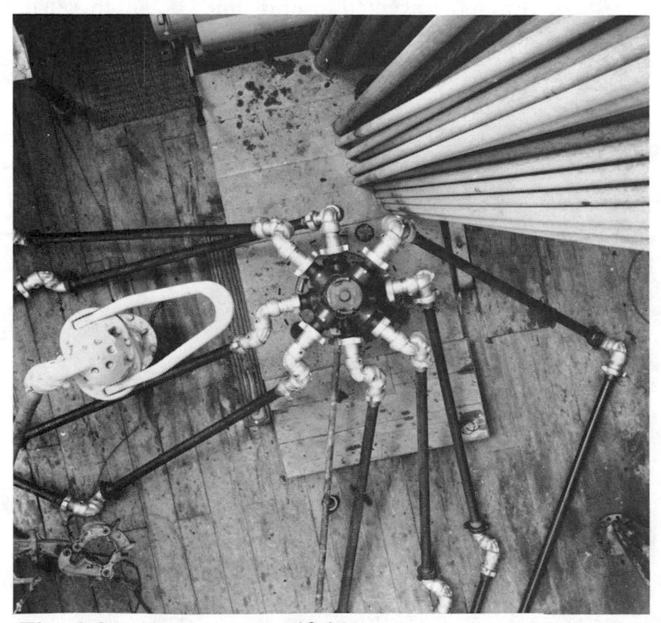

Fig. 8.28 Wellhead manifold.

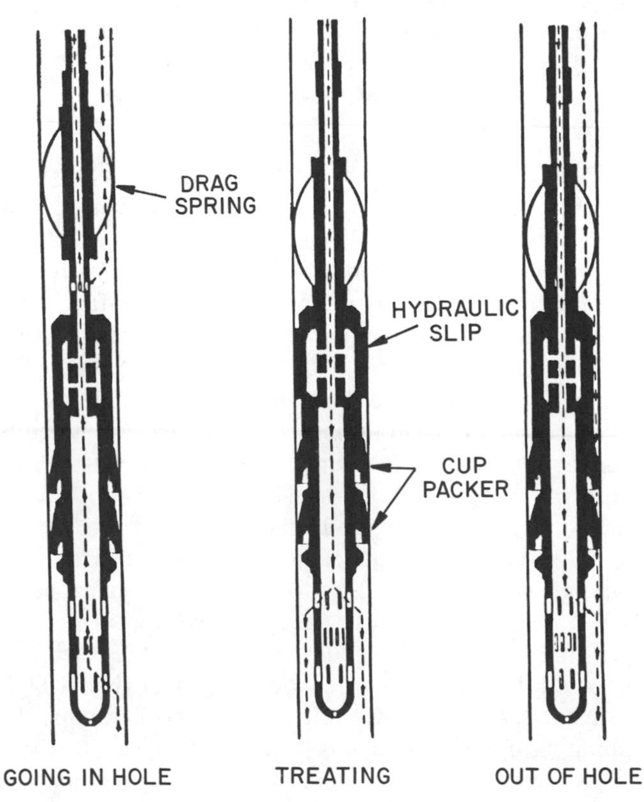

Fig. 8.30 Operational diagram for cup-type retrievable packer.

pressure testing of the entire complex of pumps, connecting lines, and wellhead manifold to determine the condition of the discharge system before the actual fracturing operations are begun. Two of the adapters typically include check valves to provide a secondary means of limiting backflow from the well.

8.7 Down-Hole Tools and Packers

Packers

One of the first packers used for fracturing was a hydraulically set straddle packer. This tool was used in the early days for both open-hole and cased-hole fracturing. The tool consisted of two packer sections, pistons to expand them, perforations between the packer sections, and a bypass tube to equalize annulus pressure above the top packer and below the bottom packer.

As fracturing in casing became more prevalent, a cup-type retrievable packer utilizing hydraulic slips gained popularity as a single packer fracturing tool (Fig. 8.30). The hydraulic slips prevented the tool from moving up the hole when the fracturing pressure was applied to the formation below the packer. This tool was eventually replaced by a full opening retrievable ring packer that utilized mechanical lower slips and hydraulic upper slips to stabilize the packer in the hole (Fig. 8.31). A "J" slot controlled the action of the mechanical slips, which were set first, after which weight applied to the packer expanded the rubber packing elements. Fracturing pressure would then set the hydraulic upper slips.

The full opening retrievable packer (Fig. 8.32) could be converted into a straddle packer by inserting a perforated section between the packers.

Multiple fracturing was accomplished by using the full opening retrievable straddle packer without moving the packer. Three or more packers were used, and again there was only one set of hydraulic slips and one set of mechanical slips. Perforated sections were provided between the packers, and a sliding sleeve valve was provided for all but the bottom set. After the bottom formation was fractured through the perforated section, a small bronze ball was dropped. This ball would pass

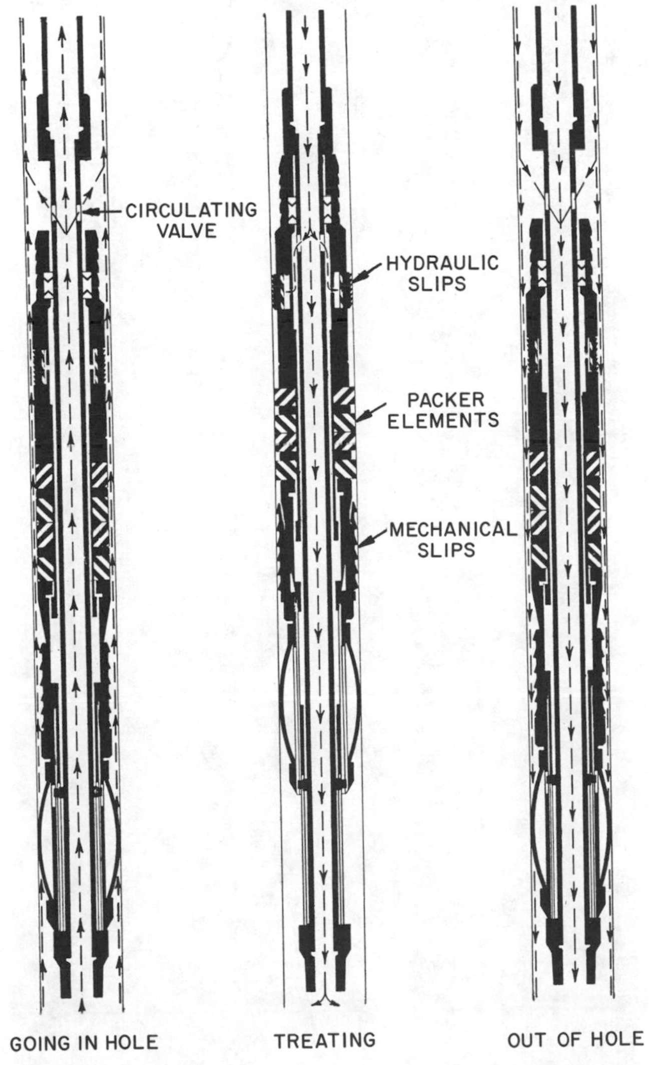

CIRCULATING VALVE

HYDRAULIC SLIPS

PACKER ELEMENTS

MECHANICAL SLIPS

GOING IN HOLE TREATING OUT OF HOLE

Fig. 8.31 Operational diagram for ring-type retrievable packer.

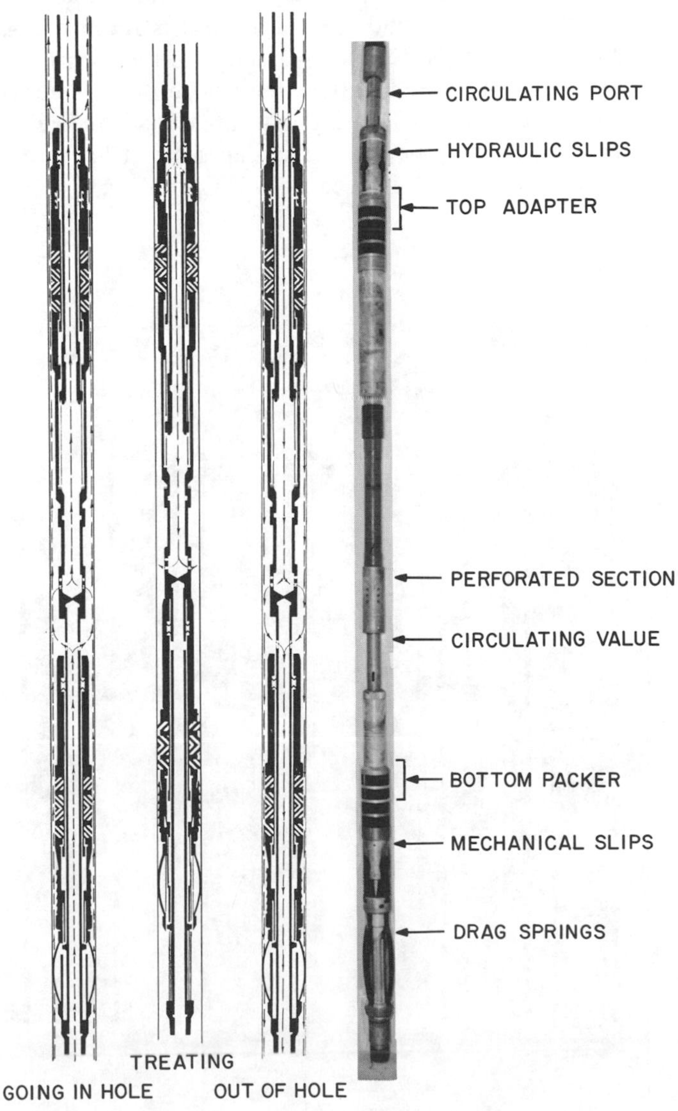

Fig. 8.32 Full-opening retrievable straddle packer.

through all upper sliding sleeve valves and land on the lower sliding sleeve, closing off the lower perforations and opening the sliding sleeve valve ports in the next higher section. After the second zone was treated, a second ball (slightly larger) was dropped to open the next sliding sleeve valve above. The tubing size was the limiting factor on how many jobs could be run because a larger ball was required for each successive treatment stage.

Sometimes, after the last zone was treated through this packer, one additional zone was treated through the annulus (between tubing and casing).

The multiple purpose retrievable packer (Fig. 8.33) is used for testing, treating and squeezing. This tool has hydraulic upper slips, mechanical lower slips, a circulating or bypass valve and ring-type sealing elements. A "J" slot controls the mechanical slips and the setting of the packer.

The retrievable bridge plug has cup sealing elements on both ends (Fig. 8.34) with two-way mechanical slips in the middle. A valve provides a bypass for well fluid when the tool is being moved. The valve is closed when the tool is set and opened when it is retrieved. A removable setting mechanism attached to the top of the tool is used to set and retrieve the bridge plug. A combination cup and packer ring that has the advantage of a double seal may be used. The cup seals at low pressure and the packer ring seals at high pressure.

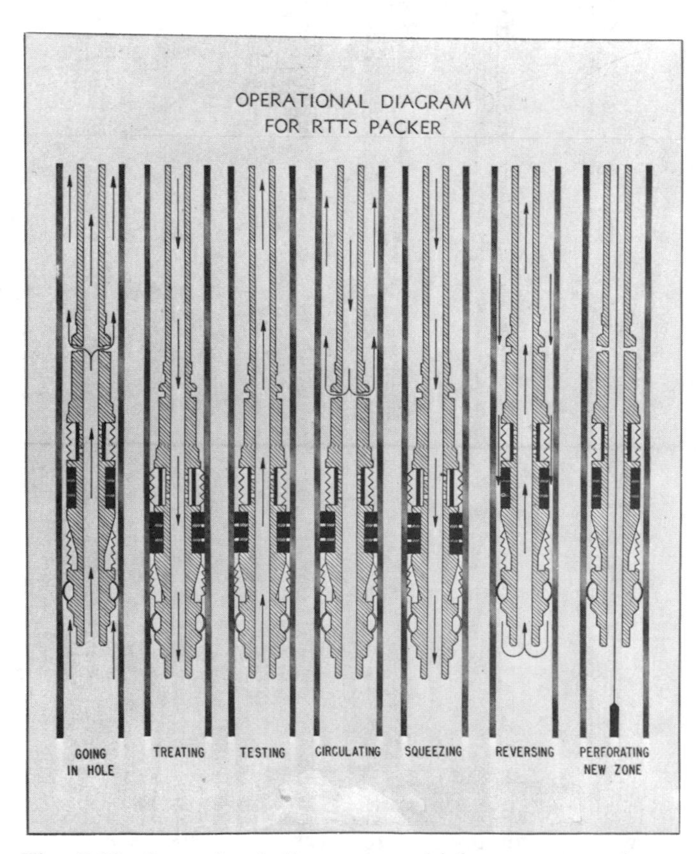

Fig. 8.33 Operational diagram—multiple purpose packer.

Two tools, the multiple purpose packer and the retrievable bridge plug, provide another means for multiple fracturing inside the casing. The bridge plug is run on the bottom of the packer and set below the formation to be fractured. Next, the packer is released from the bridge plug, pulled up, and set above the formation. The formation is then fractured through the opening in the packer. To fracture the next zone the packer is unseated and the bridge plug picked up. After the two

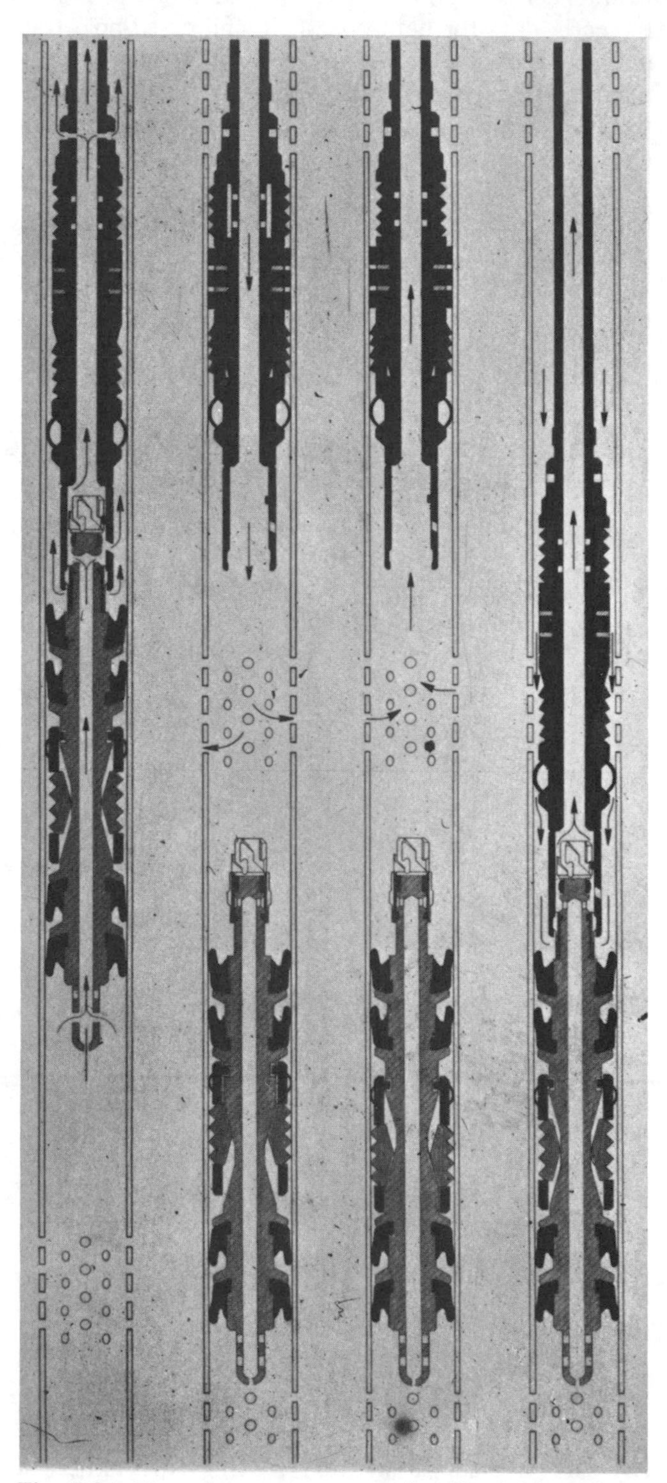

Fig. 8.34 Two-piece cup-type retrievable bridge plug.

Fig. 8.35 Hydraulic jetting tool.

tools are moved to the next zone the steps are repeated — as many times as desired.

Hydraulic Jetting

The desire to initiate fractures in certain planes or directions led to the development of hydraulic jetting (Fig. 8.35). This process is used for perforating, fracture initiation, and clean-out, and consists of pumping a sand-fluid mixture under pressure through nozzles in order to attain jet velocity. The casing, scale, or formation is cut away as the jetted sand-fluid mixture impinges upon it. Because the energy is released over a sustained period of time rather than instantaneously, cement is not shattered, junk is not left in the hole, and the formation is not compressed. Although almost any convenient fluid can be used as a carrying agent, the less viscous fluids are more desirable because they cause less friction loss in the tubular goods.

In initiating a fracture, a circular or vertical notch is cut in the formation with hydraulic jetting, thus creating a "point of weakness". Subsequent fracturing treatment pressure causes a break at the point of weakness. With hydraulic jetting, formation breakdown pressures are generally lower than when conventional well perforators are used.

Cross-Over Valves

As fracturing volumes increased, it became necessary to increase treating efficiency. One method involves pumping fracturing fluid through both casing and tubing. However, the casing cannot always withstand the breakdown pressure. Pressure records made during well stimulation treatments have shown that after the fracture is initiated, the surface injection pressure often declines and is within the working pressure of the casing throughout the remainder of the job. To help overcome this problem, a cross-over valve (Fig. 8.36) is used to permit stimulation treatments to be conducted through the manifolded tubing and casing annulus after initial formation breakdown through the tubing. This technique has greatly increased treating efficiency and has reduced costs.

The cross-over valve may be run with any retrievable packer commonly used in fracturing, and is installed as an integral part of the tubing string. It is completely pressure controlled from the surface and can be opened or closed as often as necessary during well stimulation by starting or stopping injection down the annulus. Injection down the tubing is uninterrupted during operation of the valve.

The valve is designed so that reverse circulation may be established at any time without moving the tubing. This feature permits immediate removal of fracturing fluid if a sand-out occurs in treatments where the wellhead is flanged down at the surface.

Removable Liners

Fracturing of wells that had several perforated zones presented a new problem. The zones not being treated had to be isolated so that the treating fluids could be diverted to other perforations. A removable fracturing liner (Fig. 8.37) was developed for this purpose. This liner consists of three sections: an upper sealing element, a long standard-OD tubular spacer, and a lower sealing element. Both the upper and lower sections are equipped with two flexible, cup-type sealing elements that are mounted back-to-back. In this manner, one element or the other functions as a packer when a pressure differential exists across the tool section. The upper sealing section, equipped with a drag spring assembly and a "J"-slot locking and setting arrangement, is attached to heavy duty slips that support the liner

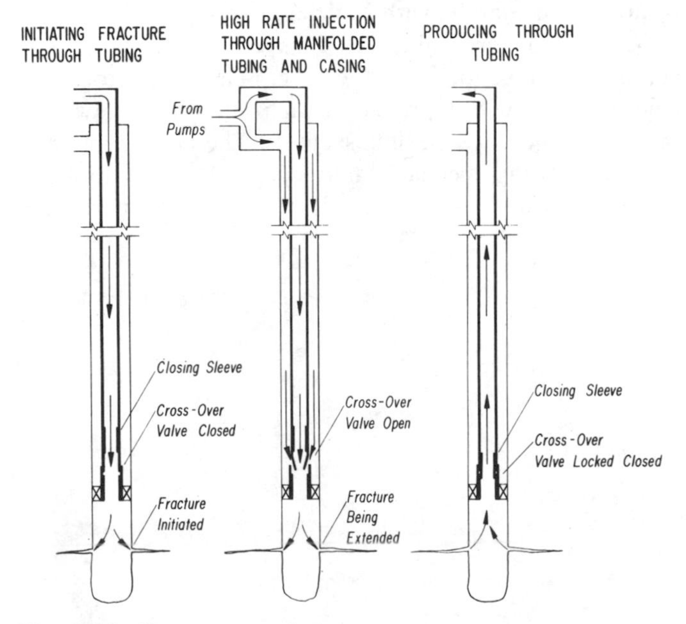

Fig. 8.36 Cross-over valve.

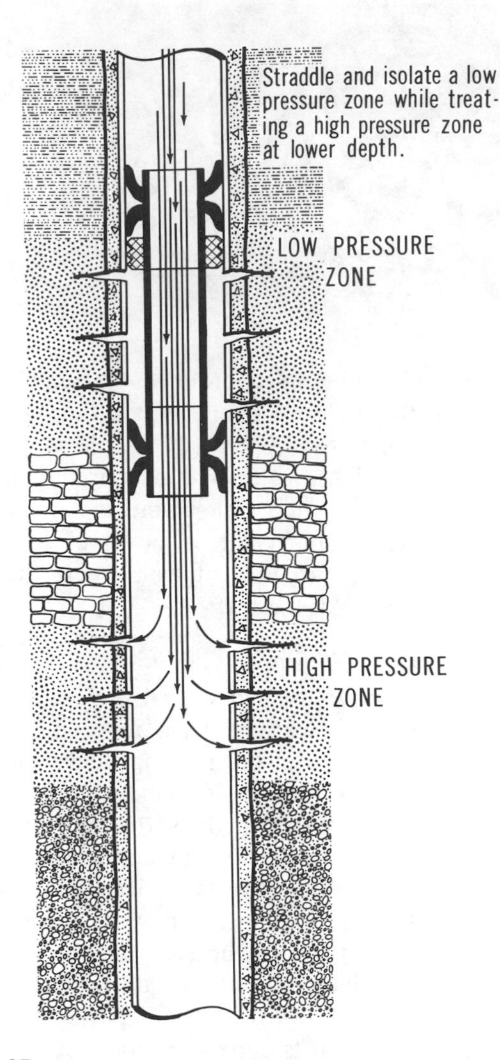
Straddle and isolate a low pressure zone while treating a high pressure zone at lower depth.
LOW PRESSURE ZONE
HIGH PRESSURE ZONE

Fig. 8.37 Removable fracturing liner.

assembly while it is set. The upper and lower sections are separated by the straddling spacer. The length of the spacer is not critical but should straddle the perforations adequately, with sufficient overlap so that both upper and lower sealing sections of the tool are in contact with good, sound casing. The liner tool is equipped also with a pressure equalization port between the upper and lower sealing sections. The port functions with the setting mechanism to open when the tool is run and close when the liner is set across the perforations.

Although many other tools are available for different applications of fracturing, the ones discussed are the ones that have been most widely used.

8.8 Summary

The mechanical equipment used to hydraulically fracture a well has evolved from a single truck-mounted pump employing approximately 100 hp, designed for cementing oil wells, to multipump units employing 20,000 hhp. Fracturing fluid pumping rates have increased from 1½ bbl to more than 550 bbl of proppant-laden slurries per minute. Surface pressure as high as 15,000 psi has been encountered. These operations demand adequately designed equipment, detailed planning, centralized control of the operation, and recording of well treatment data.

Pressure-volume and horsepower calculations have been presented, together with an example calculation to permit the determination of horsepower requirements for a hydraulic fracturing job.

The development of special equipment, such as the liquid-propping-agent proportioner and the bulk proppant transporters, have been discussed to illustrate the need for such equipment and the degree of control required during a fracturing operation.

The discussion of wellheads and safety check valves illustrates that the treating of a well with 10 or 12 pump units involved operational problems that had to be solved before fracturing could be generally accepted as a routine well stimulation procedure.

Subsurface equipment such as packers, cross-over valves and bridge plugs were required to restrain the fluids to the intervals to be treated. The use of abrasive proppants posed design problems that had to be overcome if these tools were to operate for long periods of time and were to be removable from the well at the end of the treatment.

References

1. Gras, E. H.: "Hydraulic Fracturing Mechanical Equipment", unpublished report, Halliburton Co. (Oct., 1966).

2. Clark, J. B.: "A Hydraulic Process for Increasing the Productivity of Wells", *Trans.,* AIME (1949) **186**, 1-8.

Chapter 9

Economics of Fracturing

9.1 Introduction

During the development phase of the hydraulic fracturing process, the trend was to increase treatment volumes and pump rates and to improve fluid characteristics in order to increase productivities. Although these changes resulted in increased production, the attendant increased expense made it desirable to evaluate the benefits of these processes in terms of net profit. As a result of this evaluation, a method was developed for determining the maximum profit that may be derived from hydraulic fracturing.

The procedure used for determining the best treatment design (i.e., for deriving maximum profit) from hydraulically fracturing a given well involves three steps. (1) Determine the productivity improvement that can be obtained by various fracturing treatments. (2) Establish the cost of the treatments and the relative savings effected by the increase in the post-fracturing producing rate. (3) Combine the findings of the first two steps to determine the treatment that will yield the maximum net profit. This procedure relates pump rate, treatment volume, fracturing fluid characteristics and well manipulation costs with the monetary return resulting from the treatment.

This chapter describes a method of calculating the maximum net profit from hydraulic fracturing. The profit is derived from accelerated production or, in some cases, from an increase in ultimate recovery. No attempt has been made to evaluate all the variables that affect fracture treatment design.

The examples are presented only to demonstrate a method of analysis and to show that treatment variables can markedly affect profit. The major parameters affecting fracturing treatments that have been designed to yield maximum profit are presented and a method of evaluating these variables is outlined. Examples are given of the over-all procedure, and a complete calculation is given. For the calculation it was necessary to make a number of assumptions (see Table 9.1) with regard to the well, formation, treating and cost parameters.

9.2 Method for Estimating the Present Worth of Future Production

PW, the cumulative present worth of the production, should be closely approximated by the relationship

$$\text{PW} = \sum_i N_{nv}\left(\frac{\Delta V_i}{1+at_{im}}\right) \quad \text{. . . . . (9.1)}$$

If $q_o(t)\Delta t_i$ for ΔV_i is substituted in Eq. 9.1 and the fundamental theorem of integral calculus is applied, the expression for the PW is

$$\text{PW} = N_{nv}\int_0^T \frac{q_o(t)dt}{1+at}, \quad \text{. . . . . (9.2)}$$

where $q_o(t)$ is the production rate expressed as a function of time. Integration implies that each barrel of oil

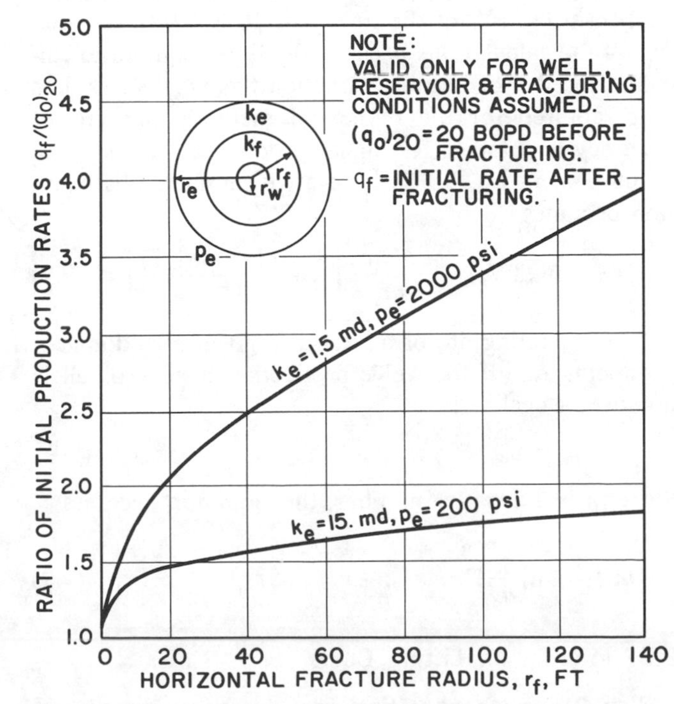

Fig. 9.1 Effect of formation permeability and fracture radius on initial production rates.

is converted to its particular discounted cash value at the time it is produced.

Eq. 9.2 may be written also as

$$\text{PW} = N_{nv}\int_0^T \{q_o(t) \cdot [1 - (at) + (at)^2 - (at)^3 + (at)^4 - \cdots]\}dt, \quad \ldots \ldots (9.3)$$

when it is recalled that

$$(1 - at)^{-1} = 1 - at + (at)^2 - (at)^3 + \ldots.$$

A necessary condition for the equivalence of Eqs. 9.2 and 9.3 is that the series expansion must converge within the limits of the integration, or that the product of $(a \cdot t) < 1$.

An integrated approximation to Eq. 9.2 may be obtained by neglecting all the at factors in Eq. 9.3 that are raised to powers greater than 1 and by substituting $q_o e^{mt}$ for $q_o(t)$. This approximation is given by the equation

$$\text{PW} \cong N_{nv} q_o \int_0^T e^{mt}(1 - at)dt, \quad \ldots \ldots (9.4)$$

or by the equation

$$\text{PW} = N_{nv} q_o \int_0^T e^{mt}dt - aN_{nv} q_o \int_0^T te^{mt}dt \quad \ldots (9.5)$$

The formula used to calculate the present worth of future production is

$$\text{PW} = N_{nv}(N_p) - \frac{aN_{nv} q_o}{m^2}[e^{mT}(mT - 1) + 1], \quad \ldots (9.6)$$

where PW is the cumulative present worth of the production in dollars.

The increased value of the well due to fracturing is obtained by subtracting the PW of the initial 20-BOPD well from the PW of the treated well. Eq. 9.6 was used for all the calculations because the time required to calculate the PW of future production this way is less than that required for the numerical method, and greater accuracy is obtained for higher productivity wells.

Since m in Eq. 9.6 is a negative number, this equation becomes

$$\text{PW} = N_{nv}(N_p) - \frac{aN_{nv} q_o}{m^2}\left[1 - \frac{(1 + |m|\ T)}{e^{|m\ T|}}\right] \quad .(9.7)$$

The operating life of a well to a given abandonment producing rate if the well's production history parallels a constant decline is

$$q_a = q_o e^{mt} \quad \ldots \ldots (9.8)$$

Since m is a negative number, this equation becomes

$$|m|t = \ln\left(\frac{q_o}{q_a}\right) \quad \ldots \ldots (9.9)$$

9.3 Interest on Lifting Costs

The lifting or operating cost rates were prorated according to the value of the initial production rates. The cumulative interest (I_c) lost on the money paid out for lifting costs was developed from the formula

$$\text{Interest lost on lifting cost money} = \sum_i aL(t - t_i)\Delta t_i = aL \int_0^T Tdt - \int_0^T tdt \quad \ldots (9.10)$$

The cumulative interest on operating expense was calculated with the following formula:

$$I_c = \frac{aLT^2}{2} \quad \ldots \ldots (9.11)$$

9.4 Productivity Improvement and Treatment Variables

The first step in calculating the optimum fracturing treatment requires that the productivity increase resulting from various fracture radii be determined. The ratio of productivity after fracturing to that before fracturing is a function of fracture radius, fracture capacity and formation characteristics. These variables are related to well production by the discontinuous permeability formula for steady-state flow.[1] With this method it is assumed that, due to the creation of a fracture, the permeability in a zone around the wellbore differs from that at a distance. In this system, illustrated on Fig. 9.1, k_e is the original permeability of the formation before treatment and k_f is the permeability of the formation from the wellbore to the fracture radius. It may be shown that

$$k_{avg} = \frac{\log(r_e/r_w)}{\frac{1}{k_f}\log\left(\frac{r_f}{r_w} + \frac{1}{k_e}\right)\log\left(\frac{r_e}{r_f}\right)}, \quad \ldots (9.12)$$

where log is to the base 10.

In applying Eq. 9.12, k_f is equal to the effective horizontal permeability of the formation lying within the radius of fracture. The value that should be assigned to this effective horizontal permeability is somewhat indefinite because the height of formation, vertical permeability, thickness of fracture, etc., all influence it. For the purpose of these calculations, however, it is believed that a sufficiently accurate estimate of its value may be determined from the following formula.

$$k_f = \frac{k_e h + k_f w}{h} \quad \ldots \ldots (9.13)$$

When all factors in Eq. 9.12 have been estimated as explained above, the average permeability of the whole producing zone (k_{avg}) is calculated. After this average permeability is obtained, the stabilized production rate following hydraulic fracturing may be estimated by:

$$q = \frac{3.07\ h\ k_{avg}\ \Delta p}{\mu \log(r_e/r_w)} \quad \ldots \ldots (9.14)$$

A plot of the data obtained from calculations using

this technique is shown on Fig. 9.1 for the reservoir conditions assumed in Table 9.1 (Cases 1 and 2).

In order to compare the effect of fracture radius on wells completed in both 1.5-md and 15-md formations, the formation pressures were set at 2,000 and 200 psi, respectively, to establish a 20-B/D prefracturing production rate in both cases. No formation damage or skin effect was assumed in either case.

It may be noted that, for the two formation conditions investigated, an increase in fracture radius effects an increase in well productivity. The increase is much more pronounced for the 1.5-md (2,000 psi) formation than for the 15-md (200 psi) formation.

Post-fracturing productivity in a given formation is affected by such things as formation damage, fracture penetration and the fluid-carrying capacity of the fracture. Since no formation damage was assumed, only the variables controlling fracture radius were investigated, and the assumed fracture capacity (300 md-ft) is considered representative of that obtained in typical field fracturing operations using a sand propping agent (see Chapter 6).

The effect of pump rate, treatment volume, and fracturing-fluid characteristics on fracture radius has been given in Chapter 4. The fracture-radius calculation method described in this chapter is used to determine the effect of the pump rate, fluid volume, formation and fracturing-fluid characteristics on the radii that may be obtained from various treatments. An example plot of data obtained from this type of calculation is shown on Fig. 9.2. The simulated fracturing programs for these example wells were varied to cover the range of injec-

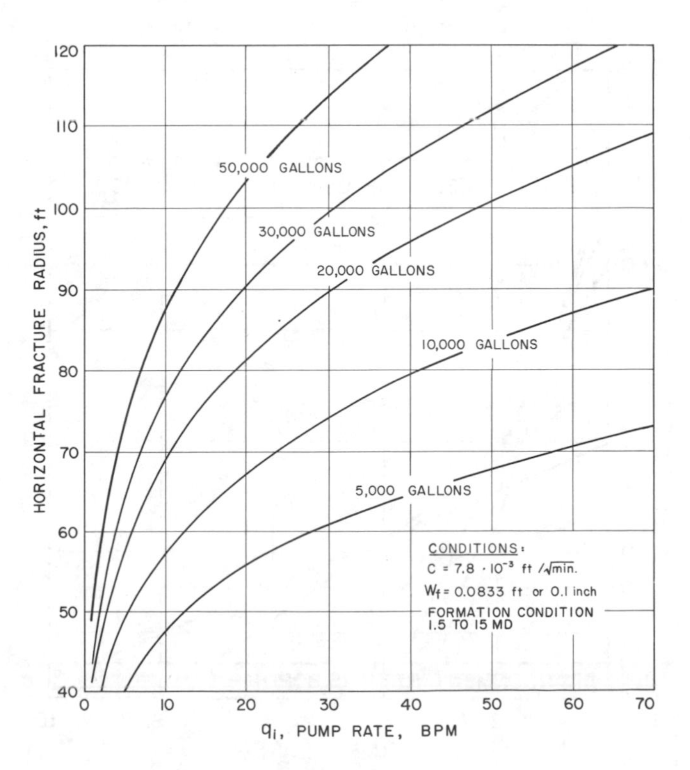

Fig. 9.2 Typical curves of fracture radius vs pump rate.

TABLE 9.1 — BASIC ASSUMPTIONS

WELL CONDITIONS	
Tubing size, OD, in.	2⅞
Casing size, OD, in.	5½
Well depth, ft	5,000
Wellbore diameter, in.	6
Abandonment rate, BOPD	5
RESERVOIR CONDITIONS	
Formation permeability (homogeneous) (k_e), md	
Case 1	1.5
Case 2	15.0
Formation thickness, ft	15
Radius of drainage (r_e), ft	660
Reservoir oil viscosity, cp	2
Reservoir oil compressibility	None
Flow conditions	Steady-state, radial throughout reservoir
Static bottom-hole pressure (p_e), psi	
Case 1	2,000
Case 2	200
Producing bottom-hole pressure, psi	
Both cases	20
Ultimate recovery, bbl	
No increase due to fracturing	36,000
Maximum increase due to fracturing	3,000
Producing rate before treatment, q_o(20), BOPD	20
Damage to formation due to well completion	None
Ratio of water-to-oil permeability, k_w/k_o	1/10
Porosity of formation, percent	15
The formation is preferentially water-wet.	
There is a constant percentage production decline.	
FRACTURING CONDITIONS	
Fracturing fluid characteristics	
Viscosity, cp	
Oil	3
Water, in tubular goods	1
Water, in formation	0.75
Fluid-loss coefficient, cu ft/sq ft/min	
Oil	0.0078
Water, 1.5-md formation	0.0058
Water, 15-md formation	0.0388
Fracture capacity, k_f, md-ft	300
Fracture clearance during fracturing (uniform), in.	0.10
Fracture shape	Horizontal pancake
Fracture orientation	Symmetrical around well
Pressure during fracturing at formation face and in fracture, psi	2,500
ECONOMIC FACTORS	
Lifting cost	
First 20 BOPD	$150/month
Each additional BOPD	$0.50
Net value of crude oil before discount and lifting cost but after taxes and royalty	$2
Pump-truck cost per truck	$500
Interest rate, percent/year	4
Cost of sand-blending equipment,	
0 to 12 bbl/min	$100
13 to 25 bbl/min	$125
26 to 40 bbl/min	$200
41 to 70 bbl/min	$250
Cost of storage tank rental	
250-bbl tank	$100
500-bbl tank	$125
Proration of production	None
Cost of propping agent, per lb	$0.016
Cost of fluid-loss additive, per lb	$0.25
Fracturing crude oil (base fluid)	No charge
Fracturing water (base fluid)	No charge

tion rates of 0 to 70 bbl/min and treatment volumes of 5,000 to 50,000 gal. It was found that for the low-fluid-loss, oil-base fracturing fluid, and for the 1.5-md and 15-md formation permeabilities used in this example, permeability was not a major factor in the fracture-radius calculations since the wall-building characteristics of the fluid controlled the loss of fracturing fluid to the formation.

Since the formation permeability range did not significantly influence the fracture-radius calculations for the example when a low-fluid-loss, 3-cp oil was used, it was possible to construct one series of curves (relating treatment volume and pump rate to horizontal fracture radius) applicable to both the assumed formations (see Fig. 9.2).

When untreated water was used as a fracturing fluid, its high fluid loss caused the formation permeability to affect the calculated fracture radius.[2] Accordingly, a different set of curves is required to relate treatment volume and pump rate for each of the 1.5-md and 15-md example formations. Plots for water, similar to those in Fig. 9.2, were made to complete the economic calculations for untreated water fracturing fluids presented as examples later in the chapter.

Once the effect of fracture radius on well productivity is determined, as well as the treating conditions necessary to obtain various radii, the cost of treatments and the relative savings effected by the increase in producing rate as a result of fracturing can be determined.

9.5 Relative Savings

In order for the operator to receive the maximum net profit from a fracturing treatment, an "economic balance" is necessary between the cost of performing the treatment and the gross savings resulting from the increased production rate following the treatment. This type of analysis permits the evaluation of pump rate, treatment volume, fracturing fluid characteristics and well manipulation costs, as related to the monetary return from the treatment. These factors are discussed separately.

Basic Considerations

To assure uniformity of meaning, the terms *relative savings* and *net profit* are defined as follows.

Relative savings: The difference between the cost of producing an unfractured well and that of producing a fractured well (both produced to the economic limit), plus the increase in present worth of the oil from the fractured well due to its more rapid recovery.

Net profit: The relative savings less the expenditures required to create the fracture.

There are three basic factors that apply in estimating the relative savings credited to a fracturing treatment.

1. The elimination of a part of the operating expense that would have been incurred had the well not been fractured.
2. The increased present worth of the oil reserves as a result of well treatment.
3. The value that accrues from the increased oil recovery resulting from the fracturing treatment.

These factors are a direct result of the increased production rate that a treated well has, and the correspondingly shorter operating life in which it must produce a given quantity of oil.

To relate the first two factors, it is necessary to assume a depletion history for each well. The life of each well in the examples was predicted to follow a straight-line plot of cumulative production vs producing rate. The actual production history of many wells has been shown to parallel closely the constant percentage decline characteristics assumed.[3] For the sake of comparison, an untreated well with an initial producing rate of 20 BOPD was used in all the example economic calculations. A plot of producing rate vs time will be a straight line when plotted on semilog paper (see Fig. 9.3).

The foregoing method of predicting production history does not take into account the pressure readjustment following a fracturing treatment. Flush production, or that production immediately following fracturing treatment, which results from production of oil from the formation adjacent to the fracture, was not considered in this analysis. This omission tends to make the example net profit calculations conservative. Flush production following fracturing can be included in this type of analysis by properly predicting the post-fracturing production-decline curve.

Reduction in Operating Expenses

Less time is required to recover a fixed reserve from

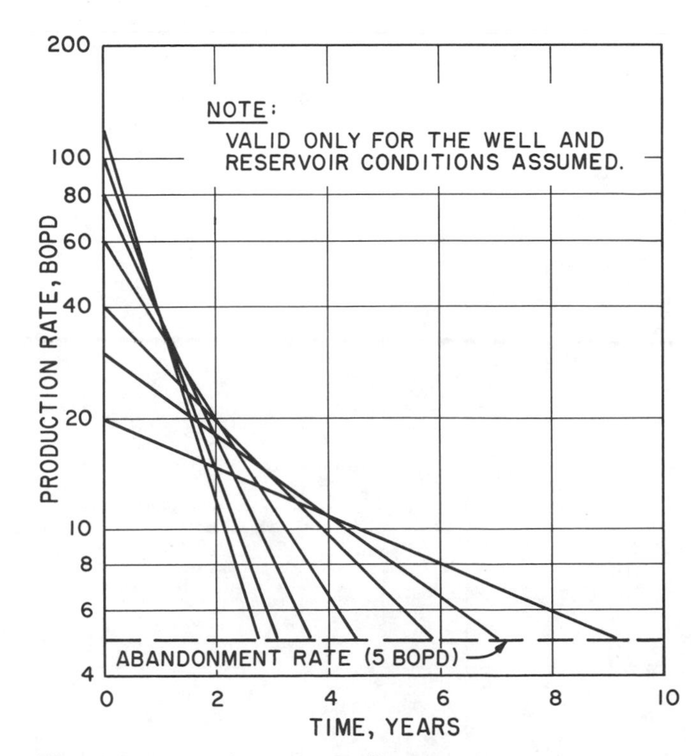

Fig. 9.3 Production rate decline curves.

a well whose productivity has been increased by a fracturing treatment than from one that has not been fractured. It is possible to calculate the effect of several different production rates on the operating life of a well with a fixed ultimate recovery, as well as on the life of a well whose ultimate recovery has been increased by the fracturing treatment. The operating expense eliminated by increasing productivity and decreasing the operating life of a well is credited to profit from the treatment. It was estimated for the example illustrated on Fig. 9.4 that monthly operating costs varied in a straight-line relationship from $150/month for a well with an initial productivity of 20 BOPD to $200/month for a well with an initial productivity of 120 BOPD. Interest on current operating expenses was also added to the final lifting cost figures. The interest on the money invested to perform the fracture treatment may be neglected because of the small sums involved for treatment costs as compared with the total value of the oil remaining in the ground and the total lifting costs involved. For the examples shown, the operating expense saved by fracturing is the difference between the total operating expense accumulated over the life of the untreated 20-B/D well, and the corresponding expenses for a well producing at rates up to 120 B/D. These data are plotted on Curve I, Fig. 9.4.

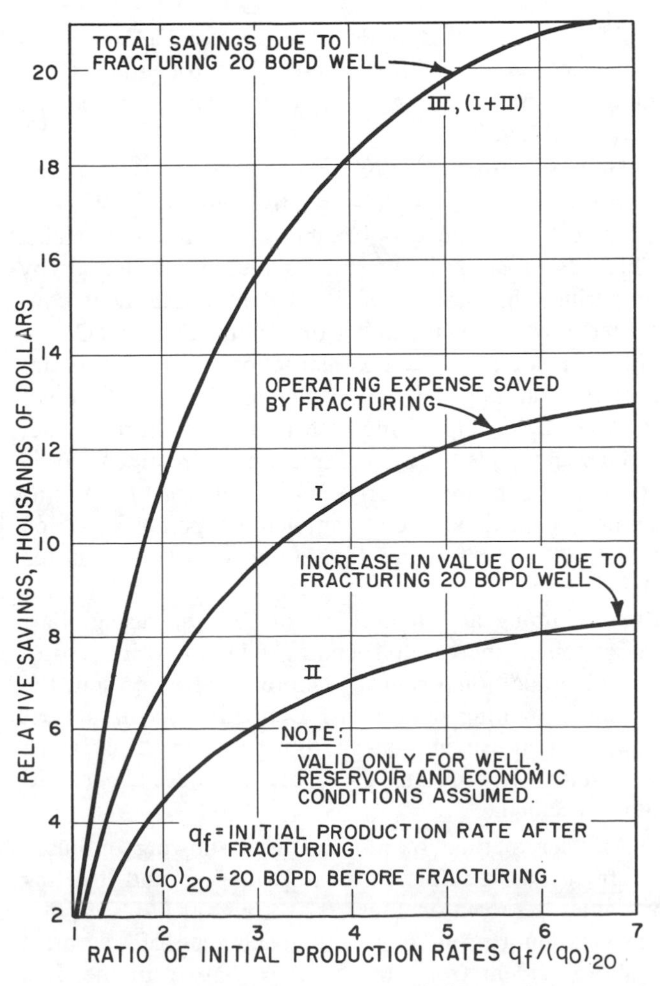

Fig. 9.4 Effect of fracturing on relative savings.

Increase in Present Worth of the Oil Reserves

The increase in present worth of the oil by reason of accelerated producing rates was calculated in the example, assuming a 4-percent discount rate. These data for various ratios of initial stabilized, post-fracturing production rates for the referenced 20-B/D oil well are plotted as Curve II on Fig. 9.4.

Curve III on Fig. 9.4 is the relative savings from increasing production by fracture-treating a 20-B/D oil well. As shown, this curve is the sum of Curves I and II and is the basis for all other economic comparisons in this chapter. A summary of these data is presented also in Table 9.2 to demonstrate the method used in making the calculations.

Effect of Increase in Ultimate Recovery

In plotting the curves for the examples on Fig. 9.4 it was assumed that the fracture treatment did not increase the recoverable reserves. Because increases in ultimate recovery are often attributed to fracturing, the effect this factor would have on net profit and optimum treating conditions must also be investigated.[4,5]

For one example in this chapter it was assumed that the increase in ultimate oil recovery is a complex function of the increase in fracture radius.[6,7] It was also assumed that the increase in ultimate recovery followed a constant percentage decline and that the cumulative production and producing life were extended by the additional recovery.

Effect of Fracture Radius on Relative Savings

To determine the effect of permeability and fracture radius on relative savings, it is necessary to combine the data obtained when the productivity from fracturing is calculated by the discontinuous permeability formula (Fig. 9.1) with the data obtained when the relative savings is determined for various production rate increases (Fig. 9.4). A cross-plot of data on Figs. 9.1 and 9.4 is shown on Fig. 9.5, where the relative savings for a range of fracture radii is shown, with the two example formation conditions as parameters.

Thus, after calculating the relative savings that will result from various fracture radii (Fig. 9.5) and the treating conditions necessary to produce these fractures (Fig. 9.2), it is then possible to calculate the net profit after subtracting the cost of the treatments.

Fracturing Treatment Costs

The factors that determine fracturing treatment costs are pump rate, treatment volume, type of fracturing fluid, and fracture propping agent. The number of pumping units (hydraulic horsepower) required to achieve a given pump rate is influenced by the length and size of tubular goods, the fluid-flow properties and the bottom-hole fracturing pressure. The friction loss down commercial tubing and casing during the pumping of untreated water and 3-cp crude oil was calculated for the examples presented in this chapter. The surface pressure (friction loss plus bottom-hole injection pres-

TABLE 9.2 — CALCULATED SAVINGS TO OPERATOR BY PRODUCING FIXED RESERVES AT ACCELERATED RATES USING 20-B/D (INITIAL RATE) WELL AS THE BASE

	Initial Producing Rate of Well, BOPD						
	20	30	40	60	80	100	120
A. Relative savings by producing oil at faster rate, dollars*	—	2,910	4,460	6,160	7,090	7,640	8,040
B. Well operating life, months**	110	84	70	54	44	37	33
C. Estimated average operating cost, dollars per month	150	155	160	170	180	190	200
D. Cumulative operating expense, dollars (B × C)	16,500	13,100	11,200	9,180	7,920	7,040	6,600
E. Cumulative interest on operating expense during well life, dollars***	3,020	1,820	1,310	830	560	435	365
F. Total operating cost chargeable to well, dollars (D + E)	19,520	14,920	12,510	10,010	8,480	7,475	6,695
G. Operating cost saved relative to 20-B/D well, dollars†	—	4,600	7,010	9,510	11,040	12,045	12,555
H. Total savings by producing reserves at faster rate, dollars (A + G)	—	7,510	11,470	15,670	18,130	19,685	20,595

*Reserves fixed at 36,000 bbl. Net value of oil after taxes and royalty, estimated at $2/bbl. Lifting cost not deducted. Increase in present worth is computed using Eq. 9.2, and relative savings are determined by deducting present worth of untreated BOPD well production from present worth of a fracture-treated well's production.
**Calculated using Eq. 9.4.
***Calculated using Eq. 9.5.
†Total operating cost of fractured well (Line F) subtracted from total operating cost of 20-BOPD well (Line F).

sure minus hydrostatic head) and pump rate dictate the number of pumps required.

The cost of renting storage tanks to accommodate the total volume of fracturing fluid must be included. No provision was made in the examples for purchasing the fracturing fluids since the 3-cp oil was assumed to be lease crude oil, and it was assumed that water would be readily available. If the base fracturing fluid is purchased, however, this factor also must be considered. The cost of sand-blending equipment must be determined for the various pump rates considered. For the examples, the cost of the fluid-loss agent and sand was determined from 1967 price schedules.

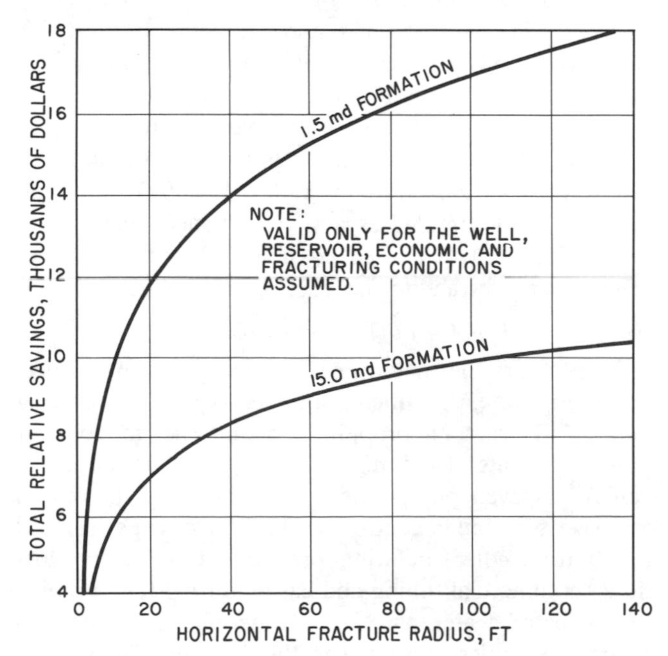

Fig. 9.5 Effect of formation permeability and fracture radius on total savings.

9.6 Relationship of Savings to Treatment Cost

After the treating costs have been calculated, it is then necessary to subtract these costs from the relative savings from fracturing to determine the net profit that may be obtained.

Optimum treating conditions can be determined by constructing plots to show relative savings and treatment cost as a function of pump rate or fracture radius. This was accomplished for the example calculations by preparing Fig. 9.6, which illustrates a treatment conducted down 2½-in. tubing or 5½-in. casing. (Calculations involved in the construction of Fig. 9.6 are presented at the end of the chapter.) Fig. 9.6 shows that when the rate of increase in savings from a fractured well equals the rate of increase in treatment costs, additional expense to create a larger fracture radius cannot be justified. For example, the points at which the slope of the fracture treating cost curve equals the slope of the savings curve are labeled A and A′ for 2½-in. tubing and B and B′ for 5½-in. casing. Past these points, increased spending will not bring corresponding additional return; therefore they represent the optimum treating conditions. Deducting the treatment cost from the savings at the optimum point gives the maximum net profit to the operator under the conditions specified.

Another method of plotting data showing optimum treatment design for various formation and treating variables is shown on Figs. 9.7 through 9.11. Note that the maximum profit region for each type of treatment is more evident from these curves than from the data presented on Fig. 9.6. However, the cause for the sharp decline in the net profit curve (a pronounced increase in treating cost) is better illustrated on Fig. 9.6.

It should be emphasized that Figs. 9.6 through 9.11 were included for illustrative purposes only. These curves are representative of the hypothetical conditions outlined in these examples and should not be used in designing fracture treatments for other reservoirs and well conditions.

9.7 Factors Affecting Profit

There are many factors that affect the profit that can be derived from a hydraulic fracturing treatment. Each of these must be carefully considered in designing a job if the optimum treatment is to be obtained.

Effect of Formation Permeability and Depletion

Based on the formation conditions assumed (see Table 9.1), the data on Fig. 9.1 demonstrate that the state of reservoir depletion (p_e) and formation permeability should be given primary consideration in designing a fracturing program. This plot shows that the productivity improvement from fracturing a 20-B/D oil well in a 1.5-md formation with 2,000 psi reservoir pressure will be much greater than that resulting from fracturing a 20-B/D oil well in a 15-md formation with 200 psi pressure. Maximum net profit obtained by fracturing a 20-B/D oil well in a 1.5-md formation (see Fig. 9.7) is about twice that obtained when fracturing a 20-B/D oil well in a 15-md formation (see Fig. 9.8). Note that lower permeability formations require larger volume and higher injection rate treatments to achieve maximum net profit than do higher premeability formations of equal prefracture productivity. Similar calculations have shown that if the fluid-carrying capacity of the fracture were increased above that assumed in the examples shown, then the economics of treating higher permeability wells would be improved appreciably.

Significance of Pump Rate

There is an optimum treatment design for the conditions assumed in preparing Figs. 9.7 through 9.11. The optimum pump rate shown on Fig. 9.6 occurs when the pump rate is at 8 bbl/min down 2½-in. tubing and 25 bbl/min down 5½-in. casing. The rapid increase in the slope of the cost-to-fracture curve is caused by in-

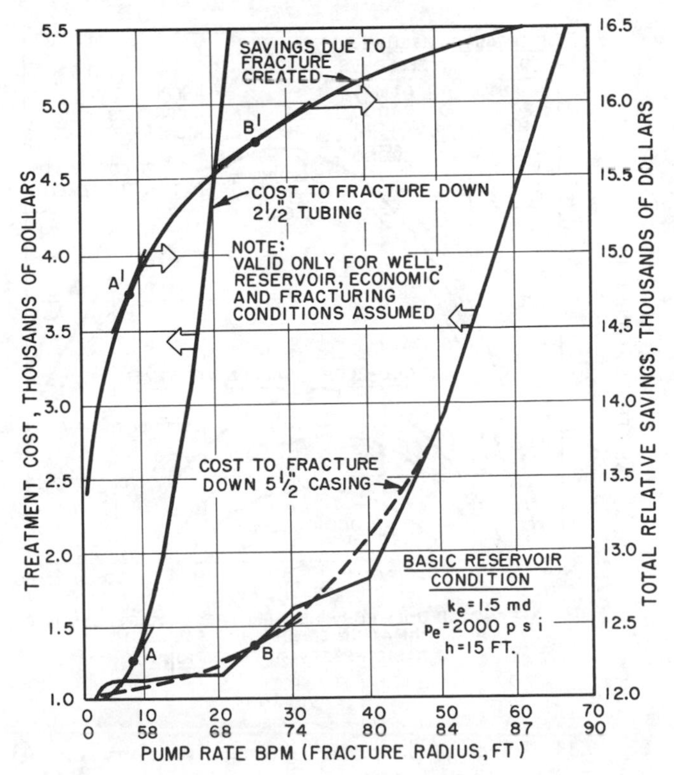

Fig. 9.6 Curves of treatment cost vs savings for a 10,000-gal treatment (fracturing fluid — sand and oil with fluid-loss agent).

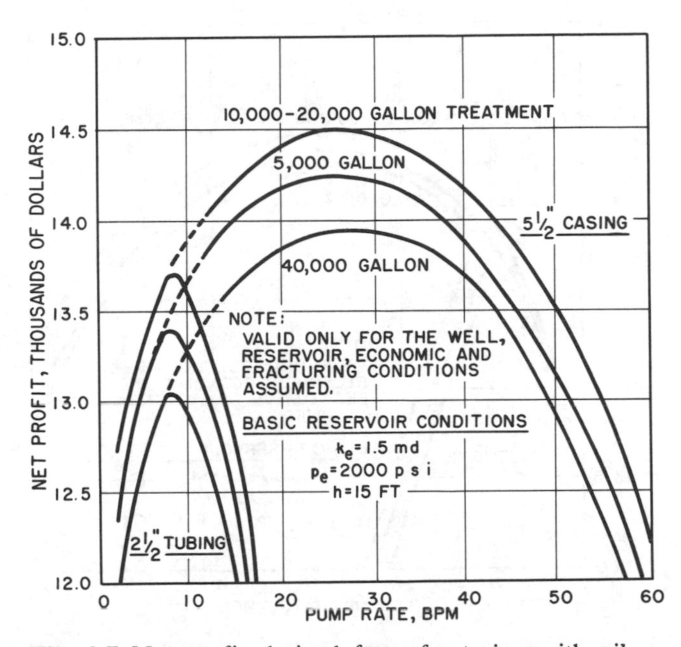

Fig. 9.7 Net profit derived from fracturing with oil.

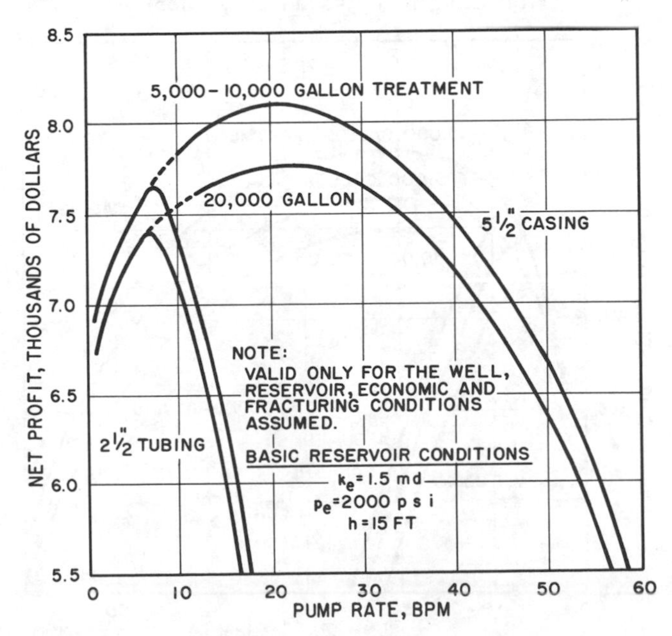

Fig. 9.8 Net profit derived from fracturing with oil (fracturing fluid — sand and oil with fluid-loss agent).

creased surface pressures at the higher pump rates. These higher pressures, caused by greater friction losses down the tubular goods, require additional pumping equipment. The over-all effect is a rapid increase in treating costs. Since a 3-cp fluid was used in constructing Fig. 9.6, the use of a higher viscosity fluid would cause an even greater increase in treatment cost for a given pump rate. Therefore, the size of the tubular goods has a major influence on the slope of the cost curve and thus on the optimum pump rate. These and other calculations have shown that in order to yield maximum profit, the pump rate is increased as the size of the tubular goods that conduct the fracturing fluid becomes larger.

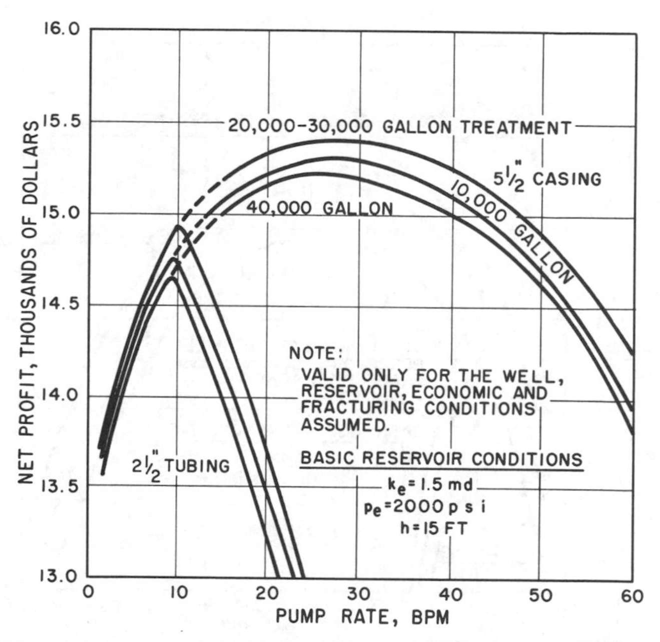

Fig. 9.9 Net profit derived from fracturing with water (fracturing fluid — sand and water).

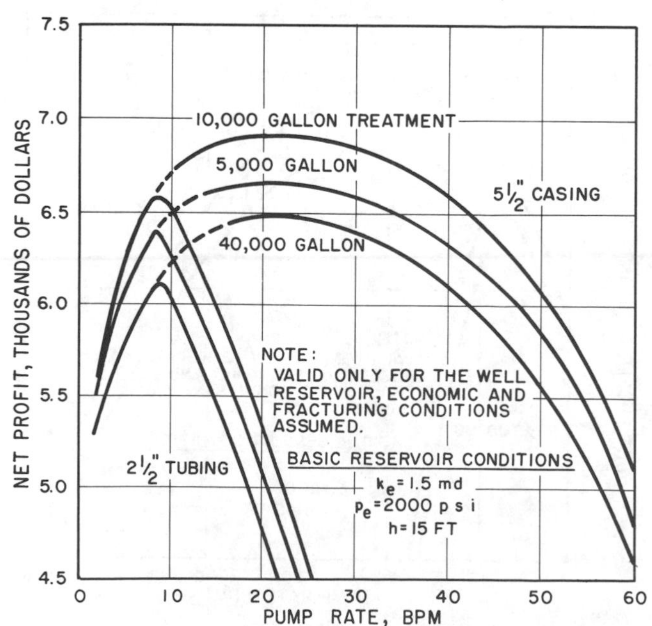

Fig. 9.10 Net profit derived from fracturing with water (fracturing fluid — sand and water).

Significance of Tubular Goods Size

Figs. 9.7 through 9.11 show that injection rates greater than 10 bbl/min down 2½-in. tubing for the assumed conditions should be avoided, since the profit from fracturing is greatly reduced because of the pump horsepower expended in overcoming friction losses at higher pump rates. The friction loss becomes even more significant as the well depth increases. These figures show that the net profit from optimum treatments down 5½-in. casing is \$300 to \$800 greater than the equivalent profit from a treatment down 2½-in. tubing for various treating conditions. When computing the profit from fracturing wells down 5½-in. casing, no cost provision was made for pulling and running the tubing, since the fracturing job can often be conducted by simultaneously pumping down the casing-tubing annulus and through the tubing. In order to determine the economic feasibility of the operation, the added expense of pulling tubing specifically for fracturing down casing must be considered in calculating fracture treatment design.

Effect of Treatment Volume

As shown by Figs. 9.7 through 9.11, the total volume of fracturing fluid injected has a significant effect on net profit. The optimum volume (for the conditions studied) varies over a range of 5,000 to 30,000 gal. The example calculations on Figs. 9.7 through 9.11 show that the optimum treatment volume for a given fracturing fluid is independent of the size of well tubular goods

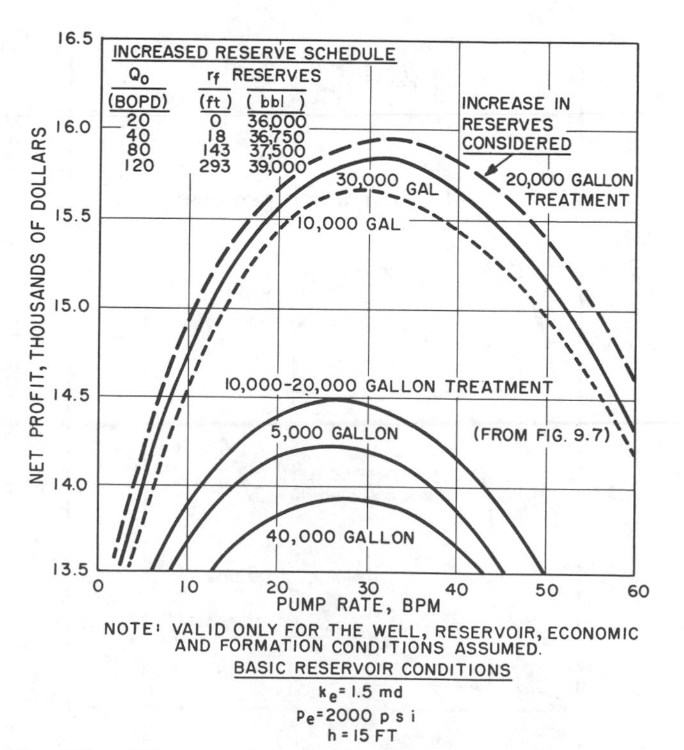

Fig. 9.11 Effect of increased ultimate recovery on net profit and treatment design (fracturing fluid — sand and oil with fluid-loss agent). Treatment down 5½-in. casing.

through which the fluid is pumped.

Effect of Type of Fracturing Fluid

The effect of the two types of fracturing fluids (fresh water or low-fluid-loss, 3-cp oil) on the net profit realized from a fracturing treatment is shown on Figs. 9.7 through 9.11. The results show that the net profit from using water as a fracturing fluid for the 1.5-md permeability formations is comparable with that obtained by using oil with fluid-loss additives. However, appreciably less net profit was realized using water as a fracturing fluid in the 15-md formations than was obtained using oil with additives. This is demonstrated by a comparison of the net profit shown in Figs. 9.7 and 9.9 for the 1.5-md formation, with Figs. 9.8 and 9.10 for the 15-md formation. It is assumed in these cases that neither the oil nor the water adversely affects the permeability.

Productivity Impairment

The increase in production rates after a fracturing treatment often is attributed to the elimination of wellbore "damage," or "skin effect," which does not extend an appreciable distance into the formation. No calculations were included in the examples to demonstrate the effects of formation damage. However, if fracturing eliminates wellbore impairment, profit will be increased.

Fracture Orientation

It was assumed in the examples used in this study that only a single horizontal fracture was involved. The effect of multiple horizontal or vertical fractures can be readily evaluated in a manner similar to the method presented here.

Proration

The maximum initial post-fracturing producing rate considered in the example calculations was 120 BOPD, which may not be compatible with proration rules in effect in some of the oil-producing states. However, it was found that, for the conditions considered, the optimum fracture-treatment design resulted in an initial stabilized production rate of approximately 60 BOPD, which would be an allowable rate in many prorated fields. In general, proration of production reduces the profit that can be obtained by fracturing.

Increased Ultimate Recovery

It has been demonstrated that hydraulic fracturing may substantially increase ultimate recovery.[4,5] With greater quantities of recoverable oil, there will be an increase in the net profit derived from the fracturing job.

If it is assumed that, as has been noted in actual field experience, the additional reserves resulting from fracturing are recovered in the first few months after the treatment, then the increase in net profit should be even greater than that shown on Fig. 9.11, primarily as a result of the increased present worth of the oil.

Example calculations were made for the formation and fracturing conditions shown on Fig. 9.7 and a comparative set of curves (Fig. 9.11) was constructed. Fig. 9.11 shows that when an increase in ultimate reserves is assumed the optimum pump rate is increased to 30 bbl/min from 25 bbl/min. The optimum treating volume is also increased. The net profit from fracturing increases from $14,500 to $15,950. Other similar calculations have shown that these general trends are true also for other conditions. That is, the pump rate and treatment volume required to achieve maximum profit from fracturing are greater for reservoirs where ultimate recovery is increased than for reservoirs where there is no increase in recovery.

9.8 Example Optimum Treatment Design Calculations

The following example summarizes briefly the calculations required for constructing Figs. 9.6 through 9.11.

At the end of each step the pertinent equations and an example of a specific case are given to illustrate the method. This example demonstrates the entire procedure and represents one point on the curves of the figures. Following are the conditions necessary to define the example.

1. Primary reserves or cumulative production (N_p), bbl — 36,000
2. Abandonment rate, BOPD — 5
3. Pressure at drainage radius of 660 ft (p_e), psi — 2,000
4. Formation permeability (k_e), md — 1.5
5. Thickness of producing interval (h), ft — 15
6. Tubular goods, casing, in. — 5.5
7. Well depth, ft — 5,000
8. Fracturing fluid — 3-cp sand-oil mixture with fluid-loss agent, and density of 8.04 lb/gal
9. Sand propping agent, lb/gal — 1.5
10. Bottom-hole pressure required to fracture, psi — 2,500
11. Injection rate during fracture extension (q_i), bbl/min — 32
12. Initial production rate before fracturing, BOPD — 20
13. Initial production rate after fracturing, BOPD — 66
14. Fracture shape and orientation — horizontal "pancake", symmetrical around the wellbore
15. Fracture clearance (w) = 0.1 in. throughout extent of fracture
16. Fracture capacity $(kh)_f$, md-ft — 300
17. Net value of crude oil (after taxes and royalty), N_{nv}, per bbl — $2
18. Annual interest rate for conservative investments a, percent — 4
19. Average lifting cost during life of well L, per month — $173
20. Open hole at producing interval

Step 1

Construct a set of production decline curves that will be most applicable to the reservoir. The curves should be chosen so that the alteration in production history due to treatment is shown. A convenient method of satisfying this requirement is to relate production decline history to the initial production rates. These curves are used to determine 1-A and 1-B described below.

1-A. The time mode of depletion, i.e., the time:rate: cumulative-production relationships for each case.

Example:
A determination of the slope m of the decline curve from Conditions 1, 2, 12, and 13 completely defines the time-mode depletion used in this chapter.

$$m=\frac{q-q_o}{N_p}=\frac{q_a-q_o}{N_p} \quad . \quad . \quad . \quad . \quad . \quad . \quad . \quad . \quad (9.15)$$

If experience shows that fracturing causes an increase in ultimate reserves, the curves should reflect that fact. This increase will change the value of m (Eq. 9.15) and the recoverable reserves from the fractured well. The depletion relationships used for this chapter are shown on Fig. 9.3.

For the initial 20-BOPD unfractured well,

$$m=\frac{5\text{ B/D}-20\text{ B/D}}{36{,}000\text{ bbl}}=-4.17\times10^{-4}\,(\text{day})^{-1},$$

and for the initial 66 BOPD fractured well,

$$m=\frac{5\text{ B/D}-66\text{ B/D}}{36{,}000\text{ bbl}}=1.695\times10^{-3}\,(\text{day})^{-1}.$$

1-B. The total operating life of the well under various degrees of well stimulation (as compared with the operating life of the well without fracture stimulation).

Example:

$$q=q_oe^{mt} \quad . \quad . \quad . \quad . \quad . \quad . \quad . \quad . \quad . \quad . \quad . \quad (9.8)$$

Since m is a negative number,

$$|m|\ t=\ln_e\left(\frac{q_o}{q_a}\right), \quad . \quad . \quad . \quad . \quad . \quad . \quad . \quad . \quad . \quad . \quad (9.9)$$

where $|m|$ is the absolute value of m.

For the initial 20-BOPD unfractured well,

$$|m|\cdot t=\ln_e\left(\frac{20}{5}\right)=1.386,$$

$$t=3{,}330\text{ days}.$$

For the initial 66-BOPD fractured well,

$$|m|\cdot t=\ln_e\left(\frac{66}{5}\right)=2.58$$

$$t=1{,}522\text{ days}.$$

Step 2

The total savings due to increasing the producing rate may now be estimated. Based on the data from sections 1-A and 1-B in Step 1, the total savings, by including 2-A, 2-B and 2-C below, can be determined.

2-A. The increased net present worth value of the treated well's reserves compared with that of the untreated well.

Example:

$$\text{PW}=N_{nv}(N_p)-\frac{aN_{nv}q_o}{m^2}\left[e^{mT}(mT-1)+1\right]. \quad .(9.6)$$

Since m is a negative number, this equation becomes

$$\text{PW}=N_{nv}(N_p)-\frac{aN_{nv}q_o}{m^2}\left[1-\frac{(1+|m|\cdot T)}{e^{|m|\cdot T}}\right]. \quad .(9.7)$$

For the initial 20-BOPD unfractured well,

$$\text{PW}=\left[\$72{,}000-(0.04/\text{yr})(\$2/\text{bbl})(20\text{ B/D})(\text{yr}/365)\div(-4.17/\text{day}\cdot10^{-4})^2\right]\left[1-\frac{2.386}{e^{1.386}}\right]$$

$$=\$72{,}000-(2.52\cdot10^{-4})(0.403)$$
$$=\$61{,}840.$$

For the initial 66-BOPD fractured well,

$$\text{PW}=\$72{,}000-(5{,}030)(0.731)$$
$$=\$68{,}330.$$

The increase in PW for the 66-BOPD well relative to that for the 20-BOPD well is

$$\$68{,}333-\$61{,}840=\$6{,}490.$$

2-B The lifting cost saved by reducing the total operating expenses.

Example:
For the initial 20-BOPD unfractured well,

Lifting Cost = (\$150/month)(month/30.4 day)(3,330 day) = \$16,450.

For the initial 66-BOPD fractured well,

Lifting Cost = (\$173/month)(month/30.4 day)(1,522 day) = \$8,660.

The net savings in lifting cost is

$$\$16{,}450-\$8{,}660=\$7{,}790.$$

2-C. The interest saved on the additional money that would have been expended to operate the well to the economic limit without fracturing.

Example:

Cumulative interest on lifting cost money $=I_c$

$$=\frac{aLt^2}{2} \quad . \quad . \quad . \quad . \quad . \quad . \quad (9.11)$$

For the initial 20-BOPD unfractured well,

$$I_c=(0.5)(0.04/\text{yr})(\$150/\text{month})(12\text{ month/yr})(3{,}330\text{ day}/365\text{ day})^2$$
$$=(\$36/\text{yr}^2)(83.3\text{ yr}^2)=\$3{,}000.$$

For the initial 66-BOPD fractured well,

$$I_c=(0.5)(0.04/\text{yr})(\$173/\text{month})(12\text{ month/yr})(1{,}522\text{ day}/365\text{ day})^2$$

or

$$I_c=\$723.$$

The net interest saved is \$3,000 − \$723 = \$2,277.

The sum of these factors constitutes the total savings (see Fig. 9.4).

Example:
The total savings due to increasing the initial rate of the 20-BOPD well to 66 BOPD is the sum of the individual savings or

$$\text{Total savings} = \text{PW} + \text{Lifting cost savings} + \text{Interest saved}$$
$$\$6{,}490+\$7{,}790+\$2{,}277=\$16{,}557 \quad (9.16)$$

The data outlined under Steps 1 or 2 may also be obtained graphically or analytically.

Step 3

Select a curve or formula relating increased production rate and assumed values of fracture dimensions for the type of fracture being considered. For this step an equation presented by Muskat[1] was used (see Fig. 9.1). This equation is

$$\frac{q_o}{q}=\frac{F_k \log (r_e/r_w)}{\log (r_f/r_w + F_k) \log (r_e/r_f)}, \quad (9.17)$$

where $F_k = k_f/k_e$, and q is the production before fracturing.

Multiplying the numerator and the denominator by $1/k_f$ yields

$$\frac{q_o}{q_{20}}=\frac{\frac{1}{k_e}\log\left(\frac{r_e}{r_w}\right)}{\frac{1}{k_f}\log\left(\frac{r_f}{r_w}+\frac{1}{k_e}\right)\log\left(\frac{r_e}{r_f}\right)} \quad (9.18)$$

Example:

$$k_f=\frac{300\text{ md-ft} + (1.5\text{ md})(15\text{ ft})}{15\text{ ft}}=21.5\text{ md.} \quad (9.13)$$

For the example considered, r_f may be calculated as follows:

$$\frac{66}{20}=\frac{1}{1.5\text{ md}}\log\frac{660\text{ ft}}{0.25\text{ ft}}\Big/(1/21.5\text{ md})(\log r_f)+(1/21.5\text{ md})(\log 4)+(1/1.5\text{ md})(\log 660)-(1/1.5\text{ md})(\log r_f)$$

or $r_f = 91.3$ ft, which is the calculated fracture radius required to produce 66 BOPD for the reservoir conditions assumed.

Step 4

Using as parameters the ratios of the initial production rate of the stimulated well to that of the untreated well, prepare a plot of fracture dimension vs total savings due to stimulation. In other words, combine the results of Steps 2 and 3 so that the ratio of initial production rates is eliminated as a variable (see Fig. 9.5).

Example:
For the example point on the curves, this step means that a horizontal fracture with a radius of 91.3 ft will result in a relative savings of \$16,557.

Step 5

Prepare plots relating fracture area and treatment conditions, e.g., type of fluid, pump rate and treatment volumes. An example is given on Fig. 9.2. The equation[2] used to prepare the plots shown on Fig. 9.2 is

$$Y=\frac{2}{\sqrt{\pi}}X-1+B \quad (9.19)$$

where

$$Y=\frac{4\pi C^2 A}{q_i W} \quad (9.20)$$

$$X=\frac{2C\sqrt{\pi t_f}}{W} \quad (9.21)$$

Chapter 4 presents the method of calculation from which can be made such plots as are shown on Fig. 9.2.

Step 6

Calculate treatment costs for the various conditions considered under Step 5. The methods outlined in Chapter 7 may be used for calculating friction losses of fracturing fluids and for estimating the pump horsepower required. It is convenient to plot treatment cost as a function of pump rate and/or fracture radius for various treatment volumes and fluids since, of all the treating parameters, pump rate has the greatest influence on cost (see Fig. 9.6).

Example:
Calculation of cost for a 20,000-gal fracture job at 32 bbl/min.

Fluid cost:

Fluid-loss additive for a volume of 20,000 gal (sand volume neglected)

$$\frac{20{,}000\text{ gal}}{42\text{ gal/bbl}}(25¢/\text{lb})\ (4\text{ lb/bbl}) = \$476$$

Sand
(20,000 gal) (1.5 lb/gal) (1.6¢/lb) = \$480

Total Fluid Cost \$956

Pump cost for a rate of 32 bbl/min:

Static head (weight of oil + sand = 8.04 lb/gal)

$$(5000\text{ ft})\ \frac{8.04}{8.33}\cdot\frac{62.4\text{ lb/cu ft}}{144\text{ sq in./sq ft}}=2090\text{ psi}$$

Head loss due to flow friction (See Chapter 7 for 3-cp mixture at 32 bbl/min.)

$$(5000\text{ ft})\cdot\frac{8.04}{8.33}\cdot(0.15\text{ psi/ft})=724\text{ psi}$$

Surface pressure (pump pressure)
2400 psi − 2090 psi + 724 psi = 1134 psi

Number of trucks and cost

Two pump trucks costing $400 each are required to deliver 32 bbl/min at 1,134 psi. This will cost $800.

$$hhp = bbl/min \times psi/40.8 = 32 \times 1134/40.8 = 890 \quad . \quad . \quad (9.22)$$

Breakdown of total treatment cost

Sand and fluid-loss additive	$ 956
Two pump trucks	800
Blender cost	200
Storage tank rental	300
Total cost of treatment	$2,256

Step 7

Replot savings as a function of pump rate for various treatment volumes and fluids as in Step 6. Results from Steps 5 and 6 may be used for this (see Fig. 9.6).

Example:

In terms of the one-point example, this step means that both the total relative savings of $16,557 and the treating cost of $2,256 can be plotted or expressed in terms of the pump rate of 32 bbl/min or of the fracture radius of 91.3 ft.

Step 8

Subtract total costs from the total savings at given values of pump rate to obtain the net profit curves as a function of pump rate and treatment volume. Figs. 9.7 through 9.11 present the new profit curves in this chapter.

Example:

The net profit derived from fracturing a 20-BOPD well so that the initial production after fracturing is 66 BOPD is $16,557 — $2,256 = $14,301, where $16,557 is the total saving (Eq. 9.16, Step 2-C), and $2,256 is the total treatment cost (Step 6-E).

As stated earlier, these calculations involved in designing an optimum fracturing treatment can be readily programmed for electronic computing machines. This greatly reduces the engineering time and cost involved in making such an analysis.

9.9 Summary

A study of the examples shown on Figs. 9.7 through 9.11 reveals that profit from hydraulic fracturing can be increased by properly designing the treatments. These plots also indicate that for a given formation there is an optimum treatment design to yield maximum profit. The principal calculations involved in this design can readily be programmed for electronic computers, which greatly reduces the engineering time involved in the analysis. The following conclusions can be drawn from the studies made of hydraulic fracturing design methods.

1. Profit from hydraulic fracturing can be increased by properly designing fracturing treatments.
2. For a given formation there is an optimum treatment design that will yield maximum profit.
3. The optimum treatment volume for a given fracturing fluid is independent of the size of well tubular goods through which the fluid is pumped.
4. To yield maximum profit, the pump rate is increased as the size of the tubular goods conducting the fracturing fluid becomes larger.
5. Formations with lower permeability require larger volume and higher injection-rate treatments to achieve maximum profit than do higher permeability formations of equal productivity.
6. The added expense of pulling tubing specifically for the purpose of fracturing down casing must be considered in the calculations so as to determine the economic feasibility of this operation.
7. The pump rate and treatment volume required to achieve maximum profit from fracturing are greater for reservoirs where ultimate recovery is increased than for reservoirs where there is no increase in recovery.

References

1. Muskat, M.: *Physical Principles of Oil Production,* McGraw-Hill Book Co., Inc., New York (1949) 242.
2. Howard, G. C. and Fast, C. R.: "Optimum Fluid Characteristics for Fracture Extension", *Drill. and Prod Prac.,* API (1957) 261.
3. Arps, J. J.: "Estimation of Primary Oil Reserves", *Trans.,* AIME (1956) **207,** 182-191.
4. Campbell, J. B.: "The Effect of Fracturing on Ultimate Recovery", paper 851-31-L, presented at the Mid-Continent District Meeting, API Div. of Production, Tulsa, Okla., April 10-12, 1957.
5. Garland, T. M., Elliott, W. C., Jr., Dolan, Pat and Dobyns, R. P.: "Effects of Hydraulic Fracturing Upon Oil Recovery from the Strawn and Cisco Formations in North Texas", RI 5371, USBM (1957).
6. Roberts, George, Jr.: "A Review of Hydraulic Fracturing and Its Effect on Exploration", paper presented at AAPG Meeting, Wichita, Kans., Oct. 2, 1953.
7. Wilsey, L. E. and Bearden W. G.: "Reservoir Fracturing — A Method of Oil Recovery from Extremely Low Permeability Formations", *Trans.,* AIME (1954) **201,** 169-175.
8. Howard, G. C., Flickinger, D. H., Fast, C. R. and Evans, R. B.: "Deriving Maximum Profit from Hydraulic Fracturing", *Drill. and Prod. Prac.,* API (1958) 91.

Chapter 10

Results of Hydraulic Fracturing

10.1 Introduction

The concept of hydraulically fracturing a formation to increase well productivity was presented to the petroleum industry in March, 1948, by J. B. Clark.[1] His technical paper summarized an extensive program in which the Hydrafrac process was developed and field tested in a six-state, eight-field testing program. During the first 20 years of its commercial use, more than 500,000 wells were treated by this process.

Detailed treating records were maintained by most service companies and by many of the oil producers during the critical 5-year industry evaluation phase. Early treating records reflected a success ratio of 75 to 80 percent. As pumping equipment improved and as the mechanics of fracturing became better understood, this success ratio climbed toward the 90-percent mark. Widespread use of fracturing has resulted in the general acceptance of this process as a necessary step in completing a well, and fewer and fewer detailed records are maintained of well production before and after fracturing.

The recognition that wellbore damage caused by drilling mud can be easily overcome by fracturing has prompted the engineer to use the drilling fluid that will effect the fastest and most economical penetration rate. Frequently, production increases observed after hydraulic fracturing are greater than those that would have been predicted from known formation characteristics. These large increases in productivity are usually a result of overcoming wellbore damage and increasing the effective permeability to produced fluids in the interwell area. The failure of a well to respond to hydraulic fracturing treatment in a formation known to contain oil and gas generally is traceable to an inability to restrict the fracture to the producing interval. The failure of a well to maintain high post-fracturing producing rates in a formation known to contain oil and gas is attributed either to the inability to maintain a propped fracture, or to the presence of insoluble solids in the fracturing or produced fluid. These solids generally plug the fracture in or near the wellbore.

10.2 Theory

Van Poollen, Tinsley, and Saunders: Electrical Model Tests

Van Poollen, Tinsley, and Saunders used electrical models and physical tests to calculate the effect of fractures on well productivity.[2] Their studies indicated that in homogeneous and isotropic formations, the productivity ratio (fractured well productivity to unfractured well productivity) is directly proportional to the ratio of fracture capacity to formation capacity (both expressed in md-ft).

These authors presented the following equation for determining the ratio of well productivity before and after stimulation by a horizontal fracture.

$$q_f/q_o = \frac{\log(r_f/r_w)}{\log(r_e/r_f) + \dfrac{\log(r_f/r_w)}{\left[1 + \dfrac{(kh)_f}{(kh)_o}\right]}} \quad \ldots\ldots (10.1)$$

Dyes, Kemp, and Caudle's Tests

A similar investigation by Dyes, Kemp, and Caudle using electrical models showed that the physical arrangement of the fracture (i.e., length, flow capacity, and orientation), controlled the increase in production obtained from fracturing.[3] It was noted that the fracture resistance to flow was negligible when fracture capacity or conductivity (defined as the permeability of the propped fracture times the fracture width) divided by the average formation permeability was over 10,000.

Craft: Fracture Capacity vs Well Damage vs Productivity Ratio

Craft *et al.*[4] combined the works of Dyes *et al.*[3] with those of Crawford and Landrum[32] to arrive at a correlation of fracture conductivity or capacity and percent penetration of the well drainage area with well productivity ratio for both horizontal and vertical fractures. The productivity ratio for a horizontal fracture is given by

$$\frac{q_f}{q_o} = \left(\frac{k_f W_f}{kh}\right)\left[\frac{\left(\frac{kh}{k_f W_f}+1\right)\ln\left(\frac{r_e}{r_w}\right)}{\left(\frac{k_f W_f}{kh}+1\right)\ln\left(\frac{r_e}{r_f}\right)+\ln\left(\frac{r_f}{r_w}\right)}\right], \quad (10.2)$$

where W_f is the width of the fracture.

The productivity ratio for vertical fractures was derived from electric analog data that correlated productivity ratios for various fracture penetrations with a factor C.

$$C = k_f W_f / k \quad , \quad . \; . \; . \; . \; . \; . \; . \; . \quad (10.3)$$

where W_f is the propped width of the fracture in inches. (This correlation is shown on Fig. 10.1.)

McGuire and Sikora: Productivity in Solution-Gas-Drive Reservoirs

McGuire and Sikora studied the effect of vertical fractures (symmetrical to the well) on the productivity of wells in solution-gas-drive reservoirs.[5] They assumed that the reservoir is homogeneous and isotropic, that the fluid is homogeneous, that the fracture extends from the top to the bottom of the reservoir and that production results from fluid-expansion drive.

Recognizing that the flowing pressure varies from point to point along the external boundary, McGuire and Sikora elected to express the productivity index in measurable terms, as follows:

$$J = \frac{q}{\bar{p} - p_{wf}} \quad , \quad . \; . \; . \; . \; . \; . \; . \; . \; . \; . \quad (10.4)$$

where q is the flow rate in barrels of oil per day.

Fig. 10.2 shows the effect of fracture length and conductivity on well productivity. The ordinate is the ratio of the productivity indexes for fractured to unfractured formations multiplied by a scaling factor. The scaling factor extends the investigation to well diameters other than the 6 in. used in these studies.

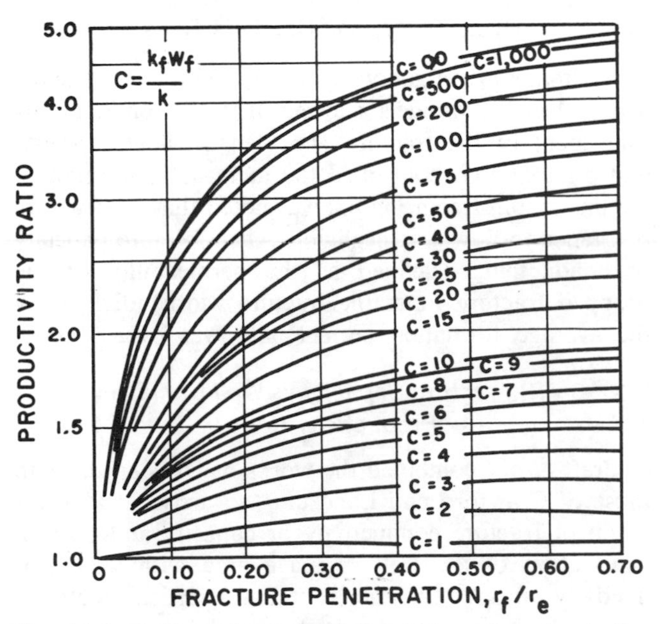

Fig. 10.1 Estimated productivity ratio after fracturing (vertical fracture).[4]

The abscissa expresses the ability of the fracture to conduct fluid relative to the ability of the formation to do so. It is the ratio of two products — fracture permeability times fracture width divided by formation permeability times the width of the drainage area (drainage radius). For simplicity, McGuire and Sikora used $\sqrt{A}$ for the drainage radius. By dividing 40 (the well spacing used in McGuire's study) by A (the well spacing to be investigated), they provided a simple means of calculating relative conductivity of a fracture system.

To illustrate the use of Fig. 10.2, assume that the following conditions exist in the reservoir under study. The matrix permeability varies from 1 to 10 md and has relative conductivities between 10^4 and 10^3, respectively. The fracture is vertical, 0.01 in. wide, packed with $-10+40$ mesh U. S. Standard Sieve sand, and has a fracture permeability of 10^5 md. Fig. 10.2 shows that a twofold to sixfold increase in well productivity can be expected. Note the small effect that fracture length has on increasing well productivity.

Prats: Sand-Filled Vertical Fracture

Prats considered mathematically the effect of a sand-filled vertical fracture of limited radial extent and finite capacity on the flow behavior of a reservoir producing an incompressible fluid through a well.[6] His work indicated that a vertical fracture is equivalent to increasing the effective well radius. Prats showed that if the wellbore radius is small and the fracture is of a high capacity the effective well radius may be assumed to be one quarter of the total fracture length. This may be expressed as follows:

$$\frac{q_f}{q_o} = \frac{\ln\left(\frac{r_e}{r_w}\right)}{\ln\left(\frac{r_e}{\frac{1}{4} r_f}\right)} \quad . \; . \; . \; . \; . \; . \; . \; . \; . \; . \quad (10.5)$$

Bearden: Symmetrical Horizontal Fractures

Bearden,[7] in his investigation of the effect of hy-

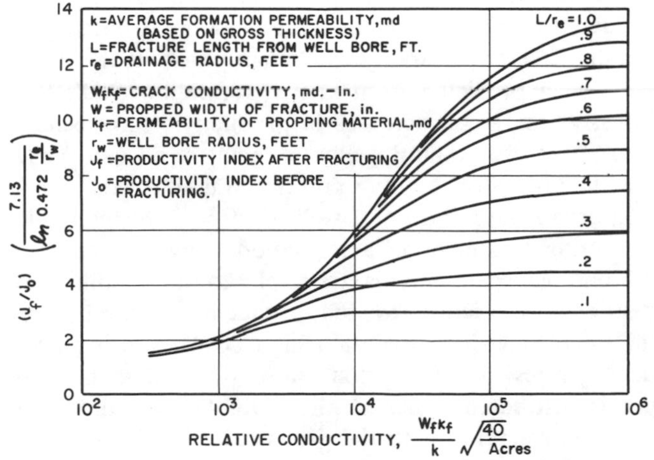

Fig. 10.2 Increase in productivity from vertically oriented fractures.[5]

draulic fracturing on well production, showed that if it is assumed that a symmetrical horizontal fracture is created in a zone of uniform permeability, then the production increase can be approximated with fair accuracy by either of the two following methods.

1. Enlarged Wellbore. The maximum stabilized production following a hydraulic fracturing treatment may be approximated by calculating the effect of enlarging the wellbore to the specific fracture radius being considered. This effect is readily calculated by setting the value of the wellbore radius equal to the fracture radius in a radial flow formula. Under these conditions, Eq. 10.1 becomes

$$q = \frac{3.07\ h\ k_h\ \Delta p}{\mu \log r_e/r_f} \quad . \quad . \quad . \quad . \quad . \quad . \quad . \quad . \quad . \quad . \quad (10.6)$$

where

q = stabilized producing rate, B/D
h = thickness of formation, ft
r_e = external boundary radius, ft

In the application of this formula, the distance, r_e, is again quite arbitrary and may be chosen as the distance to such points where a reasonable estimate of the formation pressure can be made. The use of a value of r_e equal to 500 ft is probably satisfactory unless otherwise indicated by specific well data. A plot of the calculated production increase is shown on Fig. 10.3. In this example, the term r_e was set equal to 500 ft. The extremely high permeability of even a very thin fracture makes this method of calculation reasonably accurate for thin zones.

2. Discontinuous Radial Permeability. Another method for calculating stabilized production increases from hydraulic fracturing is by considering a discontinuous and approximately radial permeability distribution. In this case, it is assumed that because of the creation of a fracture the permeability in a zone around the wellbore differs from that at a distance away from the wellbore. This system is illustrated in Fig. 10.4, where k_2 is the original permeability of the formation before treatment and k_1 is the permeability of the formation from the wellbore to the fracture radius. Under the conditions displayed on Fig. 10.4, it may be shown that

$$k_{avg} = \frac{\log r/r_w}{\frac{1}{k_1}\log\frac{r_f}{r_w} + \frac{1}{k_2}\log\frac{r_e}{r_f}}, \quad . \quad . \quad . \quad . \quad . \quad . \quad (10.7)$$

where

k_f = effective horizontal permeability of zone affected by fracture, md
k_e = original horizontal permeability of formation, md

In applying Eq. 10.7, it may be assumed that r_e is 500 ft (unless other data are available) and k_1 is the effective horizontal permeability of the formation lying within the radius of fracture. The value to assign to this effective horizontal permeability is somewhat indefinite, since the height of formation, vertical permeability, thickness of fracture, etc., all influence it. For these calculations, however, it is believed that a sufficiently accurate estimate of the permeability may be determined from the following formula.

$$k_1 = \frac{k_2\ h + k_f W_f}{h}, \quad . \quad . \quad . \quad . \quad . \quad . \quad . \quad . \quad . \quad (10.8)$$

where

k_1 = effective horizontal permeability, md

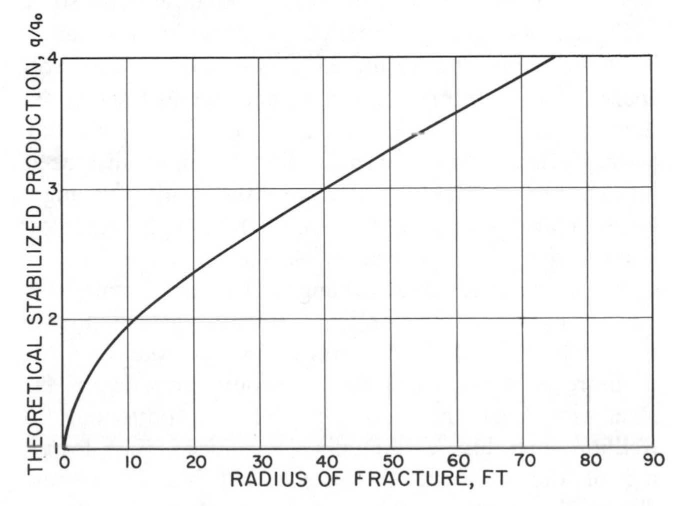

Fig. 10.3 Stabilized production increase from hydraulic fracturing.

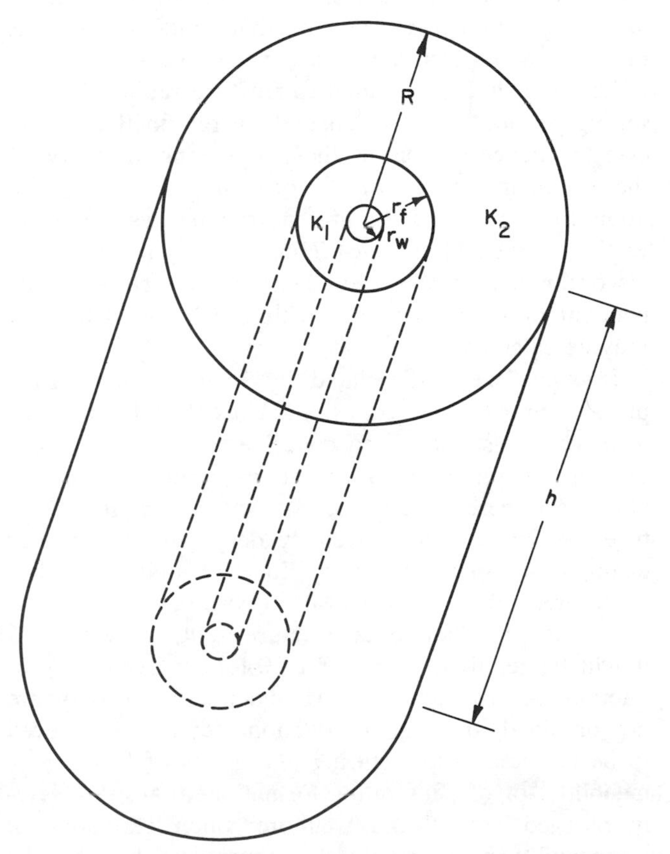

Fig. 10.4 Discontinuous permeability method for calculating production increases from fracturing.

k_2 = original horizontal permeability, md
h = height of formation, ft
w = width of fracture, ft

The flow capacity of the fracture $k_f W_f$ has been determined experimentally and has been found to vary from 100 to 40,000 md-ft, depending upon the overburden pressure, the formation hardness, and the type and concentration of propping agent used.

When all factors in Eq. 10.7 have been estimated, the average permeability of the whole producing zone is calculated. After the average permeability is obtained, the stabilized production rate following hydraulic fracturing may be determined by

$$q = \frac{3.07\ h\ k_{avg}\ \Delta p}{\mu \log r_e/r_w} \quad \text{(10.9)}$$

To illustrate this method of calculating production increases, assume that a fracture 35 ft in radius with a flow capacity of 750 md-ft is created in a formation that is 20 ft thick, has 15-md permeability, and has been penetrated by a 6-in. hole. From Eq. 10.8, k_1 is found to be 52.5 md; if this value of k_1 is applied in Eq. 10.7, the average permeability of the zone is 28.02 md. When this value is substituted in Eq. 10.9, the production increase q/q_o is 1.87. By comparing this value with that shown on Fig. 10.3, it may be seen that the production increase is approximately 66 percent of the theoretical maximum increase.

Provided that all factors can be estimated with reasonable accuracy, this method of calculating stabilized production increases from hydraulic fracturing is probably more accurate than the method that assumes wellbore radius to be equal to fracture radius. The resulting production as calculated by this method, however, is dependent upon the average permeability of the producing zone, which, of course, is dependent upon the flow capacity of the fracture, variations in vertical permeability, etc. Thus the accuracy of the discontinuous permeability distribution method is dependent upon the accuracy with which the factor k_{avg} may be calculated.

In some wells stimulated by hydraulic fracturing, production is increased as much as 20-fold. In such instances, it is difficult to conceive that the large productivity increases from formations of uniform permeability are merely the result of fracturing, as such fractures would have an extremely deep penetration. For example, it can be shown by Eq. 10.6 that if a radius of drainage of 1,500 ft and a wellbore radius of 0.25 ft are assumed, a horizontal fracture radius of some 975 ft will be required to cause a 20-fold increase in production. As fractures of this magnitude generally are not obtained, the high production increase is believed to be the result of fracturing into zones of higher permeability, or of penetrating formations that are severely plugged around the wellbore. Such variations in permeability may be natural occurrences; they may be the result of (1) drilling fluid's reducing interwell permeability, (2) carbonaceous or other foreign matter carried by water or oil, or (3) the reduction of the relative permeability to oil caused by the presence of another fluid phase in the reservoir.

The effect of nonuniform permeability on well productivity is readily calculated by Eqs. 10.7 and 10.9. Also, if the fracture created by hydraulic fracturing extends uniformly through the zone of higher permeability, an approximation of the resulting production increase may be obtained by means of either the enlarged wellbore concept, as shown on Fig. 10.3 or by the discontinuous permeability distribution method. The application of these two methods to field conditions is difficult since the permeability of the formation not immediately surrounding the wellbore is unknown. Usually, workover data from similar wells in the area will indicate whether zones of nonuniform permeability are likely to be encountered by the fractures.

To illustrate the possible effect of fracturing such a formation, assume that the rock exhibits a permeability of 5 md in a zone 2 ft in radius around the well axis, and that the surrounding formation has a permeability of 100 md. Also, assume that r_f = 500 ft, Δp = 500 psi, h = 10 ft, r_w = 0.25 ft, and μ = 2.0 cp. From Eq. 10.7 it may be calculated that the average permeability of such a formation prior to treatment is 16.13 md. From Eq. 10.9 it may be calculated that the pretreatment production will be 37 BOPD. If a horizontal fracture with a 25-ft radius is created, the post-treatment production may be estimated (Eq. 10.6) to be 590 BOPD, which is an increase of approximately 16-fold.

When extremely large stabilized production increases follow hydraulic fracturing, nonuniform permeability conditions are to be suspected. Under such conditions, the results from hydraulic fracturing may vary greatly, depending upon the magnitude of the permeability variations.

It has been observed that production is high immediately after the hydraulic fracturing treatment. From this high initial production, the productivity of some wells declines rapidly. Some of this declining productivity is certainly the result of dissipation of reservoir energy, and some may be due to changes in the characteristics of the reservoir fluid or to a gradual reduction of the formation's oil permeability. In many instances, however, the productivity decline is the result of changes in boundary conditions that occur within the reservoir immediately after the well is placed on production.

To understand these changes, let us assume that after a hydraulic fracturing treatment the pressure in the wellbore and in the formation surrounding the wellbore is stabilized at the maximum pressure of the reservoir. When the well is placed on production, the wellbore pressure is immediately reduced. The lowering of the wellbore pressure establishes a pressure differential between the wellbore and the reservoir. During the early stages of post-treatment production, the pressure differential is effective over a very short radial distance, causing the pressure gradient and the resulting post-fracturing producing rate to be high.

As the well is continuously produced, the pressure transient moves outward from the wellbore. This action reduces the pressure gradient at the wellbore, which, in turn, causes the flow rate to decline.

At the beginning of production r_e is very small; i.e., the distance at which the reservoir pressure has been affected by withdrawals is very short. This makes the value of log r_e/r_w very small, which makes the producing rate large. As r_e increases, the value of log r_e/r_w increases. This causes the production to decrease at a similar rate.

The time required for r_e to expand to its normal position may be investigated by assuming steady-state conditions. In a finite time, steady state will not actually be obtained, but it can be approximated closely enough for all practical purposes. The time required to approach steady-state conditions is a function of the permeability and porosity of the formation, the compressibility and viscosity of the fluid, and the radius of drainage. It is independent of the rate of production and the net thickness of the pay. The time is given by[8]

$$t_s = \frac{50\,\phi\,c\,\mu\,r_e^2}{k_o} \qquad (10.10)$$

If the reservoir is undersaturated and a gas phase is present, this may be taken into consideration by using a value for the fluid compressibility that reflects the condition of the total system, and by multiplying the denominator by $(\rho_o + \rho_g R)B_o\rho$. In this term, ρ_o is the density of the oil phase in lb/cu ft; ρ_g is the density of the gas phase in lb/cu ft; R is the GOR in scf/bbl; ρ is the total density in lb/cu ft; and B_o is the formation volume factor, reservoir bbl/STB.

The oil production rate after hydraulic fracturing is determined using Eq. 10.9 at several assumed values of r_e. Usually r_w is assumed equal to the radius of fracture and the assumed values of r_e are varied from

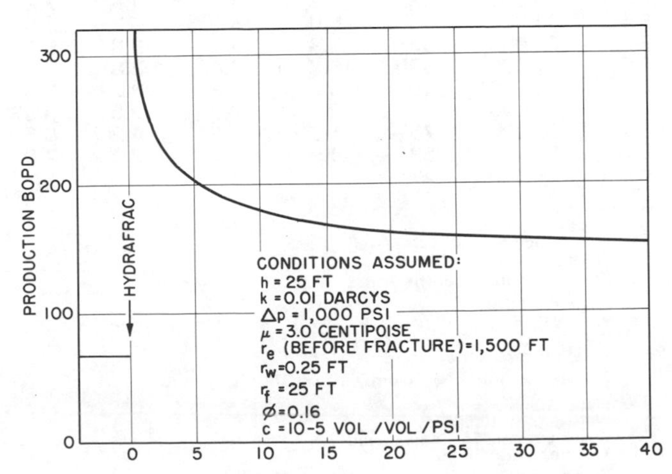

Fig. 10.5 Production decline following hydraulic fracturing.

TABLE 10.1 — THEORETICAL PRODUCTION DECLINE FOLLOWING HYDRAULIC FRACTURING

Drainage Radius, (r_e) (ft)	Producing Rate (BOPD)	Time (days)
100	424.9	0.24
200	283.3	0.96
300	237.1	2.16
500	196.6	6.0
750	173.0	13.5
1000	159.7	24.0
1500	143.9	54.0

r_w (assumed equal to r_f) to a value approaching stabilized conditions (typically 1,000 to 2,000 ft). After the producing rates are determined for several values of r_e, Eq. 10.10 is solved to determine the length of time required for r_e to advance from $r_w(r_f)$ to the value assumed. The length of time (t_s) determined from Eq. 10.10 is then plotted against the producing rate (q) from Eq. 10.9 for each assumed value of r_e.

To illustrate this type of solution, let us assume the following conditions:

$$\begin{aligned} h &= 25 \text{ ft} \\ k &= 0.01 \text{ darcy} \\ \Delta p &= 1{,}000 \text{ psi} \\ \mu &= 3.0 \text{ cp} \\ r_e \text{ (before fracture)} &= 1{,}500 \text{ ft} \\ r_w &= 0.25 \text{ ft} \\ r_f &= 25 \text{ ft} \\ \phi &= 0.16 \text{ percent expressed as a decimal} \\ c &= 10^{-5} \text{ vol/vol/psi} \end{aligned}$$

From Eq. 10.9 it may be determined that, before fracturing, this well should produce 67.7 B/D. The theoretical production of the well after the creation of a 25-ft horizontal fracture is shown on Table 10.1.

These data are also plotted on Fig. 10.5, which shows that the production following hydraulic fracturing declines quite rapidly for the first 10 days and then begins to stabilize. It should be pointed out, however, that Fig. 10.5 applies only to the conditions assumed. If either the porosity or compressibility of the formation, or the viscosity of the oil, is greater than those values assumed in the example, a longer time will be required for the post-treatment production to become stabilized. Also, formations of lower permeability will exhibit a declining production rate over a longer period of time.

As previously mentioned, the production decline is assumed to be caused by the gradual dissipation of the expanding pressure transient. Any other factors that cause declining productivity — changes in reservoir fluid characteristics, depletion of reservoir energy, decreasing permeability to oil, etc. — must be considered over and above the production decline calculated by this method.

Boriskie: Theoretical vs Observed Well Productivity Increases

Boriskie,[9] in his study of the effect of hydraulic fracturing on well productivity, compared the post-fracturing production calculated by Crittendon,[32] Mc-

Guire and Sikora,[5] and Prats[6] with the production actually obtained in the field.

To make this comparison, the production history before and after fracture stimulation of 84 wells in seven West Texas Permian Basin fields was studied. The wells varied from 1,500 to 12,500 ft in depth. Fracture area and percent penetration of the drainage area were calculated using R. D. Carter's equation[10] (Eq. 4.6).

Charts prepared by Craft *et al.*[4] and McGuire and Sikora[5] (presented on Figs. 10.1 and 10.2), were used by Boriskie.

A review of the well productivity increases due to hydraulic fracturing showed that, generally, in the shallow zones the treatment volumes were smaller, injection rates were higher, and the penetration of the resulting fractures was greater than in the deeper reservoirs. The estimated productivity increases read from the curves developed by McGuire and Sikora and by Craft *et al.*, together with the estimated increases as determined by the Prats mathematical method, are tabulated for individual wells in Table 10.2.

Figs. 10.6 through 10.8 present comparisons of the individual well productivities as predicted by McGuire and Sikora, Prats, and Craft *et al.* Fig. 10.6 shows that the method used by McGuire and Sikora generally

TABLE 10.2 — WELL PRODUCTIVITY DATA

Well	Treatment Volume (gal)	Percent Penetration[a]	Productivity Ratio			
			Field	Crittendon	McGuire and Sikora	Prats
A-1[b]	20,000	69	7.5	4.8	7.9	7.1
A-2	20,000	69	6.4	4.8	7.9	7.1
A-3	20,000	69	7.4	4.8	7.9	7.1
A-4	20,000	69	7.6	4.8	7.9	7.1
A-5	20,000	69	8.0	4.8	7.9	7.1
A-6	20,000	69	7.5	4.8	7.9	7.1
A-7	20,000	69	6.0	4.8	7.9	7.1
A-8	20,000	69	6.2	4.8	7.9	7.1
A-9	20,000	69	7.3	4.8	7.9	7.1
A-10	30,000	84	8.4	4.8	8.4	8.8
A-11	20,000	69	8.0	4.8	7.9	7.1
A-12	20,000	69	7.0	4.8	7.9	7.1
A-13	20,000	69	10.1	4.8	7.9	7.1
A-14	20,000	69	8.5	4.8	7.9	7.1
A-15	20,000	69	9.0	4.8	7.9	7.1
A-16	20,000	69	8.1	4.8	7.9	7.1
A-17	20,000	69	7.6	4.8	7.9	7.1
A-18	17,000	62	7.0	4.7	7.8	6.7
A-19	20,000	69	6.7	4.8	7.9	7.1
A-20	20,000	69	7.5	4.8	7.9	7.1
A-21	20,000	69	9.3	4.8	7.9	7.1
A-22	20,000	69	5.8	4.8	7.9	7.1
A-23	15,000	56	6.5	4.6	7.5	5.9
A-24	10,000	42	5.8	4.2	6.5	4.8
B-1[c]	60,000	84	9.5	4.9	6.8	9.4
B-2	60,000	84	10.3	4.9	6.8	9.4
B-3	60,000	84	9.4	4.9	6.8	9.4
B-4	60,000	84	8.6	4.9	6.8	9.4
C-1[c]	60,000	75	8.1	4.8	5.8	8.4
C-2	60,000	75	6.8	4.8	5.8	8.4
C-3	60,000	75	7.4	4.8	5.8	8.4
C-4	60,000	75	8.3	4.8	5.8	8.4
C-5	47,500	59	6.8	4.8	5.7	6.7
C-6	60,000	75	7.4	4.8	5.8	8.4
C-7	60,000	75	8.7	4.8	5.8	8.4
C-8	60,000	75	8.3	4.8	5.8	8.4
C-9	60,000	75	7.9	4.8	5.8	8.4
C-10	60,000	75	8.2	4.8	5.8	8.4
C-12	47,500	59	5.5	4.7	5.7	6.7
C-13	60,000	75	7.5	4.8	5.8	8.4
C-14	60,000	75	7.8	4.8	5.8	8.4
D-1[d]	35,000	51	7.4	4.4	6.9	5.5
D-2	35,000	51	6.7	4.4	6.9	5.5
D-3	35,000	86	9.2	4.9	7.8	7.7
D-4	60,000	70	8.9	4.8	7.6	7.3
D-5	40,000	55	7.0	4.6	7.0	5.8
D-6	40,000	55	7.5	4.6	7.0	5.8
D-7	40,000	55	8.1	4.6	7.0	5.8
D-8	40,000	55	7.1	4.6	7.0	5.8
D-9	40,000	55	6.6	4.6	7.0	5.8
D-10	25,000	42	5.7	4.2	6.2	4.7
D-11	60,000	70	8.2	4.8	7.6	7.3
D-12	50,000	65	7.0	4.7	7.5	6.7
E-1[e]	20,000	36	5.1	4.1	5.2	5.0
E-2	10,000	21	2.9	3.4	3.7	3.8
E-3	40,000	62	9.8	4.8	7.3	7.3
E-4	35,000	55	7.8	4.7	6.7	6.7
E-5	35,000	55	8.2	4.7	6.7	6.7
E-6	15,000	29	4.5	3.8	4.5	4.4
E-7	20,000	36	6.4	4.1	5.2	5.0
E-8	20,000	36	5.7	4.1	5.2	5.0
E-9	20,000	36	5.3	4.1	5.2	5.0
E-10	35,000	55	7.0	4.7	6.7	6.7
E-11	55,000	79	8.8	5.0	7.8	9.2
E-12	45,000	66	10.1	4.9	7.5	7.7
E-13	40,000	62	8.4	4.8	7.3	7.3
E-14	45,000	66	7.7	4.9	7.5	7.7
E-15	45,000	66	9.3	4.9	7.5	7.7
E-16	45,000	66	9.0	4.9	7.5	7.7
E-17	45,000	66	6.0	4.9	7.5	7.7
E-18	45,000	66	7.2	4.9	7.5	7.7
E-19	20,000	36	4.0	4.1	5.2	5.0
E-20	20,000	36	5.3	4.1	5.2	5.0
F-1[f]	150,000	67	7.5	4.7	3.8	7.2
F-2	100,000	55	6.3	4.5	3.8	6.7
F-3	150,000	67	8.1	4.7	3.8	7.2
F-4	60,000	42	5.0	4.2	3.7	5.5
F-5	100,000	55	5.8	4.5	3.8	6.7
F-6	80,000	49	6.6	4.4	3.8	6.1
F-7	72,000	46	5.8	4.3	3.7	5.8
G-1[g]	4,500	17	2.5	2.9	2.8	3.5
G-2	35,000	75	5.5	4.6	4.2	8.6
G-3	25,000	59	4.9	4.4	2.1	7.0

[a]Penetration of the drainage radius, r_e, percent

[b]Field A
Depth: 1,600 ft
Well spacing: 10 acres/well
Treating rate: 40 bbl/min
Fracturing fluid coefficient: 6.4×10^{-3} ft/$\sqrt{min}$

[c]Fields B and C
Depth: 2,900 ft
Well spacing: 40 acres/well
Treating rate: 40 bbl/min
Fracturing fluid coefficient: 2.1×10^{-3} ft/$\sqrt{min}$ (Field B)
2.5×10^{-3} ft/$\sqrt{min}$ (Field C)

[d]Field D
Depth: 3,500 ft
Well spacing: 20 acres/well
Treating rate: 35 bbl/min
Fracturing fluid coefficient: 5.0×10^{-3} ft/$\sqrt{min}$ and 2.5×10^{-3} ft/$\sqrt{min}$

[e]Field E
Depth: 8,000 ft
Well spacing: 160 acres/well
Treating rate: 30 bbl/min
Fracturing fluid coefficient: 1.04×10^{-3} ft/$\sqrt{min}$

[f]Field F
Well depth: 12,000 ft
Well spacing: 160 acres/well
Treating rate: 35 bbl/min
Fracturing fluid coefficient: 3.0×10^{-3} ft/$\sqrt{min}$

[g]Field G
Well depth: 12,500 ft
Well spacing: 160 acres/well
Treating rate: 30 bbl/min
Fracturing fluid coefficient: 3.1×10^{-3} ft/$\sqrt{min}$

predicts post-fracturing production lower than that actually experienced in the field. These inaccuracies are apparently the result of incorrect fracture capacity (conductivity) estimates. Prats' prediction method (Fig. 10.7) shows good average agreement in the field tests. The predictions of Craft *et al.* show the poorest correlation, exhibiting as much as a twofold variance.

10.3 Effect on Productivity and Oil Recovery

By 1969 hydraulic fracturing was the most widely used and most successful well stimulation process yet devised. It was used extensively as a workover procedure and as a vital part of initial well completion.

Production Behavior After Fracturing

The effect of fracturing on both short and long term well productivity has been studied by many investigators, most of whom conclude that, regardless of the kind of treatment, four basic patterns of production behavior have been observed.

Type A — Sustained increase in well production accompanied by a flattening of the production decline curve following treatment.

Type B — Sustained increase in production with the well's higher rate of production after the treatment declining essentially at the same rate established before treatment.

Type C — Transitory increase in production lasting from a few months to several months, after which the well continues to follow the production decline trend that was observed prior to treatment.

Type D — No increase in production after treatment, with the well continuing to follow its established, normal production history.

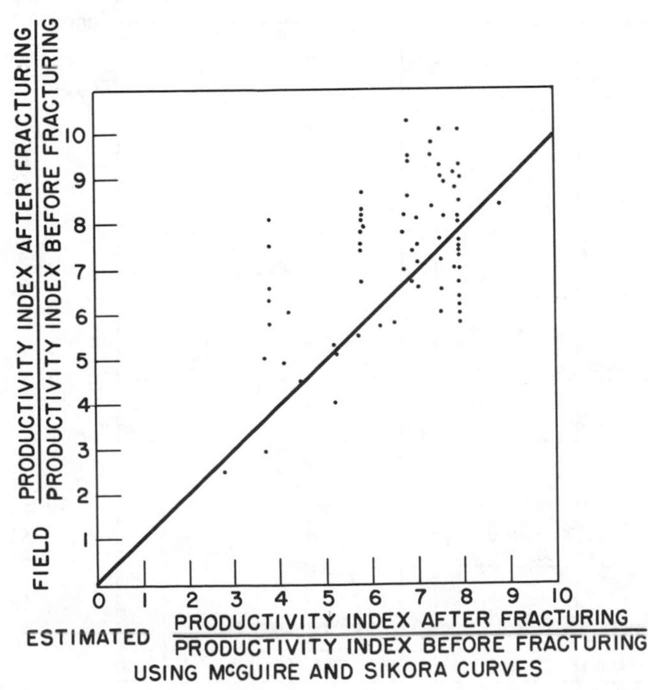

Fig. 10.6 Comparison of field productivity increases with estimated values.[5]

In no case did a treatment have a discernibly detrimental effect on the production performance of a well.

Ghauri,[11] in his study of the Brea-Olinda and Esperanza fields of California, presented four wells (Fig. 10.9) whose production decline illustrates the four types given above. The productive area is 2,500 to 8,000 ft deep and is a steeply dipping monocline. The oil production rate following treatment is the net rate of total injected fluids after recovery. The water production and GOR curves (not shown on the figure)

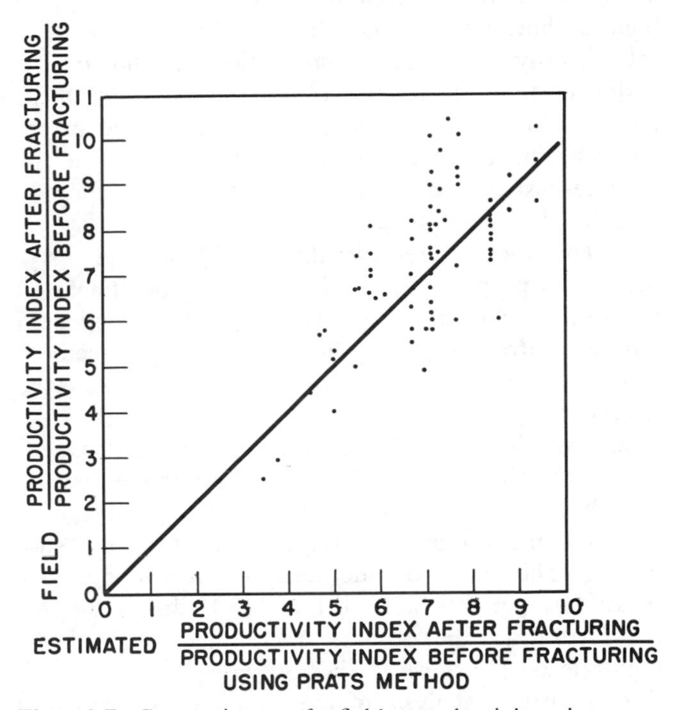

Fig. 10.7 Comparison of field productivity increases with estimated values.[6]

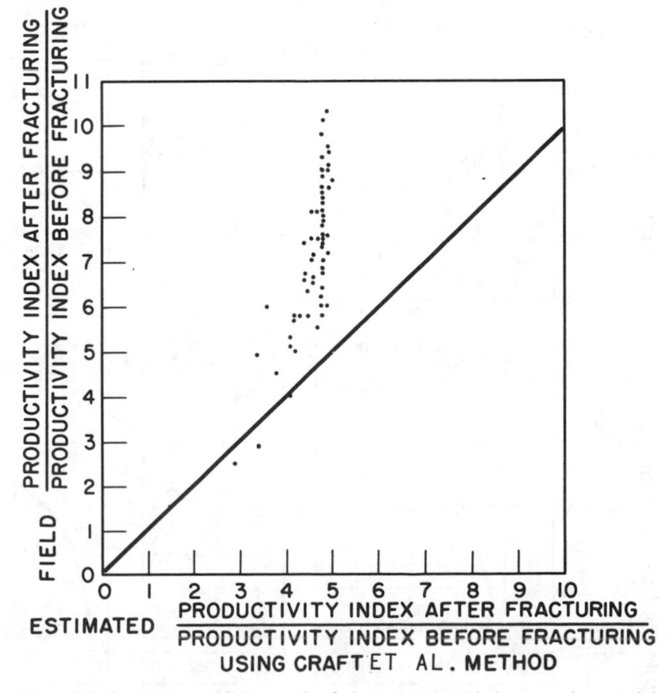

Fig. 10.8 Comparison of field productivity ratios with estimated values.[4]

showed no noticeable effects.

Individual Well Decline-Curve Analysis

Ghauri determined the additional oil by analyses of individual well decline curves. Additional oil for each well was the estimated net ultimate oil (after recovery of injected fluids) realized as a result of the fracturing treatment in excess of the well's extrapolated total production without the treatment. Production performance curves were extrapolated on the basis of the hyperbolic decline where the exponent equals 2.[12] The hyperbolic decline, fitted empirically to the production data, satisfactorily described the production behavior over a major portion of the life of the reservoirs in these fields. Table 10.3 is a summary of the range of recoveries, the average additional oil per job, and the average cost of each hydraulic fracturing operation.

Table 10.3 shows the following:

1. The fracture created with an oil-base fluid carrying a sand propping agent yielded higher additional oil recoveries than either the viscous gel fracturing fluid job (Hydrafrac) or the high-rate oil-backflush method. The recoveries attributable to the last method were the lowest.

2. The high-rate oil backflush was the least expensive of the three methods; the costs for the Hydrafrac and the oil-base fracturing job were comparable.

3. On the average, all three treatments were successful. The ratios of additional oil recovery to expenditures per job were 1:1.9, 1:5.1, and 1:3.8 for the Hydrafrac, the oil-base fracturing fluid, and the high-rate oil backflush, respectively.

A statistical analysis of the additional oil recovery data leads to the following observations regarding these stimulation treatments.

TABLE 10.3 — COMPARISON OF RESULTS OF THREE FRACTURING METHODS

Type of Treatment	Number of Jobs	Range of Additional Oil Recovery (1000 bbl)	Average Additional Oil Per Job (1000 bbl)	Average Cost Per Job
Hydrafrac	14	0 to 90	30	$16,100
Oil-Base Fracturing	14	0 to 135*	42*	$15,400
High-Rate Oil Backflush	25	0 to 90**	14**	$ 3,700

*Excludes one highly successful oil-base fracturing job, which yielded an estimated 550,000 bbl of additional oil. In 14 oil-base fracturing treatments, the additional oil recovered ranged from 0 to 550,000 bbl, and the average additional oil recovered per job amounted to 78,000 bbl.

**Excludes six recent high-rate oil backflushes to which no additional oil is assigned because of insufficient production history following treatment.

1. The increase in recoveries for 46 jobs ranged from 0 to 135,000 bbl, with an arithmetic mean of 27,000 bbl.

2. Thirty-nine percent of the jobs yielded increased recoveries of less than 5,000 bbl.

3. Although the distribution was arrayed heavily to the right of the standard deviation, 68 percent of the well stimulations yielded recoveries of less than 62,000 bbl/job.

4. The majority of the recoveries in excess of 62,000 bbl (arithmetic mean plus one standard deviation) are attributable to the sand-oil fracturing treatment.

Insofar as the Hydrafrac and the sand-oil fracturing methods are concerned, the use of fluids other than crude oil and of solids other than Ottawa sands did not enhance the recovery of additional oil. These data indicate that sand-oil fracturing is the preferable means of production stimulation for achieving maximum additional oil recovery. The larger volumes of fracturing fluid and higher injection rates associated with

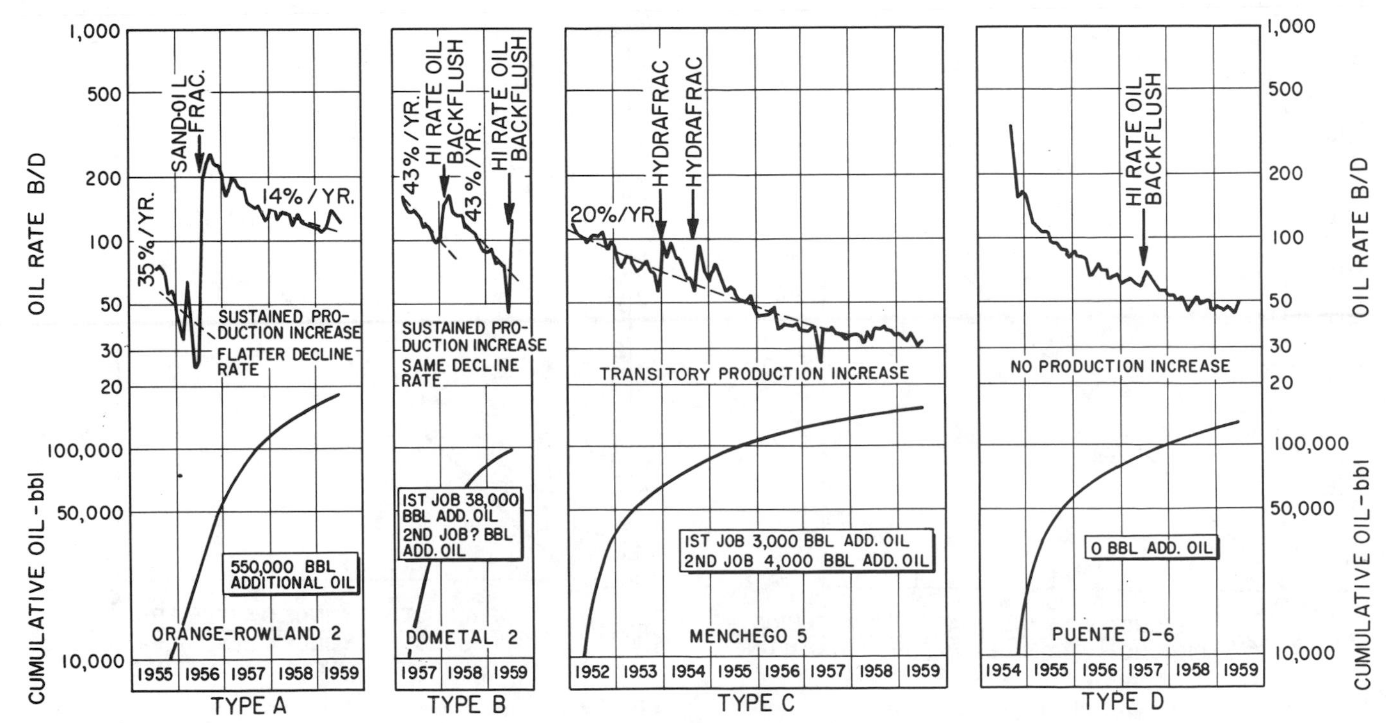

Fig. 10.9 Well decline curves showing production response, Brea-Olinda and Esperanza fields, Calif.[11]

sand-oil fracturing effect deeper penetrating fractures.

Composite Well Decline-Curve Analysis

The average production response (composite decline curves for the Brea-Olinda and Esperanza fields) for the Hydrafrac, sand-oil fracturing, and high-rate oil backflush treatments is presented in Fig. 10.10. In these curves, jobs for the respective treatments are referred to a common time datum, with the actual production shown on the graph for a period of 1 year prior to and following the treatment. Production rate following treatment is the net production rate after the recovery of total injected fluids.

Oil wash jobs with insufficient production histories were not included in the composite curve for the high-rate oil backflush, whereas all of the Hydrafrac and sand-oil fracturing jobs were included in their respective curves. The initial hyperbolic decline ratio used to extrapolate the production to the economic limit both before and after the treatment is shown on the figure, as is the additional oil calculated per job. The additional oil recovered from these fields was 29,000, 76,000, and 15,000 bbl for the Hydrafrac, the sand-oil fracturing, and the high-rate oil backflush, respectively.

The estimated production decline rate after the Hydrafrac and oil-base sand-oil fracturing treatments (Fig. 10.10) is lower than the production decline rate before the treatments. In the case of the high-rate oil backflush, the production decline rate is higher than that observed prior to treatment. This suggests that the production increase realized as a result of the backflush was due primarily to a decrease in the skin effect, with a relatively minor contribution coming from undrained portions of the reservoir. By contrast, fractures created and kept open by the propping agent in the Hydrafrac and sand-oil fracturing treatments resulted in drainage of relatively undepleted portions of the reservoir, adding significantly to the existing well drainage pattern.

Ultimate Recovery—Semi-Depleted Reservoirs

Campbell investigated the effect of hydraulic fracturing on ultimate recovery and concluded that in the Mid-Continent region of the U. S. — in fields where fracturing is applicable and judiciously applied — an increase in ultimate oil recovery can be expected.[13] Campbell's study was limited to wells in semi-depleted reservoirs with sufficient production history before and after treatment to give a relatively accurate evaluation of ultimate recovery.

Fig. 10.11 presents decline curves for two typical fractured wells that produce from the Earlsboro sand of the Keokuk pool (Oklahoma). The curve of the rate of production decline after fracture treatment is approximately parallel to the curve of the primary production decline, although at a higher level. This indicates that additional oil may be recovered before the well reaches its economic limit. These wells have been produced on a capacity decline, with only minor interruptions because of proration. On an average, for the 19 successful fracturing treatments evaluated in this 3,400-ft sand, the daily oil production per well increased from 7 to 63 bbl, and it is estimated that about 10,700 additional barrels of oil per well were recovered. The $2,800 cost of the hydraulic fracturing workover amounted to about 26¢ for each barrel of increased oil, or $50 for every barrel of increased daily

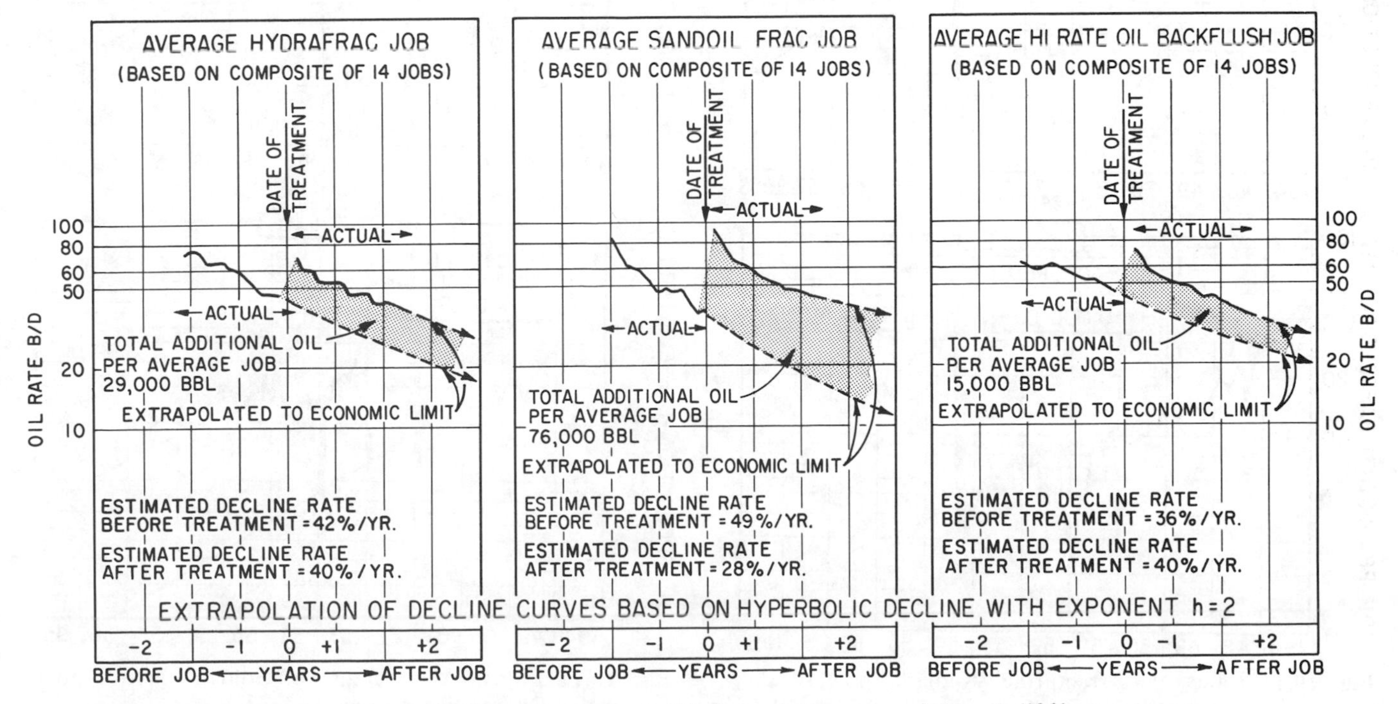

Fig. 10.10 Composite production decline curves, Brea-Olinda and Esperanza fields, Calif.[11]

oil production immediately following the fracturing treatment.

Another Oklahoma area, the Golden Trend pool, is represented in the two typical decline curves on Fig. 10.12. The production decline curve after fracturing is about parallel to the primary oil production decline curve. These two curves show the increase in ultimate recovery effected by the fracturing treatment. Well C has had only minor interruptions resulting from proration, but Well D was restricted for about 3 years after treatment. Successful hydraulic fracturing treatments in the 7,200-ft Gibson sand increased daily oil production, on the average, from 6 to 157 bbl for the 47 fracturing treatments evaluated. Ultimate increase in oil production per treatment is estimated to be 12,800 bbl. The average cost of the hydraulic fracturing jobs was $5,200, or 41¢ for each barrel of additional oil.

Typical decline curves before and after fracturing in the Illinois basin area are presented in Figs. 10.13 and 10.14. The oil production before and after treatment again indicates an increase in ultimate oil recovery. On Well E, the future trend of oil production may prove to be more nearly parallel to the prefracturing decline curve than is shown. The same is true of Well F, particularly in view of the short production history subsequent to the fracture job. In 11 hydraulic fracturing treatments in the 2,700-ft Mississippian sands, average oil production increased from 6 to 107 B/D. Increase in the ultimate production is estimated at 7,400 bbl/well. The $3,900 cost of the average fracturing job is equal to 53¢ for each barrel of increased oil recovery.

Fig. 10.15 is the decline curve from a 4,300-ft well in the Red Fork sands of Oklahoma's Mt. Vernon pool, and is an illustration of an outstanding example of increased ultimate oil recovery. The well, originally nonproductive, was fractured upon initial completion, and refractured several months later with a larger treatment. Increased oil recovery from the second fracture treatment is estimated at 34,500 bbl. The total cost of this hydraulic fracturing workover operation was $2,500, or 7¢ for each barrel of additional oil recovered.

Table 10.4 presents the experience of a major oil producer operating in the Mid-Continent area, using fracture treatments successfully on properties with long production histories. The 142 successful treatments were from 28 oil provinces in eight states. An average successful treatment increased daily oil production from

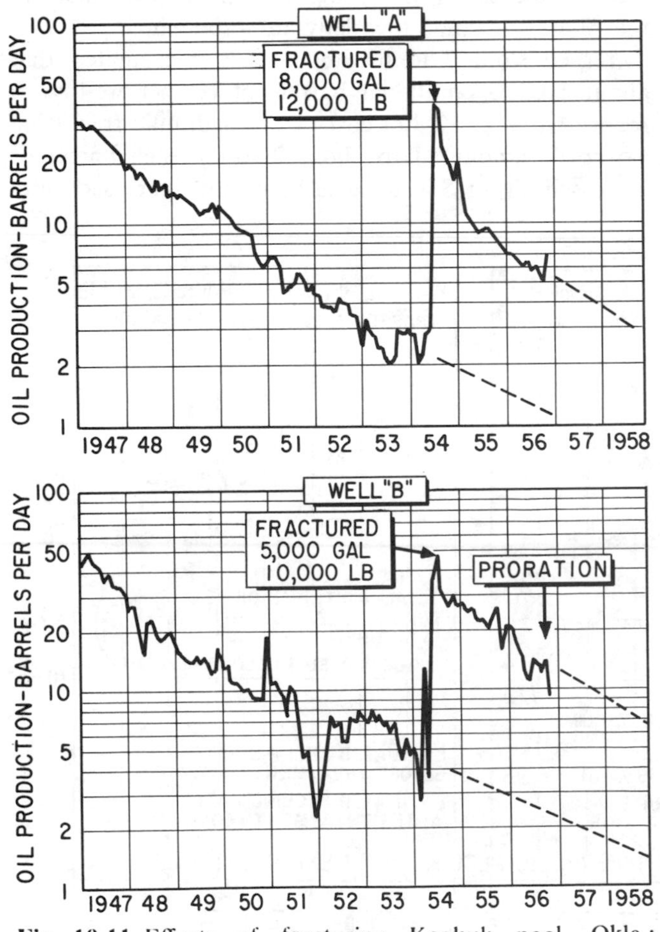

Fig. 10.11 Effects of fracturing Keokuk pool, Okla.; Earlsboro sand.[13]

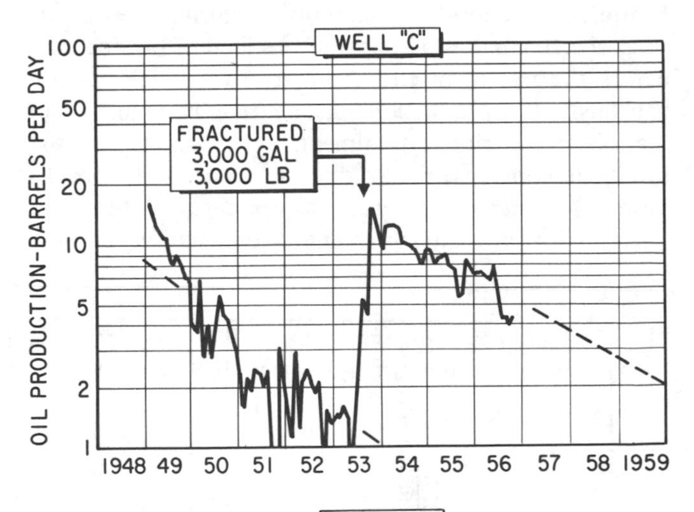

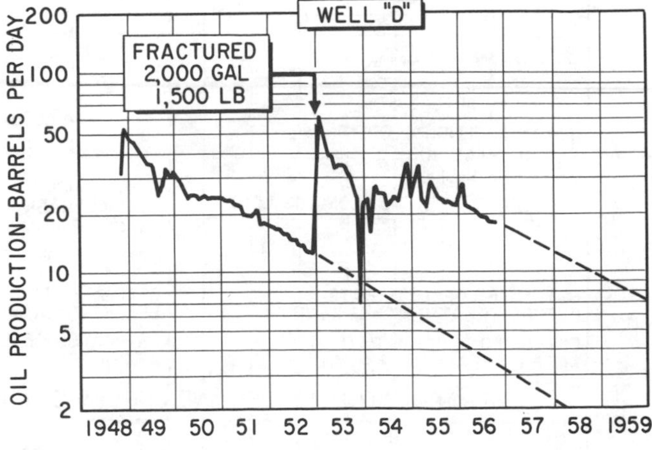

Fig. 10.12 Effects of fracturing Golden Trend pool, Okla.; Gibson sand.[13]

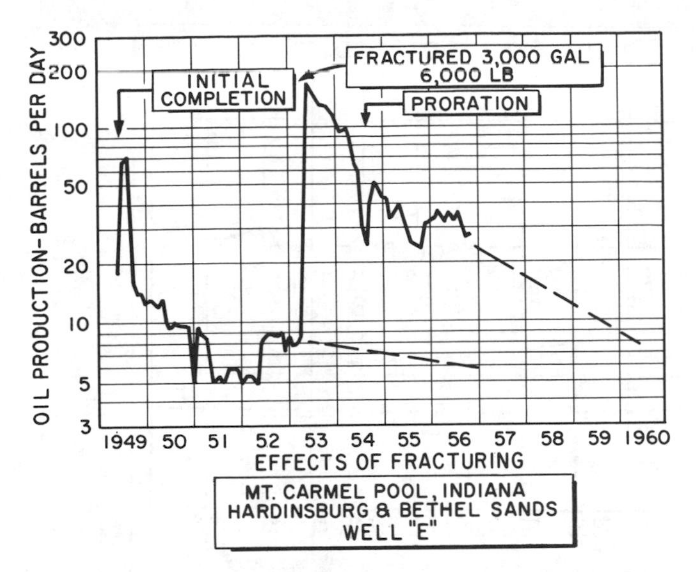

Fig. 10.13 Effects of hydraulic fracturing on well productivity.[13]

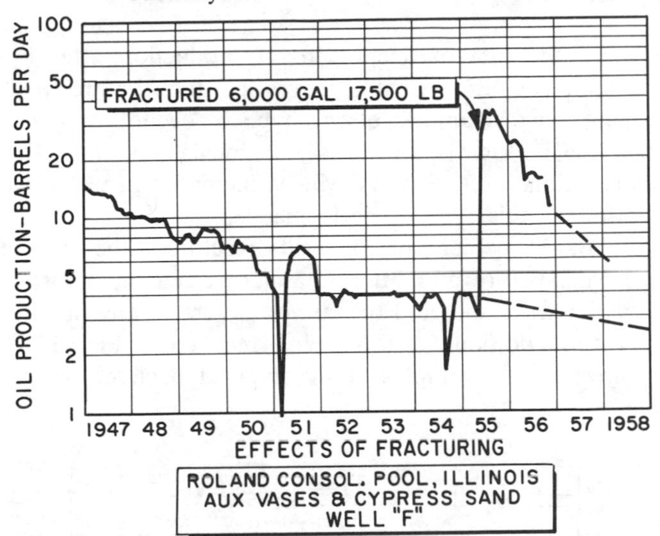

Fig. 10.14 Effects of hydraulic fracturing on well productivity.[13]

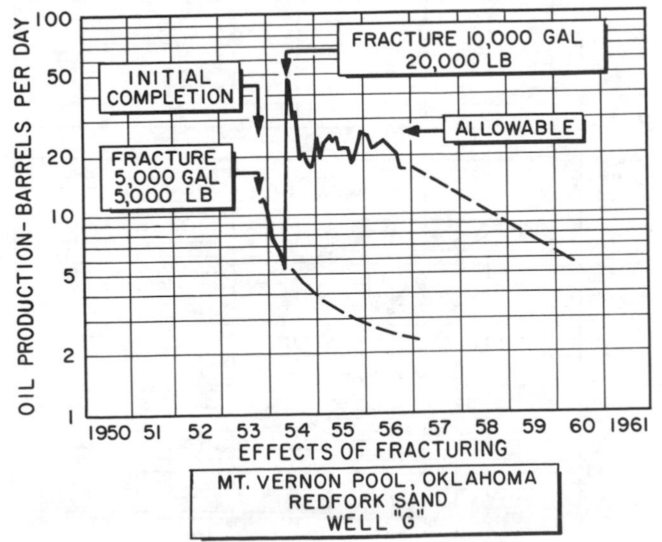

Fig. 10.15 Effects of hydraulic fracturing on well productivity.[13]

7 to 92 bbl and increased the recoverable oil approximately 12,100 bbl/well. The typical fracturing treatment cost about $3,900 or 32¢ for each barrel of increased recoverable oil. Assuming a success rate of 80 percent, the cost of fracturing, adjusted for unsuccessful jobs, would be increased to $4,875, or 40¢ for each barrel of additional oil. Thus, it is evident that when a hydraulic fracturing treatment is properly applied, it can increase the ultimate recovery of oil in semidepleted reservoirs.

Sallee and Rugg[14] and other investigators [15] have reported the results of two or more fracturing jobs conducted on a well over an extended period of time, with each fracturing job resulting in substantial and sustained increases in production and ultimate oil recovered. Fig. 10.16 illustrates such a well in a North Texas field.

North Texas Ultimate Primary Recovery

Garland, Elliott, Dolan and Dobyns concluded, in their report of investigations, that hydraulic fracture treatments have substantially increased ultimate primary recovery in North Texas.[19] Within a 2-year interval after fracture treatments, 1,278,000 bbl of oil were recovered (from 182 wells) that would not have been recovered by primary means if the wells had not been treated. Increased oil recovery during the 2-year period averaged 7,000 bbl/well for those completed in the Strawn formation and 4,600 bbl/well for those completed in the Cisco formation.

Oil Recovery from Strawn Formation Wells. Fig. 10.17 shows the composite oil production history of 69 wells producing from the Strawn formation. These wells are in 18 fields in Archer, Clay, Cooke, Jack and Montague Counties, Tex.

As a result of hydraulic fracturing treatments, the total average monthly oil production was increased from 18,400 to 45,100 bbl for the first 6 months following the fracturing of the wells, an increase of 145 percent or 2.45 times the monthly production before treatment. Extrapolating the normal rate of decline of oil production before treatment, the monthly oil production would have been 15,700 bbl if the wells had not been treated. Therefore, the net increase in rate of oil production during the first 6-month period after treatment was 29,400 bbl/month.

TABLE 10.4 — EFFECT OF FRACTURING ON ULTIMATE RECOVERY, AVERAGE SUCCESSFUL TREATMENT, MID-CONTINENT AREA

Number of successful treatments evaluated	142
Average depth, ft	4,400
Gross production before, B/D	7
Gross production after B/D	92
Production on Oct. 1, 1956, B/D	11
Gross accumulated increase to Oct. 1, 1956	7,500
Gross estimated ultimate production, bbl	12,100
Cost of fracturing	$ 3,900
Cost of fracturing per barrel of increased oil, assuming 80 percent success ratio	$ 0.40
Cost of fracturing per barrel of increase in potential, assuming 80 percent success ratio	$ 57

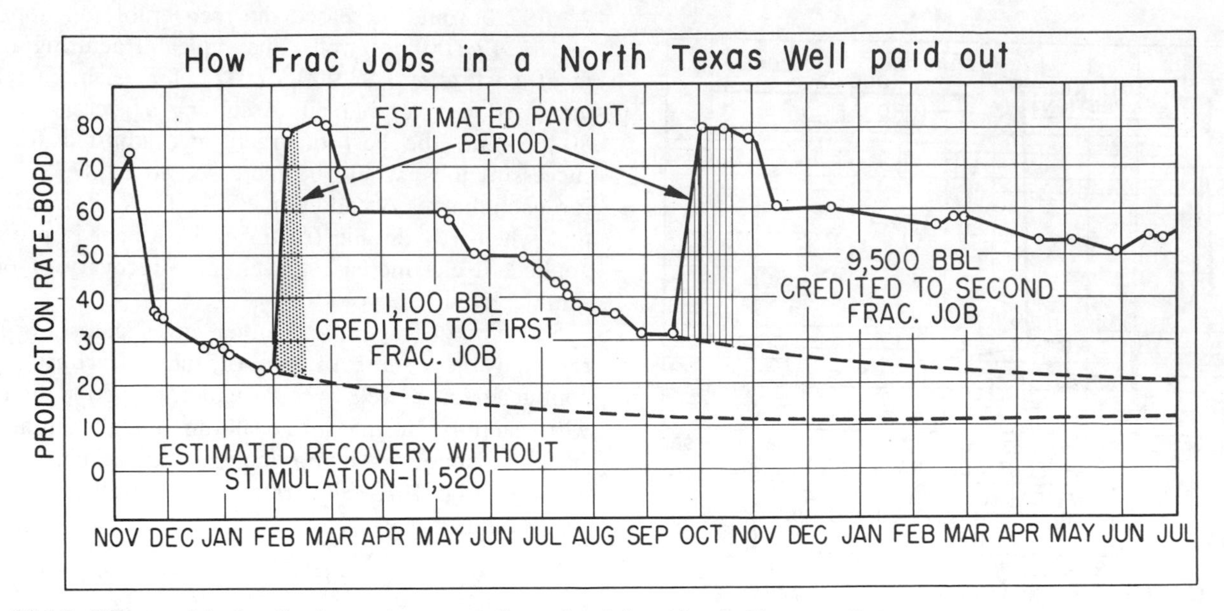

Fig. 10.16 Effects of hydraulic fracturing on well productivity, North Texas well.[14]

The monthly oil production during the fourth 6-month period after treatment was about 23,400 bbl, or 5,000 bbl more per month than the wells were producing during the 6-month period immediately before treatment. From the oil production decline curve, the estimated monthly production during the fourth 6-month period would have been about 9,700 bbl of oil if the wells had not been treated. Thus, an estimated net gain in production of 13,700 bbl of oil per month during the fourth period after treatment is indicated.

Graphic calculations show that about 484,200 bbl of additional oil were recovered from the 69 wells during the 24 months after the hydraulic fracture treatment, an average increase of about 7,000 bbl of additional oil per well.

Production Decline After Fracture Treatment. The rate of oil production decline before and after fracture treatments was compared for individual wells completed in the Strawn formation. For 30 percent of the wells, the rate of decline in oil production after hydraulic fracturing treatment was less than before treatment. Two examples of this type of decline are illustrated in Fig. 10.18. These decline curves show that the ultimate oil recovery will be increased in wells that evidence this type of performance.

For 41 percent of the wells fractured, the rate of decline before and after treatment remained approximately the same, and the curves are essentially parallel. Again, additional oil is being recovered. Fig. 10.19 shows three examples of this type of decline.

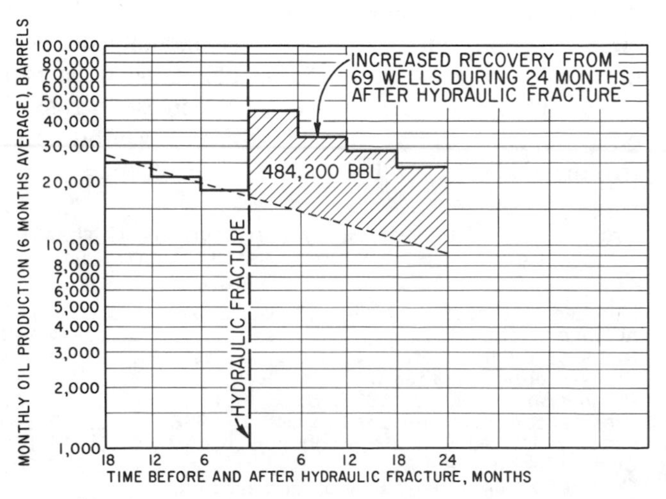

Fig. 10.17 Composite oil production history and estimated additional oil recovery, Strawn formation, North Texas area.[19]

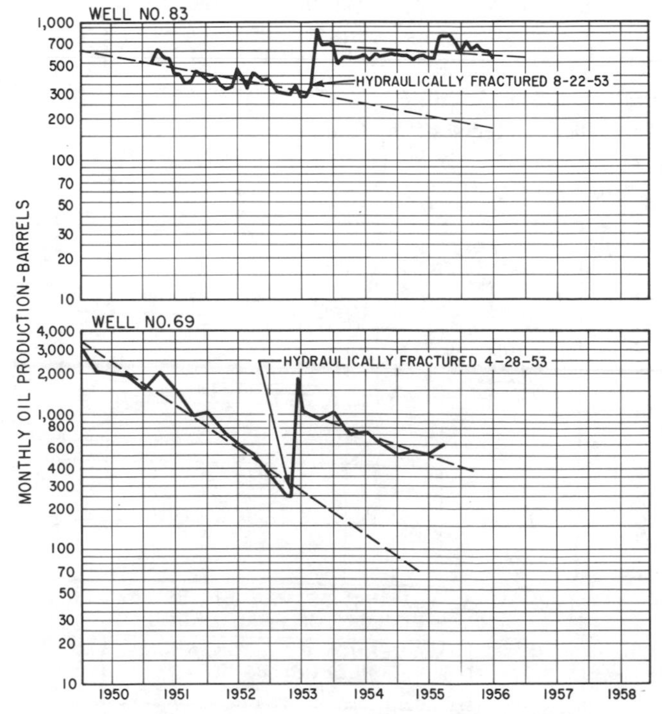

Fig. 10.18 Production history, showing rate of decline after treatment less than rate of decline before treatment.[19]

For 29 percent of the wells the rate of oil production decline after fracture treatment was greater than that before treatment. For 24 percent of the wells, the decline curve will intersect the pretreatment decline curve at a point below an estimated economic limit of 3 BOPD, which indicates that the ultimate oil recovery from these wells will be increased as a result of hydraulic fracturing. Two examples of this type of decline are illustrated in Fig. 10.20.

The well production histories in 5 percent of the wells showed that the two decline curves intersect at a point above the estimated economic limit. Despite this, the ultimate oil recovery is estimated to be approximately the same as if the wells had not been treated. Therefore, this type of result may be considered profitable because production costs are reduced by recovering the oil within a shorter period of time.

Data obtained from Strawn wells were used to plot the total monthly production after fracture treatment (expressed as a percent of the total prefracture production) against time in months after treatment (Fig. 10.21). The curve indicates that the rate of oil production from these wells did not decline to the pretreatment level until approximately 30 months after treatment. Thus, increases in production rate as a result of hydraulic fracturing were not temporary, but were sustained increases.

Production increases from hydraulic fracturing treatments performed on formations that had previously been shot with nitroglycerin were compared with production increases resulting from fracturing formations that had not been shot. (Some of the wells had more than one hydraulic fracturing treatment.) Results obtained by successive treatments were also compared.

Twenty-four Strawn wells were shot with 20 to 150 qt of nitroglycerin at their original completions. On the average, these wells had a production history of 5 to 10 years before they were fractured. The average rate of oil production of the 24 wells that had previously been shot with nitroglycerin was boosted from 241 to 563 bbl a month after the hydraulic fracturing treatments. The average rate of oil production of 41 wells that had not been shot was raised from 333 barrels a month to 804. Before treatment, the average oil

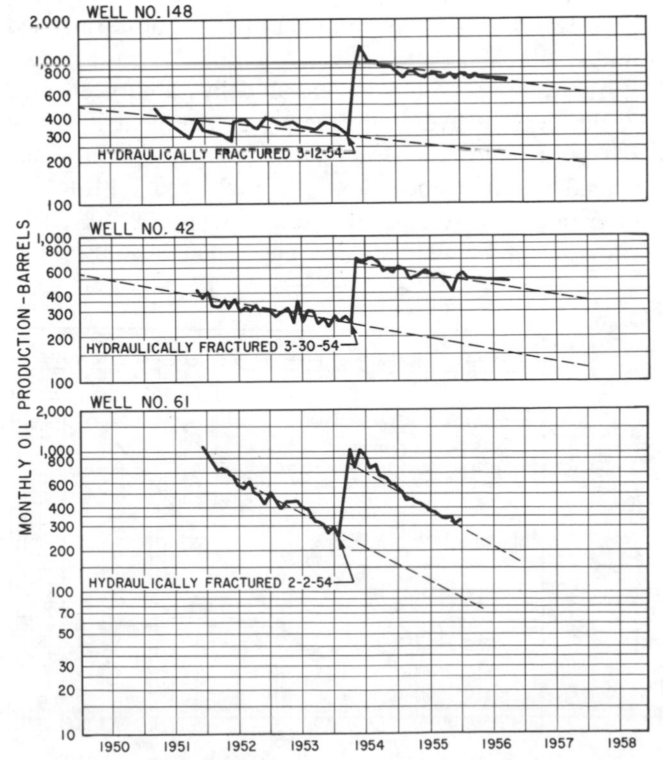

Fig. 10.19 Production history, showing rate of decline after treatment same as rate of decline before treatment.[19]

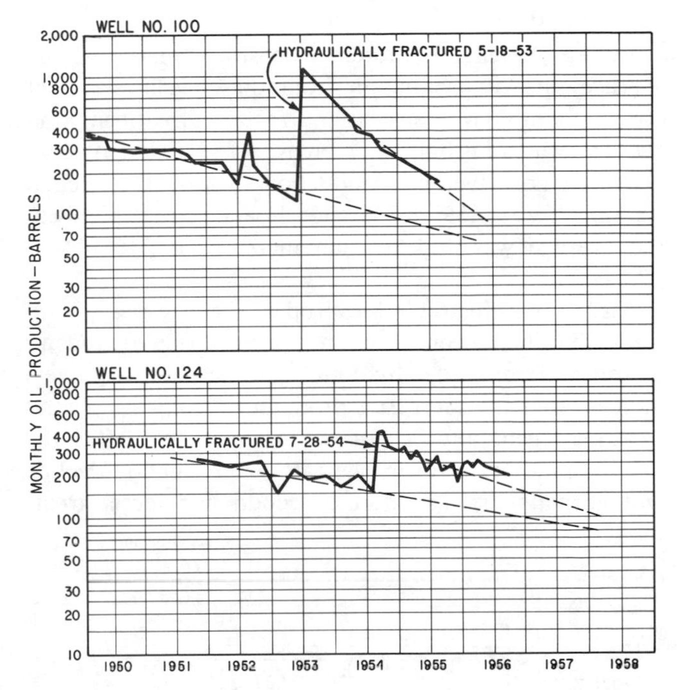

Fig. 10.20 Production history, showing rate of decline after treatment greater than rate of decline before treatment, but additional oil being recovered.[19]

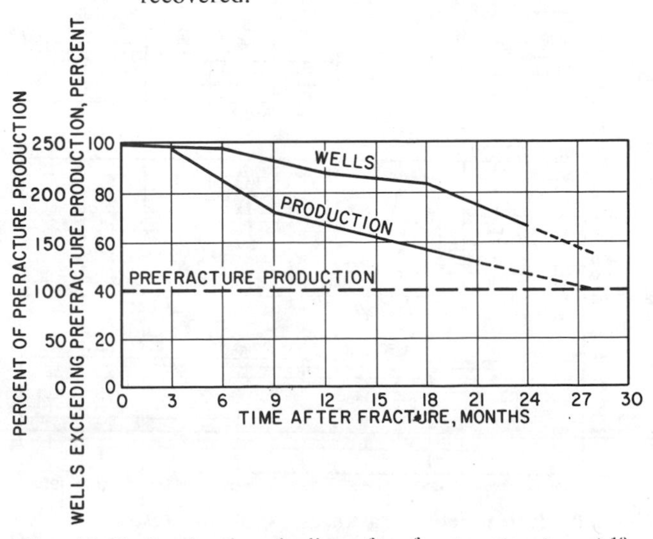

Fig. 10.21 Production decline after fracture treatment.[19]

production rate of the wells that had not been shot was 92 bbl a month more than that of the "shot" wells. After treatment, this difference was increased to 241 bbl a month, indicating that the wells that had not been shot responded better to treatment than did those that had been shot.

Eleven of the 65 Strawn wells had been subjected to more than one hydraulic fracturing treatment. Generally, the initial hydraulic fracturing treatments used less than 1,000 gal of fracturing fluid and 1,000 lb of sand proppant. In the second and third treatments the volume of fracturing fluid ranged from 1,500 to 15,000 gal and that of sand proppant from 1,200 to 30,000 lb. Three wells were re-treated within 21 months after their first hydraulic fracturing treatment, and eight wells were re-treated within a period of 29 to 55 months.

Before the first hydraulic fracturing treatment, the average rate of production from these wells was 199 bbl a month, whereas after fracture treatment the average was 711 bbl, an increase of 512 bbl or 257 percent. The average production before the second treatment was 259 bbl a month and after treatment the average was 474 bbl, an increase of 215 bbl or 83 percent.

Fig. 10.22 illustrates the production histories of two wells that had been given more than one treatment. Results of multiple hydraulic fracturing treatments indicate that the increased production from the second treatment generally is not so large as that after the first treatment; however, the rate of production decline will normally be less after a second and larger treatment.

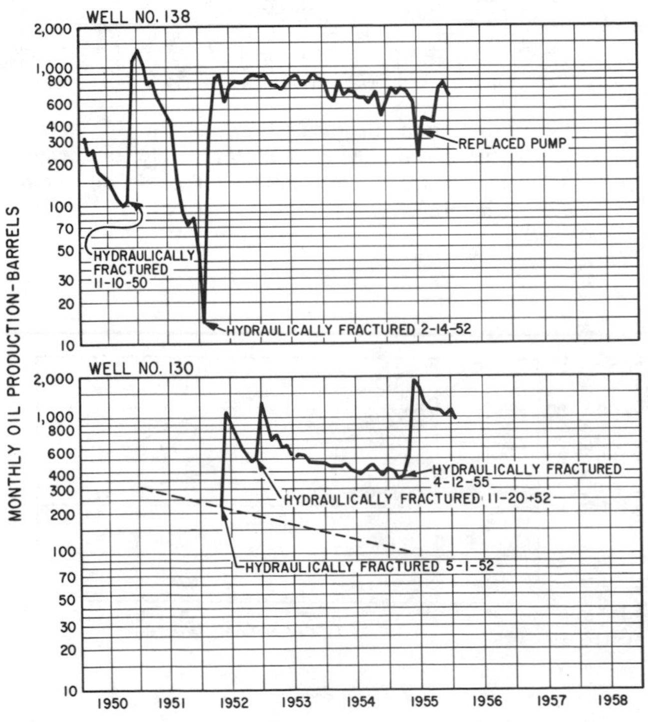

Fig. 10.22 Production histories, showing results of two or more fracture treatments.[19]

Gas Well Deliverability

Deliverability from thousands of gas wells has been improved by hydraulic fracturing, this being reflected in both flowing rate and wellhead pressure. In many cases the gas production following treatment has been several-fold greater than that obtained upon initial completion.[16] Fracturing has proved to be quite beneficial in the gas storage fields. The Oriskany formation encountered at 6,400 ft in north central Pennsylvania is a good example. Results of these fracturing tests are shown on Fig. 10.23.

The hydraulic fracturing treatment shown on Fig. 10.23 illustrates a fourfold increase in well deliverability and injectivity after fracturing. A typical injection rate after fracturing was 2.12 MMcf/D at 2,853 psi surface pressure against a formation pressure of 2,588 psi. A typical deliverability test after fracturing and after the first injection cycle was 8.0 MMcf/D with a formation pressure of 3,040 psi and a bottom-hole flowing pressure of 1,194 psi. Twenty wells in this field were treated in a similar manner with approximately the same results.

The Vicksburg formation (10,000 ft) encountered in the Lower Rio Grande Valley, Tex., is composed of a fine-grained sand containing clay and a calcareous filler. Depending upon location, there is 100 to 300 ft of the heterogeneous zone. Average permeability is about 0.25 md, with porosity averaging 16.5 to 18.5 percent. The formation contains large quantities of gas. Wells completed in this formation were fractured with 100,000 to 150,000 gal of gelled water, transporting 1 to 2 lb of $-20+40$ mesh U. S. Standard Sieve Series round sand proppant per gallon of injected fluid.

An example of the increased deliverability effected by this type of hydraulic fracturing treatment is a well that produced 4.35 MMcf/D at 4,130 psi with a calculated absolute open flow potential of 6.4 MMcf/D. After a large volume treatment, it produced 5.056 MMcf/D at 8,442 psi for a calculated absolute open

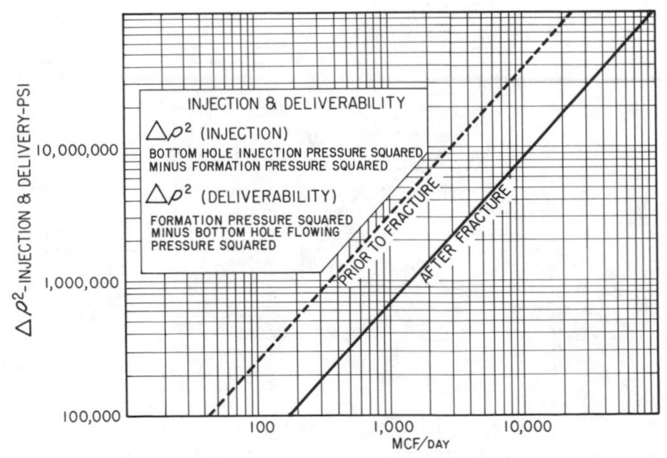

Fig. 10.23 Performance of Well A, gas storage field, North Central Pennsylvania.[16]

flow of 89.0 MMcf/D. Gas deliverability after 14 months of production was 7.0 MMcf/D at 4,700 psi. For another well subjected to a medium-sized treatment, deliverability increased from 0.7 MMcf/D at 2,700 psi to 4.4 MMcf/D at 6,300 psi.

In their study of the effect of proppant spacers on well productivity, Fast, Nabors and Mase not only demonstrated the improvement in gas well productivity to be expected from fractures, but also provided an excellent example of the effectiveness of an increased fracture flow capacity on a well's response to hydraulic fracturing.[17]

The results of these spaced-proppant treatments are shown in Fig. 10.24. The average initial potential for the 15 wells fractured with the resin proppant-spacer tail-in technique was 5,358 Mcf/D, which is 19 percent higher than that for conventionally treated offset wells. Eleven of the 15 wells had higher potentials than did their conventionally hydraulically fractured offsets. It is interesting to note that three of the four spacer wells (E, H, I, and J) did not respond so well as their offsets and produced abnormally high quantities of condensate, which restricted the inflow of gas.

Evaluation of the Hydraulic Fracturing Process

To determine the results of hydraulic fracturing, a study of 1,250 fractured wells was conducted.[20] These wells produced from 240 different formations and were spread geographically over 192 counties in 10 states.

The production data from these 1,250 wells before and after hydraulic fracturing are plotted on Figs. 10.25 and 10.26. Fig. 10.25 is a plot of the percent

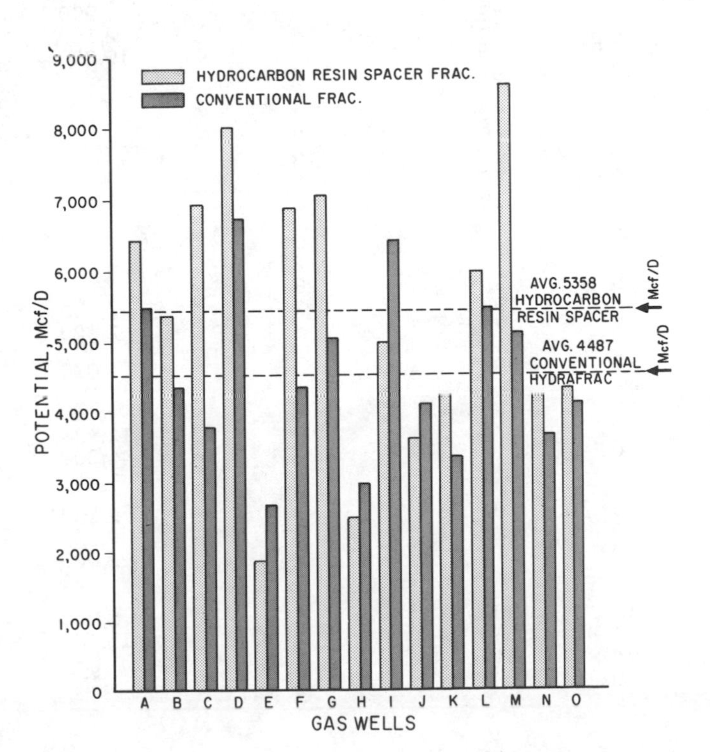

Fig. 10.24 Field test results, gas condensate hydraulic fracturing, Basin Dakota field, N. M.[16]

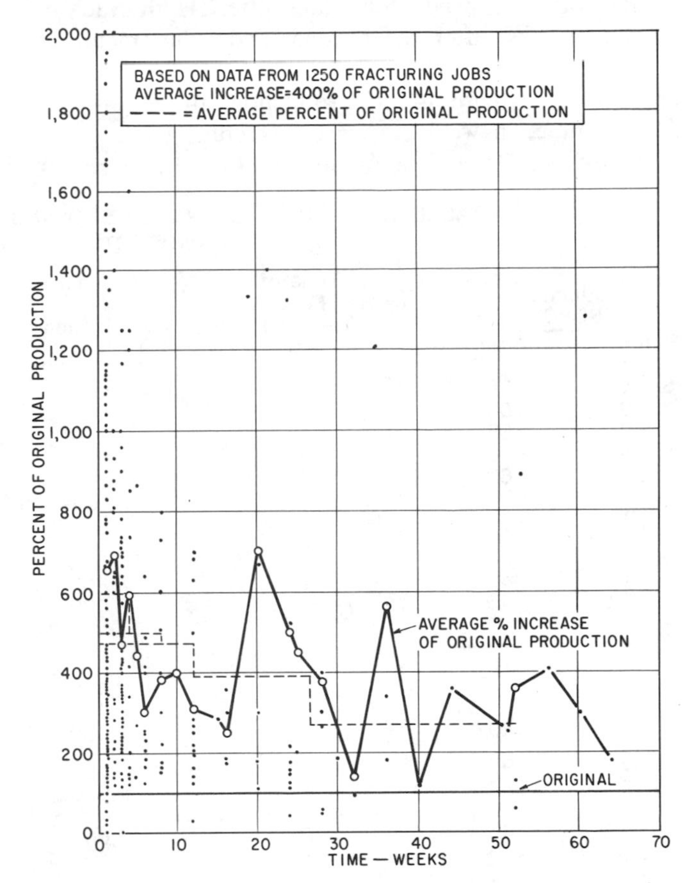

Fig. 10.25 Effect of hydraulic fracturing on well produc-

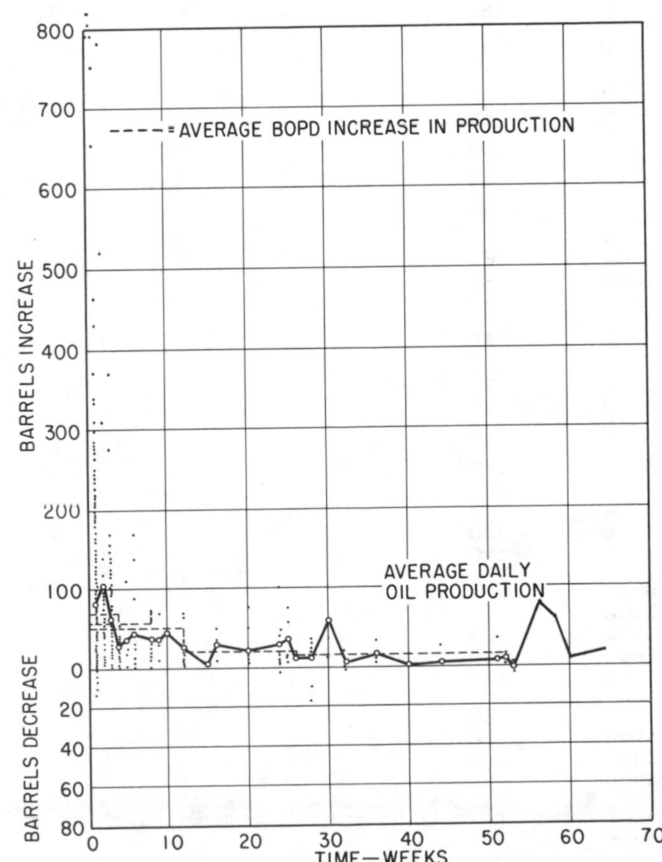

Fig. 10.26 Effect of hydraulic fracturing on well producing rate.

of the original production rate after Hydrafrac vs time, in weeks, after the original hydraulic fracturing operation. Fig. 10.26 is a plot of the same data expressed as increase in oil production, in barrels per day, vs time, in weeks, after the fracturing job.

A review of the data plotted on Fig. 10.25 shows that 126 of the 1,250 wells, or approximately 10 percent of the wells treated, failed to show an increase in well production. The success figure of 90 percent is based on a higher oil production rate after fracturing than before, and does not take into consideration whether or not the job was economically successful.

TABLE 10.5 — DATA ON WELLS SHOWING GREATER THAN 2,000 PERCENT INCREASE IN PRODUCTION FOLLOWING HYDRAFRAC

Production (BOPD)		Time of Post-Treatment Test*	Increase in Production (percent)	Production (BOPD)		Time of Post-Treatment Test*	Increase in Production (percent)
Before	After			Before	After		
Show	10	26		37	905	1	2,446
3	150	1	5,000	Show	24	-	
0	15	1		Show	33	-	
Show	75	-		0	174	-	
Show	60	4		2	75	2	3,750
Show	50	-		Show	10	-	
Show	16	-		Show	4	-	
Show	8	-		Show	45	2	
1	48	1	4,800	Show	15	20	
1.5	45	2	3,000	30	840	-	2,800
0	35	-		2	86	2	4,300
Show	15	-		2	70	9	3,500
7	168	-	2,400	2	80	3	4,000
0.3	7.5	-	2,500	2	71	12	3,550
3	120	-	4,000	1	65	3	6,500
Show	192	-		1	50	12	5,000
3	87	1	2,900	1	121	1	12,100
0	0.1	4		1	66	4	6,600
0.5	40	2	8,000	0	72	1	
Show	9	-		0	104	30	
2.5	55	4	2,200	0	61	58	
6	288	-	4,800	2	90	-	4,500
6	144	1	2,400	25	960	2	3,840
Show	50	2		1	90	1	9,000
2	49	-	2,450	Show	20	8	
0	108	-		5	125	1	2,500
0	76	-		5	145	-	2,900
0	210	-		3	65	3	2,167
0	50	8		0.5	96	-	19,200
0	35	8		0	97	3	
Show	30	-		0	50	-	
0	8	20		0	16	3	
Show	60	2		0	18	12	
Show	35	28		0	16	52	
Show	8	20		0	12	2	
0	15	-		0	12	30	
Show	12	-		0	15	2	
2	75	1	3,750	0	15	28	
0	8	-		7	196	1	2,800
0	40	-		7	177	6	2,529
Show	150	1		10	307	-	3,070
Show	50	4		Show	209	-	
Show	12	28		0	15	2	
6	130	1	2,167	3	373	-	12,433
1	120	2	12,000	0.5	210	-	45,000
Show	15	-		0	72	8	
Show	75	-		5	108	3	2,160
Show	185	-		6	128	3	2,133
Show	50	20		0.5	110	-	22,000
5	144	6	2,880	Show	110	-	
0	35	16		10	800	1	8,000
Show	58	8		4	656	1	16,400
Show	8	2		Show	312	1	
Show	10	9		0	76	-	
2.5	52	-	2,080	0	75	20	
Show	278	-		0	45	-	
1	212	-	21,200	3	120	-	4,000
1	75	25	7,500				

*Weeks after hydraulic fracturing treatment

A check of Fig. 10.25 reveals that the over-all average increase in oil production was 400 percent of the original production. This figure is the numerical average of the percent of original oil production of 1,136 hydraulically fractured wells. Data were not included from 114 wells in which the final post-treatment production was greater than 2,000 percent of the original production. The production data and percentage of original oil production obtained from these wells are presented in Table 10.5. This information was not plotted on Fig. 10.25 because these large percentage increases tend to mask data from other wells.

Continued study of the data on post-fracturing oil production showed that the time at which the post-fracturing production tests were run had a marked effect on the percent of production increase observed. It was noted, for example, that approximately 1,100 wells had data for only the first 3 months following treatment. The gain, in terms of percent of original production, effected by fracturing must therefore be stated for a definite interval of time following the fracture treatment. Because such information is needed, the percentage increase in production resulting from Hydrafrac for finite periods of time is given in Table 10.6.

A plot of the increase or decrease in production rate effected by hydraulically fracturing the 1,250 wells appears in Fig. 10.26. An average of all production figures for the 64-week period in which the post-fracturing oil production data were accumulated reflects an increase of 32 BOPD.

Since most of the data were obtained in the first 12 weeks, a more accurate evaluation of the Hydrafrac process can be obtained by calculating the average increase in production for varying periods of time following Hydrafrac. (See Table 10.7.)

In applying these data to predict future production increases, one should carefully consider whether the proposed fracturing treatment is in a new or an old area. The former would be expected to show much larger increases than indicated by Table 10.7. The response of old wells may be less.

Table 10.8 gives "typical" results from the hydraulic fracturing of wells completed in 39 producing formations. The production data selected at random from well files show that fracturing is effective in most producing geological formations. This tabulation also shows that hydraulic fracturing has been employed successfully in every major oil producing province in the continental U. S. Historically, Hydrafrac is most effective in the medium depth wells (2,500 to 7,500 ft) producing from relatively strong formations (embedment pressures ranging from 75,000 to 100,000 psi). Fracturing is least effective in the incompetent, poorly consolidated formations (where embedment pressure is less than 50,000 psi) characteristic of the regions surrounding the Gulf of Mexico.

Failure of a well to respond to fracturing (presuming the oil is in place) is usually attributed to the inability to maintain the hydraulically created fracture in an open or high flow capacity state.

In most of the wells that have been fractured, the proppant has been flint shot Ottawa sand. A normal maximum usable depth for the sand proppant in single layers is 7,500 ft, which accounts for the depth range mentioned (2,500 to 7,500 ft). Fracturing operations have been successfully performed in wells deeper than 20,000 ft. However, to maintain maximum fracture flow capacity, special high strength proppants are usually required. Failure to recognize the basic fact that proppants must possess the strength required to hold a fracture open causes most of the difficulties encountered in deep well fracturing.

10.4 Effect on Secondary Recovery

Vertically Oriented Fractures

Crawford, Pinson, Simmons and Landrum, using potentiometric models, studied the effect of vertical fractures on the sweep efficiency of the five-spot well pattern.[21] They assumed that all fractures originated at the well and extended out into the reservoir for various distances and orientations. The sweep efficiencies were found to vary from near zero to 73 percent, depending upon the length and orientation of the fracture. They observed that if only the center well in the five-spot pattern were fractured, the sweep efficiency would never exceed the unfractured sweep efficiency of 72.3 percent.

For example, if the length of a vertical fracture at the center well was 20 percent of the well spacing, the sweep efficiency was between 50 and 71 percent, depending upon the orientation of the fracture. For a

TABLE 10.6 — EFFECT OF TIME ON PERCENT GAIN IN PRODUCTION FOLLOWING HYDRAFRAC

Time Interval	Percent Gain in Production (original production = 100 percent)	
0 to 1 month	500	87 percent of the data available are on production observed in first 3 months
0 to 2 months	400	
0 to 3 months	375	
3 to 6 months	295	
6 to 12 months	185	

TABLE 10.7 — EFFECT OF TIME ON BOPD INCREASE IN PRODUCTION FOLLOWING HYDRAFRAC

Time Interval	Increase in Production (BOPD)
0 to 1 month	72
0 to 2 months	59
0 to 3 months	52
3 to 6 months	23
6 to 12 months	17

fracture length of 50 percent of the well spacing, a sweep efficiency from 18 to 62 percent was observed. If the center well and a corner well of the five-spot pattern were fractured, it was possible to increase the sweep efficiency by creating a third fracture at a corner well. The fracturing of a diagonally opposite well might or might not be more effective for this purpose than fracturing an adjacent corner well, depending upon the particular existing fractures.

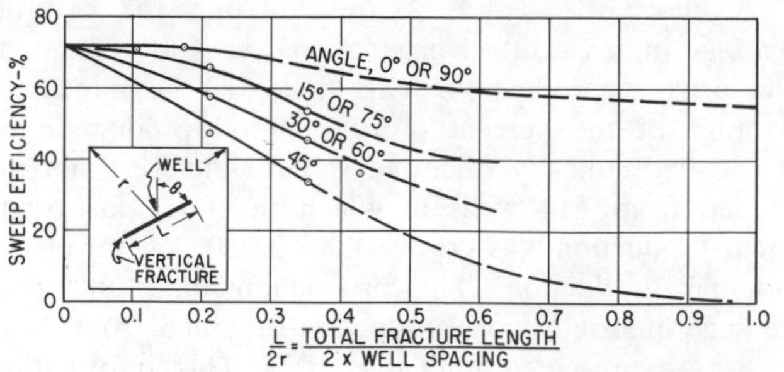

Fig. 10.27 Effect of vertical fracture length and orientation on five-spot sweep efficiencies.[21]

Fig. 10.27 shows the effect vertical fracture length and orientation have on sweep efficiencies of the five-spot pattern with unit mobility ratio when only the center well is fractured. The abscissa shows the total fracture length divided by two times the well spacing; the ordinate shows the sweep efficiency. If the pattern is unfractured, a sweep efficiency of 72.3 percent would be expected. If a vertical fracture exists at the center well and is oriented at either zero or 90° from the *y*-axis (see insert), then a sweep efficiency of about 72 percent would be expected for half fracture lengths up to 25 percent of the well spacing. If the fracture length were increased indefinitely, the sweep efficiency

TABLE 10.8 — TYPICAL RESULTS OF HYDRAULIC FRACTURING

State	Formation	Production (BOPD or Mcf/D) Before	After
Alaska	Kenai sand	1128	1584
California	Stephens sand	20	120
	Vaqueros sand	10	70
Canada	Beaverhill Lake limestone	0	75
	Cardium sand	50	204
	Villeneuve Basal quartz	0	4 MMcf/D
Colorado	Weber sand	20	110
Illinois	Cypress sand	4	35
	McClosky limestone	4	50
Kansas	Arbuckle limestone	10	40
	Herrington Krider limestone	670 Mcf/D	2050 Mcf/D
	Kansas City limestone	6	30
Louisiana	Annona chalk	5	20
	Hackberry sand	30	115
	Marginulina Texana sand	15	104
Michigan	Richfield limestone	22	150
New Mexico	Grayburg limestone	15	110
	San Andres limestone	20	60
	Mesa Verde	490 Mcf/D	5000 Mcf/D
North Dakota	Madison limestone	111	295
Ohio	Berea sand	0	60
Oklahoma	Bartlesville sand	25	275
	Basal sand	80	350
	Granite wash	120	550
	Mississippi limestone	Show	168
	Pennsylvanian	130	475
Pennsylvania	Bradford sand	6	55
Texas	Brown dolomite	6	65
	Camerina sand	26	230
	Canyon Reef limestone	50	130
	Reklaw sand	72	113
	Strawn sand	10	35
	Wilcox sand	15 Mcf/D	1100 Mcf/D
West Virginia	Benson sand	3.5 MMcf/D	66.1 MMcf/D
Wyoming	Dakota sand	25	125
	Frontier sand	16	70
	Madison limestone	473	715
	Phosphoria limestone	1050 Mcf/D	3050 Mcf/D
	Tensleep sand	20	65

would approach 56 percent. It should be noted that at an orientation of 45° and a half fracture length equal to well spacing, a sweep efficiency of near zero is to be expected. Similar effects may be noted for other orientations.

Crawford *et al.* observed also that the creation of short (less than 20 percent of the well spacing), vertically oriented fractures in the corner wells of a five-spot well pattern would result in increased sweep efficiencies.[21] For a line-drive pattern the situation is considerably different. Experimentation based on flow and potentiometric models in which an oriented vertical fracture connects two or more wells demonstrated displacement efficiencies approaching 100 percent.[22] This is shown in Fig. 10.28, which deals with a vertical fracture perpendicular to the direction of the flood (Curve A). Note that as the length of the fracture is increased, the areal sweep efficiency is increased up to fracture lengths of about 80 percent. At fracture lengths near 100 percent, the sweep efficiency is slightly less than 90 percent. The marked difference in sweep efficiency caused by fracture orientation may be noted by comparing Curve A, which shows fracture perpendicular to direction of flow, with Curve B, which shows fracture parallel to direction of flow.

In a similar study, Dyes *et al.* showed that sweep efficiency is influenced by the direction in which a vertical fracture extends into a formation, and by the length and capacity of the fracture.[3]

For a favorite direction (i.e., extending into the area between wells) a fracture length up to three-fourths of the distance between the wells has little effect on the sweepout behavior. Throughput volume has a greater influence on displacement efficiency than does fracture orientation. This is shown in Fig. 10.29, which indicates that fracturing can be used to stimulate injection rate without damage to the flood.

In the event the fracture is unfavorably oriented (toward an offset producing well), the breakthrough recovery is markedly reduced. The recovery after breakthrough, however, is increased so that the total recovery approaches that of an unfractured pattern (see Fig. 10.30). A fracture as long as half the distance to the offset well does not harm ultimate recovery. However, the injection volume must be increased. For example, the volume injected for a fracture length of one-half is 20 percent greater than that for the unfractured pattern. The throughput rate for the operation using fractures, however, would be about twice that for the unfractured case; thus, more oil recovery will be obtained at any given time.

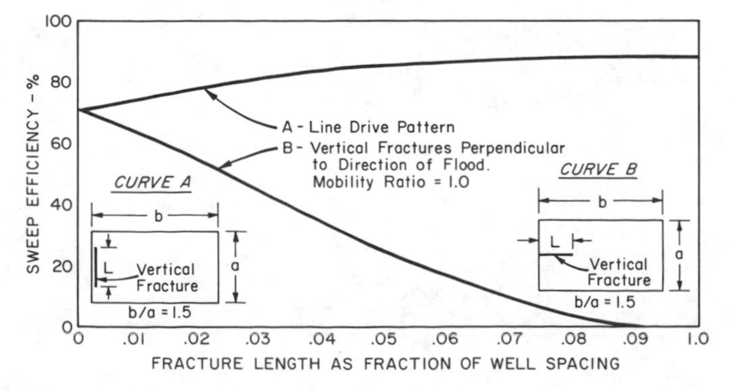

Fig. 10.28 Areal sweep efficiencies for vertically fractured systems.[21,22]

It is concluded that when one or more wells of a five-spot well pattern are fractured, a study of that pattern — taking into consideration orientation and areal extent of natural and created fractures — is required if maximum oil is to be recovered.

Horizontally Oriented Fractures

Landrum and Crawford observed in their potentiometric model studies that short-radius, horizontal fractures oriented symmetrically on the axis of an anticlinal structure greatly increase well productivity without affecting the sweep efficiency of a waterflood or gasflood.[23]

Other investigators have shown that a horizontal

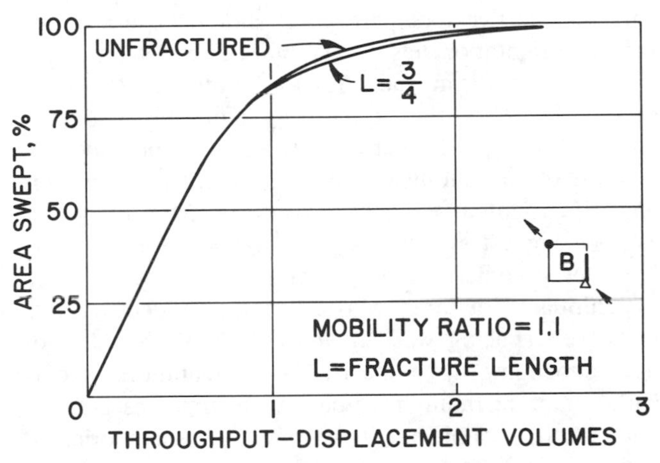

Fig. 10.29 Sweepout with fracture of favorable direction.[3]

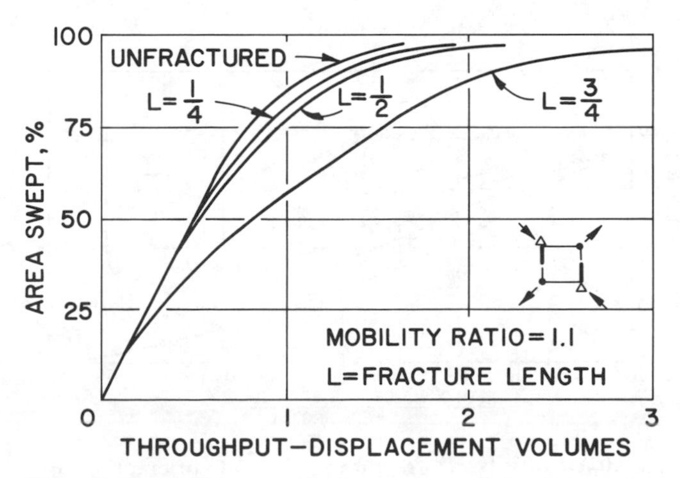

Fig. 10.30 Sweepout with fracture of unfavorable direction.[3]

fracture located in the center of the vertical section of a productive formation, particularly in a thick reservoir, will have a major effect on pattern conductivity. Horizontal fractures improve conductivities more effectively in thin reservoirs than in thick reservoirs.

Similar studies were directed at determining the sweep efficiencies of a reservoir fluid injection program in which the center well of a five-spot well pattern was fractured with an elliptically-shaped horizontally-oriented fracture.[24] The sweep efficiencies varied widely, depending upon size and orientation of the fracture system. It was observed that the orientation of an elliptical fracture with regard to surrounding wells has a greater effect on the sweep efficiency than does the areal extent of the fracture. The same size fracture could result in a sweep efficiency of 35 to 60 percent, depending upon its orientation. The most unfavorable condition was found to be that in which the major and minor axes of the ellipse coincided with lines connecting injection and producing wells. Pinson *et al.*[24] stated that maximum sweep efficiency was noted when the major axis of the fracture rotated 45° from a line connecting injection and producing wells. Fig. 10.31 shows the effect of fracture size and orientation on sweep efficiencies in elliptically fractured thin reservoirs for the five-spot pattern. It should be noted that fracture orientation has a very marked effect on sweep efficiency. Landrum and Crawford noted that if $L/2r$ is equal to 0.25, then the sweep efficiency will vary from about 48 to 65 percent, depending upon the orientation of the elliptical fracture. Fractures with their long axis oriented to either zero or 90° from the y-axis give the highest sweep efficiency. Notice, for example, that a sweep efficiency of 60 percent could be obtained for elliptical fractures whose length divided by two times well spacing was equal to 0.15, 0.25, 0.27, or 0.40, depending upon the fracture orientation. A particular fracture radius may be several times as great as other fracture radii, but if the major axes are oriented at zero or 90° with the y-axis, the sweep efficiency may be greater than that which would occur for a poorly oriented, small fracture. The data show that for fractures oriented at zero or 90°, a sweep efficiency as high as 50 percent may be expected, even for tremendously long fractures.

The sweep efficiency or area swept at water or gas breakthrough is not the sole criterion for rating the performance of pattern-type displacements. The area swept at various WOR's must also be considered. Some patterns give a rapid increase in WOR after breakthrough, others have long periods of production at low WOR's. For this reason the over-all performance should be considered.

10.5 Gravity Drainage Into Horizontal Fractures

As a means of withdrawing oil by gravity drainage, Morrisson and Henderson studied the feasibility of using a very high capacity horizontal fracture with large areal extent, placed at the base of a producing formation that had an acceptable vertical permeability.[25] Such a configuration is shown on Fig. 10.32. It should be noted that energy in the reservoir gas is not a necessary condition for this recovery process. With a sufficiently large fracture, attractive production rates will result from gravity flow alone, and the producing GOR need only be the solution GOR. Concurrent production of free gas may actually hinder oil production by reducing the relative permeability of the oil and increasing its viscosity. Since gas preferably should not be produced, the method is applicable both to reservoirs that contain oil having large quantities of associated gas and to those that contain essentially "dead oil". Fluid level in the well should be kept below the fracture depth to insure the maximum gravity flow rate, and if gas is not to be produced, the casing pressure must be at or above the bubble-point pressure of the oil.

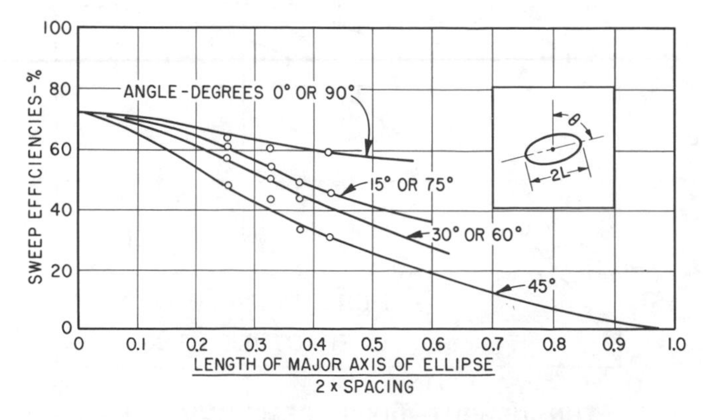

Fig. 10.31 Effect of fracture size and orientation on sweep efficiencies of ellipetically fractured thin reservoirs. (five-spot pattern[24])

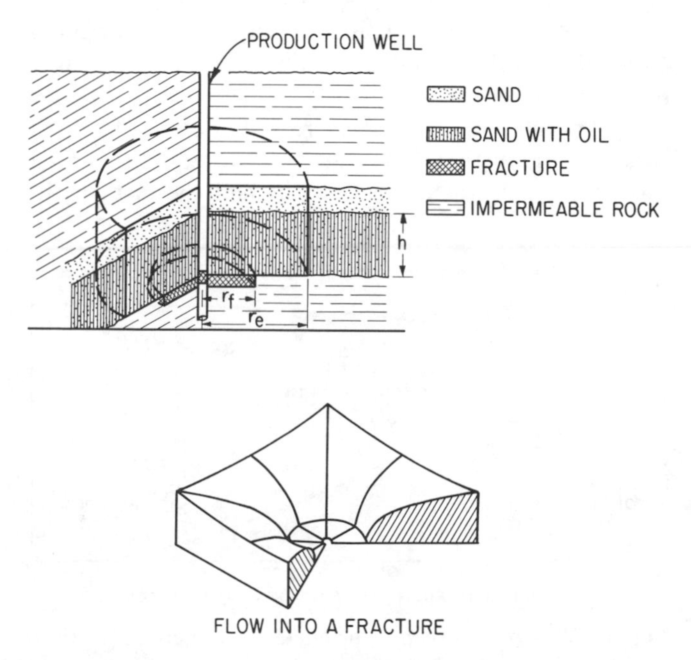

Fig. 10.32 Fracture gravity drainage recovery method.[25]

A large horizontal fracture system is defined as one whose radius is approximately 1,000 times greater than the radius of a conventional wellbore. The term "large" also implies that an appreciable fraction of the total oil contained in the drainage area of the well will lie immediately above the fracture; e.g., 15 percent of the oil in a 20-acre spacing will lie above a 200-ft-radius fracture. There is no available analytical solution to the free boundary problem posed by gravity drainage into a horizontal fracture. However, if a vertical cylindrical surface of radius r_f and height h is visualized, the production into the fracture can be divided into the production of oil from within that boundary (early production) and the production of oil that must flow across the cylindrical surface to reach the fracture (late production). If the production from one zone does not interfere with the production from the other, each could be defined with considerable confidence. The vertical flow accompanying production from the inner cylinder (radius r_f) can be defined by Darcy's law, provided proper relative permeability and capillary pressure corrections are made. The essentially horizontal flow into the inner cylinder from the region outside r_f may be defined by approximate gravity drainage solutions, which bracket the true solution.

If compensation is made for the mutual interference, two segments of a production-rate vs recovery curve are obtained that agree rather well with the production history observed in scaled model experiments. It is this agreement with the model results that led Morrisson and Henderson to justify using the model method to predict reservoir performance.[25] From their study it may be concluded that production of oil by force of gravity into a high flow capacity, long radius (hundreds of feet) horizontal fracture is a recovery method characterized by high efficiency and acceptable production rates.

Application of Fracturing to Waterflooding

Fracturing in a waterflood project offers several advantages. The injection rate may be increased at a near-constant injection pressure, or if necessary, injection pressure may be reduced while a constant injection rate is maintained. With increased water input, the floodout time for a particular reservoir would be reduced. Fracturing of production wells permits increased oil or gas recovery rates by reducing wellbore damage and increasing the effective permeability of the drainage area. A fracture may also interconnect a number of small, isolated, oil-productive lenses that were not previously in communication with the well.[26]

The possibility exists, however, that a poorly planned or poorly controlled fracture will create or aggravate channeling of the injected fluid, which will lead to an early breakthrough of the injected fluid, by-passing of oil, and lowering of sweep efficiency. In addition, a vertical fracture that extends through the cap rock would permit gas and oil to escape into the overlying, permeable barren zones. The orientation of a vertical fracture can be predicted if it is realized that the fracture will form in a direction normal to the minimum horizontal principal stress. That stress is the lesser of the two mutually perpendicular principal stresses in the horizontal plane. For instance, in an area where tectonic relaxation is caused by down-warping, as in a basin, the fractures would lie in a plane perpendicular to the regional dip. Where regional stresses are fairly uniform in orientation, all fractures should be approximately parallel. Therefore, in geologically simple and tectonically relaxed areas, fractures should be vertical and parallel.

Before planning a fracturing operation, it is necessary to be aware of wellbore blockage (any reduction in formation permeability adjacent to the wellbore that would restrict the production of oil or the injection of a displacing fluid). If damaged matrix permeability exists in the vicinity of the wellbore, a very short fracture could permit a substantial increase in productivity.

Hartsock and Slobod, in their investigation of the effect of mobility ratio on waterflood sweep efficiency, concluded that when the mobility ratio is favorable (less than 1.0), a greater sweep efficiency is attained if the fracture is located in the injection well, regardless of the fracture extent.[27] If the mobility ratio is unfavorable (greater than 1.0), the effect of fracture location is less discernible. It has been observed, however, that when the mobility ratio is greater than 1.0, and when the fracture length is less than 75 percent of the distance between wells, fractures located in the production well will effect more favorable sweep efficiencies than will fractures in the injection well. When the mobility ratio is 1.0, the effect of fracture location is negligible.

Field Results of Fracturing in Waterflood Projects

Guerrero has studied the effects of fracturing in a large number of waterflood projects in most of the oil producing regions in the U. S.[28] These projects covered a wide range of conditions of formation depth and thickness and well spacing. The over-all success-to-failure ratio was favorable, with a successful fracture treatment being defined as one in which the payout time was reasonable and some profit was realized. About 80 percent of the production well fracture treatments were considered successful, as were about 95 percent of the injection well treatments.

The principal reasons for failure in fracturing jobs include (1) fracturing into overlying free gas or underlying water, which results in high producing GOR's or WOR's; (2) incorrect location of the fracture because of a poor cement job; (3) incorrect placement of perforations; and (4) leaking packers.

Powell and Johnston discuss in detail a number of field case histories in which hydraulic fracturing treatments were successful in improving waterflooding operations.[29] To illustrate fracturing in waterflood operations, they cited the results of two treatments performed at different stages of depletion of the reservoir in two adjacent waterflood operations: Flood A, with 4,616 acre-ft and Flood B, with 3,040 acre-ft of floodable sand. These tracts were in the same geographical area, and produced from sandstones having similar characteristics.

Fig. 10.33 shows the production history of Flood A, where wells were fractured 18 months after a peak rate of oil production of 17,200 bbl/month was achieved. After a fracturing treatment, production declines at a rate approximately parallel to the estimated pretreatment rate of decline, indicating the recovery of additional oil before the flood production reached an economic limit. This increase in ultimate recovery over normal waterflood production was approximately 179,000 bbl.

Fig. 10.34 shows the production history of Flood B, the wells of which were fractured 36 months after a peak rate of oil production of 30,300 bbl/month. Again, the decline after fracturing is essentially parallel to the estimated normal decline, indicating an increase of approximately 55,000 bbl in the ultimate recovery. On the assumption that the wells of Flood B had been fractured 18 months after peak production, a hypothetical curve was drawn (Fig. 10.34) to indicate the additional oil that might have been recovered by fracturing earlier in the life of the flood. The position of the hypothetical curve was based upon the production of Flood A (Fig. 10.33) multiplied by the ratio of the productive sand volumes of the two floods.

From the information in Fig. 10.34, it is estimated that an ultimate 118,000 bbl, or an additional 63,000 bbl, of oil would have been recovered from Flood B if the wells had been fractured 18 months earlier. The earlier and additional oil recovery would have hastened the payout for the project and could have been an important economic factor.

The data presented in Figs. 10.33 and 10.34 show that fracturing improves waterflood injection-production performance.

10.6 Effect of Fracturing on the Oil Industry and on World Reserves

Since its commercial introduction to the oil industry in March, 1948, hydraulic fracturing has affected every facet of the oil industry. It has altered pipeline construction, changed production practice, killed the nitroglycerin oilwell shooting business, upset the drilling business, and revolutionized the service companies.

In the few years that hydraulic fracturing has been used, geologists and engineers have had to re-evaluate many pay sections that previously had been bypassed. Oil operators are now entering areas formerly considered nonproductive and are making productive oil or gas wells out of those that had pay formations with too little permeability to support commercial production. Areas that were formerly considered dead, having produced all the recoverable oil, are being rejuvenated and, in many instances, are now producing more oil than they did at the time of original completion.

Service companies, accustomed to having their pumping equipment last a long time, have been forced to redesign and rebuild pumps, blenders, tanks, storage bins, etc., to meet the tremendous demands of the oil industry for more and better equipment. It is not uncommon on a single fracture operation to use 10 to 20 trucks, each equipped with two pumps and requiring as much as 10,000 to 20,000 total hhp. New companies were organized to supply the special products required for the fracturing industry. Fracturing soon became the major service offered in the oil field — exceeding cementing in dollar value. The "fracture era" is one of continuous change, with hundreds of innovations being tried every year.

Oil operators have come to accept fracturing as a way of life. Most wells are fractured on initial completion while the drilling rig is still in place so that the productive interval may be properly evaluated and the well produced at its maximum capacity. More than 500,000 fracturing jobs had been completed by 1968. During the peak usage year of 1955, fracturing jobs were being performed at the rate of 4,500 every month. After 17 years of commercial use, the rate leveled off to 2,000 jobs a month. Fig. 10.35 is a plot, by year, showing the history of the use of this process.

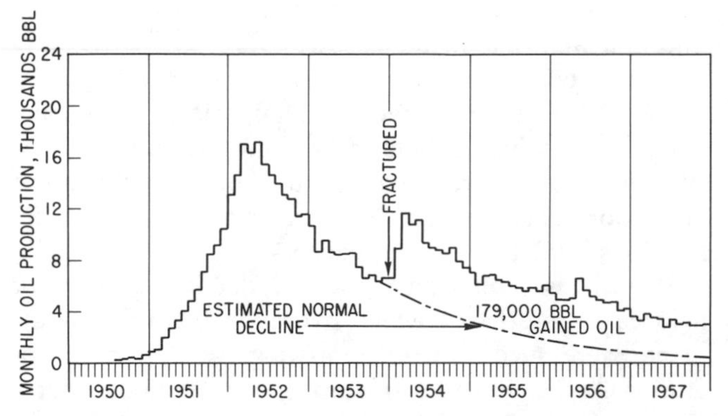

Fig. 10.33 Production history of Flood A, showing results obtained from fracture treatments 18 months after peak rate of production.[29]

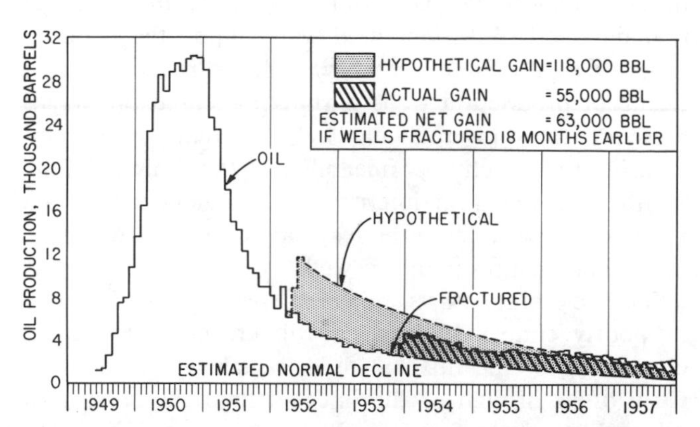

Fig. 10.34 Production history of Flood B, showing results obtained from treatments 36 months after peak rate of production and estimated results from fracturing 18 months earlier.[29]

Reserves

No one knows the total effect fracturing has had on recoverable reserves in the U. S. and Canada. Some fields such as Spraberry Trend of West Texas; Pembina field of Alberta, Canada; Pine Island field in the Louisiana extension; Morrow sand in Oklahoma; the San Juan basin of New Mexico; and the Denver Julesburg basin of Colorado are typical of hundreds of fields and areas that would be classified as *dry* were it not for hydraulic fracturing. A study of well records and hydraulic fracturing reports indicates that the average successfully fractured well increased its recoverable reserves by an estimated 17,750 bbl. This brings to approximately *7 billion bbl,* as of mid-1966, the estimated increase in recoverable reserves due to hydraulic fracturing.

Opening of Low Permeability Areas to Production

Hydraulic fracturing got its first real test in late 1949 in low permeability formations of the Spraberry area of West Texas. In this area the pay sands are so impermeable that oil operators had almost written them off as impractical to produce. Fracture treatments were so successful in increasing oil production that the hydraulic fracturing technique quickly spread to other low-permeability areas.

Western Canadian fields were hydraulically fractured with good results. Today, operators are drilling areas formerly skipped, and, with fracturing as a completion practice, are bringing in good wells.

Revival of Old Producing Areas

Many fields classified as "dead" were re-entered and fractured. These areas have added significantly to the over-all increase in recoverable reserves. The Luling field of Texas and the Red Oak field of central Oklahoma are examples. Illinois has many examples. In one instance, 88 old holes were re-entered and fractured, with many of the wells responding to fracturing by producing 150 to 200 BOPD.

Production in the eastern Kansas waterflood area was substantially increased by fracturing. In one group of fields, oil produced by waterflooding increased 85 percent when fracturing was applied.

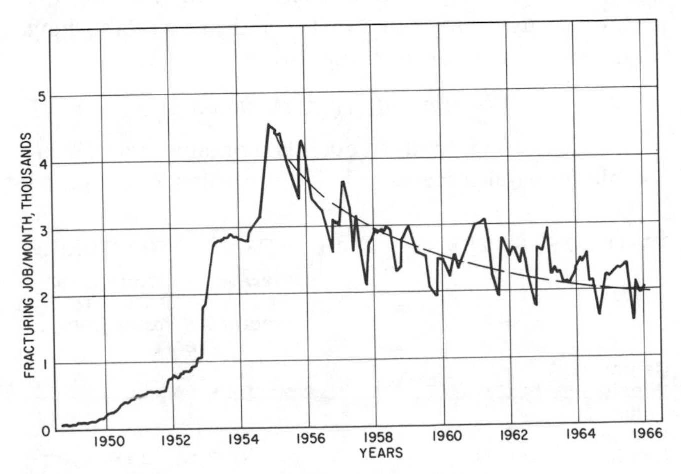

Fig. 10.35 Hydraulic fracturing activity.

Oil Company Reaction

Hundreds of companies have seen their own oil production markedly increased by the hydraulic fracturing technique. Practically without exception, all oil companies have tried fracturing.

One large Texas company states that the criterion for success in fracturing is whether, after fracturing, the company can expect an increase in oil or gas production that will enable the cost of the treatment to be returned within a reasonable time. Another company feels a typical "success" is a well that has its production increased fivefold following a fracture. One firm, which is achieving about 90 percent success on its fracturing jobs, says it will fracture wherever it can, as long as the process can yield a profit. (The main reason for this company's 10 percent failure is misapplication — fracturing where it will do no good, such as in the wrong part of the formation, or applying the method in a last-ditch attempt to salvage a well on which everything else has been tried and has failed.)

10.7 Permanence of Fracture Treatment

In any evaluation of fracture stimulation treatments three factors must be considered: (1) the prefracturing production of the well being treated; (2) the flush production immediately following hydraulic fracturing operations; and (3) the volume of oil to be recovered during the post-fracturing period.

Fig. 10.36 shows three typical responses to fracturing. Well A is a "stripper" that has produced for 10 to 20 years at a nearly constant rate, yielding a very nearly flat production decline curve. Post-fracturing analysis shows a few weeks of flush production, with the decline curve settling down to a somewhat steeper decline, but still remaining flat enough to predict a major extension in well life. It is not uncommon for several years to pass before production declines to the prefracturing level. Frequently, however, the operator is no longer content to accept the low prefracturing production and he refractures before productivity declines to its previous low level.

Well B on Fig. 10.36 shows a well whose production dropped below the economic producing limit prior to fracturing — probably because of severe wellbore plugging or isolation of the well from the more productive sections of the field. In such instances, fracturing either breaks through bore damage or reconnects the well to a more productive portion of the reservoir. The net result of this kind of treatment is that the well produces at relatively high rates for extended periods.

The third example is Well C on Fig. 10.36. Typically, in an area such as West Texas, a fracturing treatment would result in a flush production period of 30 to 60 days, with production declining in 2 to 3 years to its prefracturing level.

A study was made of various areas to determine how long productivity increases from fracturing would be sustained. (See Table 10.9.)

From this study it is concluded that the average stabilized post-fracturing production period in which production will exceed the prefracturing rate will be 3 to 4 years.

10.8 Re-treatment of Previously Fractured Wells

No discussion of hydraulic fracturing should ignore re-treatment of previously fractured wells. A review of well treatment data shows that an estimated 35 percent of fracturing operations are re-treatments. Wells that responded to the first fracturing job historically respond to re-treatment, often with production rates equal to or greater than those observed after the original stimulation. Figs. 10.9, 10.16, 10.22, 10.33, and 10.34 illustrate the benefits of refracturing operations.

Successful re-treatment of a previously fractured producing interval is due to one or more of the following: (1) extension of the existing fracture system, (2) reopening of previously formed fractures, (3) washing of fracture faces, (4) replenishing of embedded proppants, and (5) opening of new fractures in previously unfractured intervals.

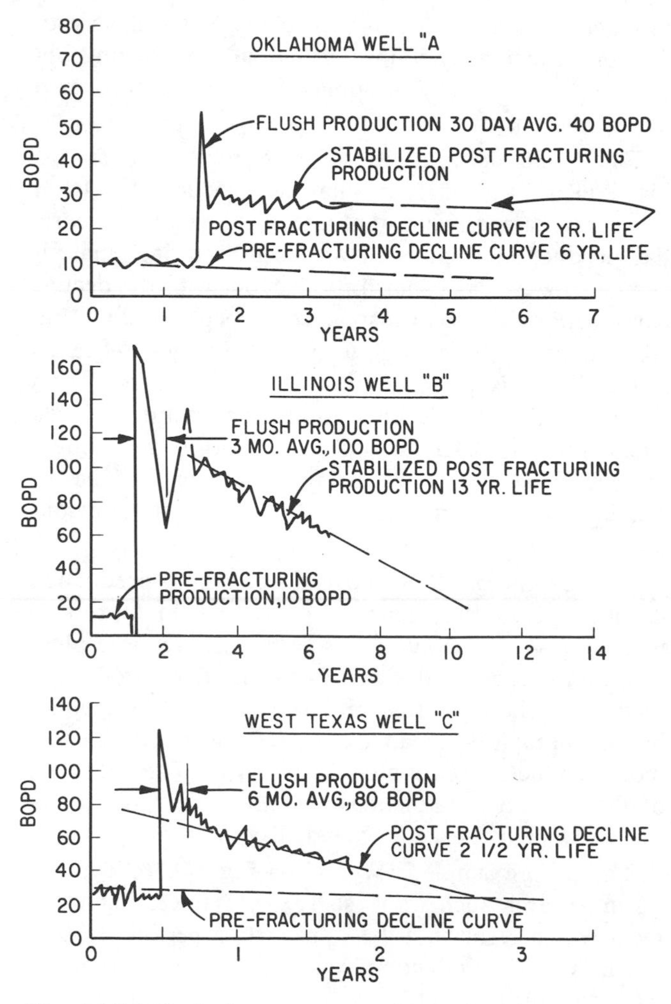

Fig. 10.36 Typical responses to fracturing.

Extension of the Existing Fracture Systems

The early years of fracturing were characterized by relatively low injection rates (4 bbl/min) and small volumes (20 bbl) of fracturing fluid. During the period in which the value of the propping agent was being established, many wells were fractured without proppant.

The wells fell into two broad categories: (1) those in which the initial fracturing treatment was unsuccessful (i.e., no increase in production was observed); and, (2) those that showed a marked increase in production as a result of fracturing but afterwards experienced a severe decrease in production.

The majority of wells requiring workovers would be included in the second category. In this instance, the initial fracturing job was a success. Most of the recoverable oil in the area surrounding the fracture has been produced. Low effective permeability either adjacent to the well or in the interwell area has prevented inflow of oil. Historically, extension of the existing fracture, exposing the fracture faces to larger areas of oil-producing rock, has effected an improvement in well productivity.

The successful refracturing job requires careful selection of fracturing fluid and isolation of the interval to be treated. Volume of fracturing fluid, pump equipment, and weight of proppant should be substantially increased over that used on the first fracturing job. Typically, fracturing fluid volumes will be two to four times larger than the original volumes used, and pump rates may be increased similarly.

Reopening of Previously Formed Fractures

In formations that have been either fractured without proppant or fractured with proppant that completely embeds in the formation, the wells often return to their original prefracturing production level in a few months following fracturing. Reopening of the fractures will normally restore production to a higher level, but unless steps are taken to insure that the fracture is kept open, the gains in production will be short-lived. Frequently, either using techniques that will cause multiple layers of proppant to be deposited in the fractures, or using an acid-base fracturing fluid, will aid both in restoring the fracture to its original open state and in re-establishing oil or gas production at a high level.

Washing of Fracture Faces

The fracturing fluid frequently contains enough insoluble material to cause a buildup of filter cake on the

TABLE 10.9 — PERMANENCE OF HYDRAULIC FRACTURING

Area	Time After Fracturing for Well to Return to Pre-Fracturing Production (years)
California	2, 6, 1
Illinois	3, 5
Indiana	4
Kansas	10, 4
Oklahoma	3, 7, 4, 4, 9
Texas	3, 5, 2, 5
Wyoming	1, 3, 5

exposed faces of the fracture. Generally, backflow at high rates removes most of this material.

In some instances, very high pressures are encountered during fracturing, extension of the fracture, or fracturing of a low pressure reservoir, and there is not enough pressure to wash away the precipitated material. There are two solutions to this problem. First, inject a treating agent that will either dissolve or remove the filter cake. For example, if an oil soluble spacer has been used, a post-fracturing wash of kerosene injected at a high rate will help remove this permeability block. Enzymes may be used to remove or break down the plugging or blocking agents formed by starch or guar-gum-base fracturing fluid. Second, when there is no means of chemically removing the plugging agent, or when the cause of "blocking" is not known, an attempt should be made either to flush or to backwash the face of the fracture. This may be accomplished by rapidly injecting water or oil and then backflowing the wash fluid into the wellbore at as fast a rate as practical. The procedure should be repeated three or four times to insure maximum cleaning of the fracture. If the fracture has been formed in a carbonate reservoir, the use of an acid wash will frequently aid in restoring the well to maximum productivity.

Replenishment of Embedded Proppants

In a formation known to be relatively soft (embedment pressure less than 50,000 psi), embedment of the proppant and resulting loss of fracture flow capacity are common causes of low production in fractured formations.

Re-treatment of soft formations is accomplished best by high-rate injection that "packs" the fracture with a proppant capable of effecting high flow capacities at the depth of treatment. Multiple layers of proppants satisfactorily maintain an open fracture in the "soft" formations.

When packing a fracture, care must be exercised to avoid flushing the proppant from the area immediately around the well, the most critical part of the fracture flow channel.

Opening of New Fractures in Previously Unfractured Intervals

In wells characterized by long producing intervals or by a producing zone with limited or restricted vertical permeability, the operator may direct a fracturing retreatment to sections of the well that have never been fractured. This permits all portions of the producing formation to contribute equally to the well's productivity.

The ability to isolate the interval to be hydraulically fractured insures proper placement of the new fracture and is the main prerequisite for a successful refracturing treatment.

10.9 Damaged Formation Permeability

Van Poollen found in his study of the effect of permeability damage on well productivity that production is reduced more in a radial-flow system than in a linear-flow system.[30] In a fractured formation, this effect will be of intermediate magnitude. Figs. 10.37 and 10.38 illustrate the extent of the damage that may be expected.

Fig. 10.39 shows the effect of formation damage on the productivity of a well with a horizontal fracture, the

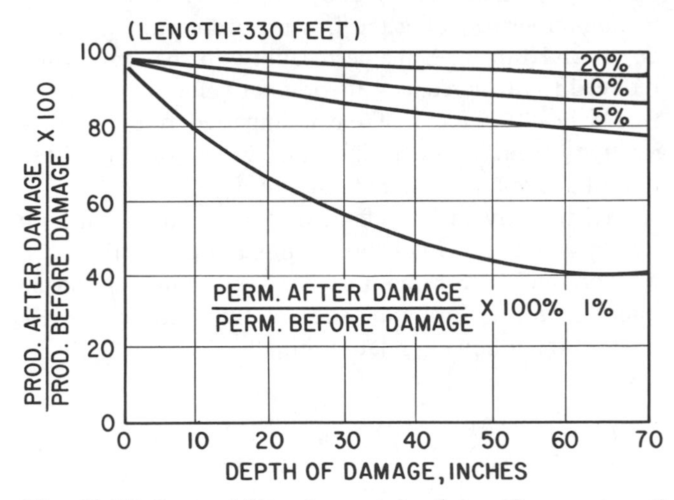

Fig. 10.37 Permeability damage in linear flow system.[30]

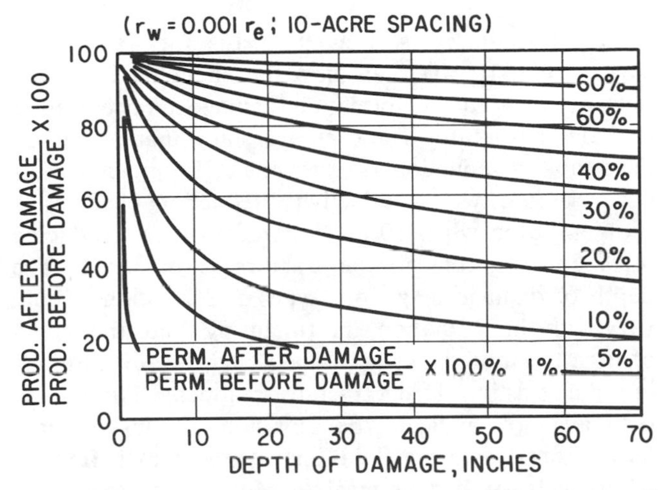

Fig. 10.38 Permeability damage in radial flow system.[30]

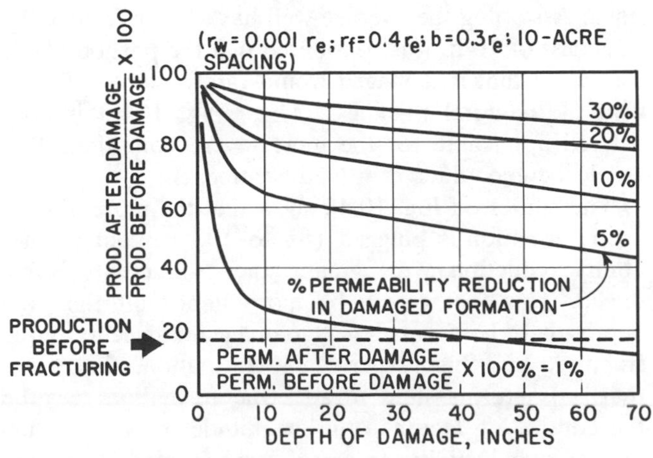

Fig. 10.39 Productivity reduction caused by formation damage around a horizontal fracture.[30]

length of which is 0.4 times the radius of drainage. The bed thickness is 0.3 times the radius of drainage, and the flow capacity of the fracture is assumed to be infinite. The dotted horizontal line indicates the productivity of the well before fracturing and without permeability damage around the wellbore. Under these conditions an undamaged fracture would increase the productivity from 17 to 100 percent. However, if the formation surrounding the fracture were damaged to a depth of 10 in., and the permeability in that zone were reduced to 10 percent of its original value, productivity would be 80 percent of the unimpaired fracture. Productivity from a damaged fracture will normally exceed that of a formation without a fracture.

In his study of the effect of formation damage on well productivity, Flickinger compared production of a well producing from both a "damaged" interval and an interval with no formation permeability impairment. He devised the following relationship:[31]

$$\frac{q_u}{q_d} = 1 + X\left(\frac{k_2}{k_1} - 1\right) \quad . \quad . \quad . \quad . \quad . \quad . \quad . \quad (10.11)$$

where k_2 is the permeability of the undamaged formation and k_1 is the permeability of the damaged formation.

This formula is used as the basis for a series of curves (Fig. 10.40) that show the effect on well productivity caused by formation damage adjacent to a horizontal fracture. These curves show that a marked permeability reduction (greater than 50 percent) must occur before well productivity reduction becomes a problem, even when 10 to 15 percent of the producing formation adjacent to the fracture is damaged. The depth of damage in various types of formations would vary with the formation, the fracturing fluid, the length of time the fluid is being pumped, and the pressure differential existing during fracture treatment. The limited data available indicate that invasion of plugging material from fracturing fluids may vary from a fraction of an inch for low permeability formations to several inches for highly permeable and porous sand formations. Assuming the average well has a net effective pay thickness of 10 ft, it is considered that the portion of the formation that is damaged would rarely exceed 10 percent of the total thickness. Thus, Fig. 10.40 is considered applicable for the most severe conditions that would be encountered in field treatments.

The curves on Fig. 10.40 show that the more a damaged formation is plugged (85 to 100 percent permeability reduction), the greater the productivity is reduced. This factor may be more important than the depth that the formation is damaged adjacent to the fracture. In a highly permeable formation, the extent that a fracturing fluid invades the formation may be the controlling factor in the magnitude of post-fracturing well productivity. In areas where formation damage is suspected, tests conducted on cores from a well to be fractured should evaluate both the amount of permeability reduction and the extent the formation is invaded during a fracturing treatment.

In Table 10.10 van Poollen illustrates the effect on productivity for the different fracture systems.[30] The tabulated depths of damage are 1 in., 2.5 in. and 5 in. into the matrix adjacent to the fracture faces; the permeability in the damaged zone is reduced to 1, 10, and 30 percent of the original permeability. Assuming productivity of an unfractured, clean well to be unity, the respective increases and decreases in productivity after fracturing and/or damaging the formation are shown as multiples or fractions of one.

Fig. 10.41 illustrates by bar graph the effect on productivity of various degrees of damage to a fractured formation. Each set of bars represents a certain depth of damage (1 in., 2.5 in. and 5 in.), where each bar in each set represents a certain degree of permeability damage to the producing formation (30, 10, and 1 percent of original permeability). This illustrates that if fluids that cause permeability damage are used, the fractured formation can be harmed.

In Fig. 10.42 the effect of formation damage on productivity is shown for systems with fractures of varying flow capacities. For example, a well with a clean fracture of infinite flow capacity will have 4.5 times the productivity of an unfractured well, and the influence

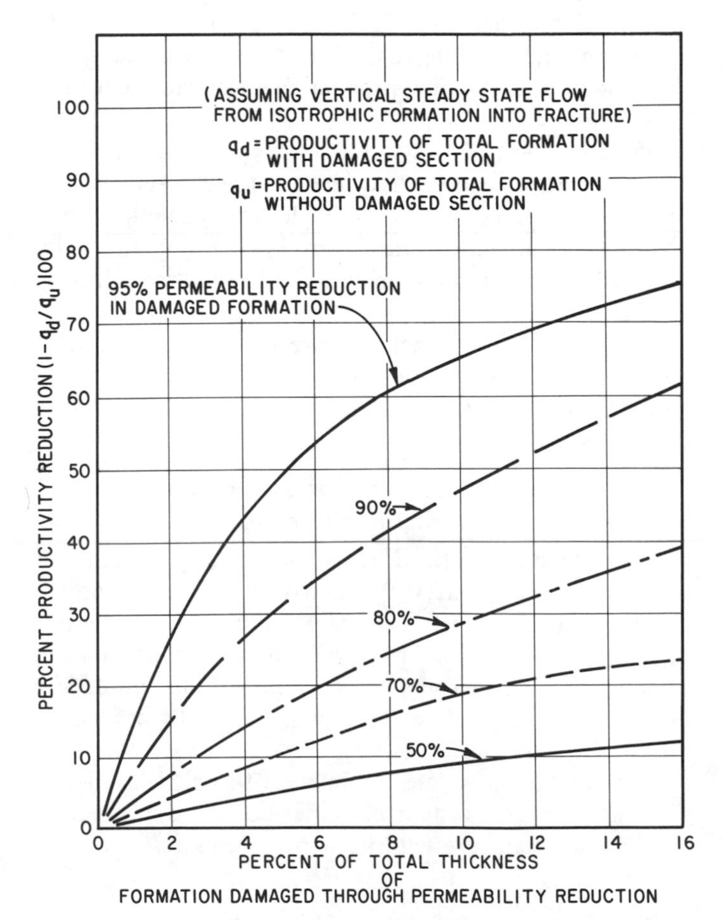

Fig. 10.40 How formation damage adjacent to a horizontal fracture affects well productivity.[31]

TABLE 10.10 — EFFECT OF PERMEABILITY DAMAGE ON WELL PRODUCTIVITY, 10-ACRE SPACING

		Productivity of Clean Well, No Fracture (q_o)	Productivity of Well with Fracture (q_f), No Damage	Percentage of Permeability Left in Damaged Zone								
				1 in.††			2.5 in.††			5 in.††		
				1	10	30	1	10	30	1	10	30
No Fracture		1	—	0.32	0.82	0.96	0.17	0.71	0.91	0.11	0.59	0.85
Linear Flow		1	—	0.96	1.00	1.00	0.93	0.99	1.00	0.87	0.98	1.00
r_f*	f**		Vertical Fracture									
0.1		1	2.4	2.0	2.4	2.4	1.6	2.4	2.4	1.3	2.3	2.4
0.2		1	3.5	3.1	3.4	3.5	2.6	3.4	3.4	2.1	3.3	3.4
0.4		1	4.5	4.1	4.5	4.5	3.5	4.4	4.5	2.9	4.3	4.5
0.6		1	6.1	5.5	6.0	6.1	4.7	5.9	6.1	4.1	5.9	6.1
0.4	4.2	1	1.6	0.6	1.3	1.5	0.3	1.1	1.4	0.2	0.8	1.3
0.4	29.8	1	2.6	1.7	2.4	2.5	0.9	2.2	2.5	0.5	1.9	2.4
0.4	84.0	1	3.3	2.0	3.1	3.3	1.2	2.8	3.2	0.8	2.6	3.1
0.4	1400.0	1	4.3	3.2	4.1	4.3	2.1	3.9	4.2	1.2	3.5	4.0
0.4		1	4.5	4.1	4.5	4.5	3.5	4.4	4.5	2.9	4.3	4.5
r_f	h†		Horizontal Fracture									
0.1	0.2	1	2.2	1.4	2.0	2.1	0.7	1.7	2.0	0.3	1.4	1.9
0.2	0.2	1	3.3	1.9	3.0	3.2	1.1	2.6	3.1	0.5	2.2	2.9
0.4	0.2	1	6.2	5.1	6.0	6.1	3.9	5.8	6.1	2.5	5.5	6.0
0.6	0.2	1	10.6	8.7	10.2	10.5	7.2	9.7	10.5	5.4	9.1	10.4
0.4	0.05	1	6.9	6.3	6.9	6.9	5.7	6.8	6.9	5.0	6.7	6.9
0.4	0.1	1	6.6	5.3	6.5	6.6	4.5	6.3	6.5	3.7	6.1	6.5
0.4	0.2	1	6.2	5.1	6.0	6.2	3.9	5.8	6.1	2.5	5.5	6.0
0.4	0.3	1	5.9	4.5	5.7	5.8	3.4	5.3	5.7	2.2	5.0	5.7

*r_f = extent of fracture expressed as a fraction of the radius of drainage (r_e).
**f = flow capacity of *horizontal* fracture is ratio of conductance of 1 sq unit of fracture to conductance of 1 cc of formation.
flow capacity of *vertical* fracture is ratio of conductance of 1 cm of fracture to conductance of 1 sq unit of formation.
†h = bed thickness expressed as a fraction of the radius of drainage.
†† = depth of penetration.

of formation damage will be minor, even if only 1 percent of the original permeability remains. However, if the fracture flow capacity decreases, the productivity of the well decreases drastically and the effect of damage is severe. It is then possible for the productivity to fall below that of a clean unfractured well.

10.10 Summary

The results of hydraulic fracturing may be summarized as follows:

1. Ninety percent of the wells treated have shown increases in productivity.

2. The ultimate recovery of oil by primary means is increased by the fracturing treatments.

3. The rate of decline of oil production after a successful hydraulic fracturing treatment is generally equal to or less than the rate of decline before treatment.

4. Results obtained from wells that have had two or more treatments indicate that the rate of production is stimulated by subsequent treatments.

5. Fracture stimulation has been effective even in gravity drainage reservoirs.

6. Fracturing of formations to be waterflooded will permit increased injection rates, increased production rates, reduced floodout times, and lower operating costs.

7. Breakthrough sweep efficiency in a line-drive pattern will be lower for fractures in an unfavorable orientation than for unfractured systems, and higher for fractures in a favorable orientation.

8. In a five-spot flood pattern, a fracture in a favorable orientation with a radius equal to 75 percent of the drainage area will have little effect on the sweep efficiency; breakthrough efficiency will be reduced by a fracture in an unfavorable orientation, but ultimate

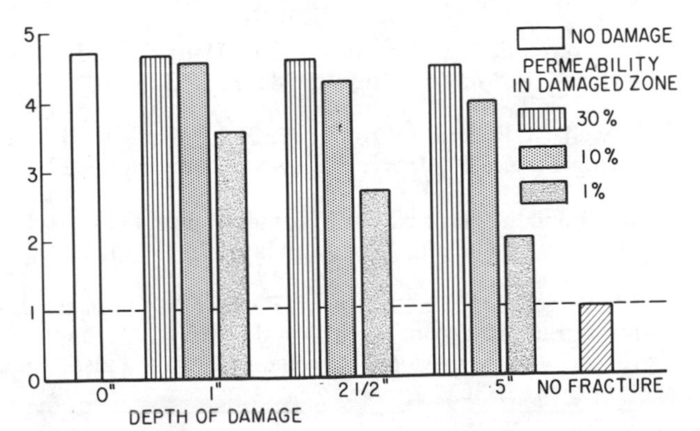

Fig. 10.41 Effect of usual depth and degree of permeability damage on average productivity for all fracture systems.[30]

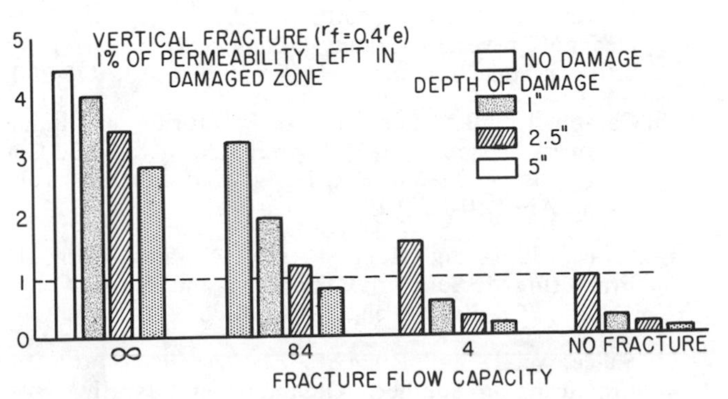

Fig. 10.42 Drastic reduction of productivity caused by decreasing flow capacity of fracture.[30]

recovery will not be significantly reduced.

9. The effect of formation damage on well productivity is more pronounced in radial systems than in linear systems.

10. Low fracture flow capacities are more detrimental to post-fracturing well productivity than is damage to formation permeability.

References

1. Clark, J. B.: "A Hydraulic Process for Increasing the Productivity of Wells", *Trans.,* AIME (1949) **186,** 1-8.
2. van Poollen, H. K., Tinsley, John M. and Saunders, Calvin D.: "Hydraulic Fracturing: Fracture Flow Capacity vs Well Productivity", *Trans.,* AIME (1958) **213,** 91-95.
3. Dyes, A. B., Kemp, C. E. and Caudle, B. H.: "Effect of Fractures on Sweep-Out Pattern", *Trans.,* AIME (1958) **213,** 245-249.
4. Craft, B. C., Holden, W. R. and Graves, E. D., Jr.: *Well Design: Drilling and Production,* Prentice-Hall, Inc., Englewood, Cliffs, N. J. (1962) 494.
5. McGuire, W. J. and Sikora, V. J.: "The Effect of Vertical Fractures on Well Productivity", *Trans.,* AIME (1960) **219,** 401-403.
6. Prats, M.: "Effect of Vertical Fractures on Reservoir Behavior — Incompressible Fluid Case", *Soc. Pet. Eng. J.* (June, 1961) 105-118.
7. Bearden, W. G.: "Technical Information on the Hydrafrac Service", unpublished report, Pan American Petroleum Corp., Tulsa, Okla. (Jan., 1953).
8. Crawford, P. B. and Landrum, B. L.: "Effect of Unsymmetrical Vertical Fracture on Production Capacity", *Trans.,* AIME (1955) **204,** 251-254.
9. Boriskie, R. J.: "An Investigation of Productivity Increases from Hydraulic Fracturing Treatments", MS thesis, Texas A&M U., College Station (Aug., 1963).
10. Howard, G. C. and Fast, C. R.: "Optimum Fluid Characteristics for Fracture Extension", *Drill. and Prod. Prac.,* API (1957) 261.
11. Ghauri, W. K.: "Results of Well Stimulation by Hydraulic Fracturing and High Rate Oil Backflush", *J. Pet. Tech.* (June, 1960) 19-27.
12. Lefkovits, H. C. and Matthews, C. S.: "Application of Decline Curves to Gravity-Drainage Reservoirs in the Stripper Stage", *Trans.,* AIME (1958) **213,** 275-280.
13. Campbell, J. B.: "The Effect of Fracturing on Ultimate Recovery", paper 851-31-L presented at the Mid-Continent District Meeting, API Div. of Production, Tulsa, Okla., April 10-12, 1957.
14. Sallee, W. L. and Rugg, F. E.: "Artificial Formation Fracturing in Southern Oklahoma and North Central Texas", *Pet. Eng.* (Feb., 1954) B-75.
15. Sallee, W. L. and Rugg, F. E.: "Artificial Formation Fracturing in Southern Oklahoma and North Central Texas", *Bull.,* AAPG (Nov., 1953) **37,** No. 11, 2539-2550.
16. Waters, A. B. and Tinsley, John M.: "Gas Well Deliverability Improved by Planned Fracturing Treatments", paper SPE 1275 presented at SPE 40th Annual Fall Meeting, Denver, Colo., Oct. 3-6, 1965.
17. Fast, C. R., Nabors, F. L. and Mase, G. D.: "Propping-Agent Spacer Effective in Gas-Well Fracturing", *Oil and Gas J.* (Sept. 27, 1965) 97.
18. Mallinger, M. A., Rixe, F. H. and Howard, G. C.: "Development and Use of Propping Agent Spacers to Increase Well Productivity", *Drill. and Prod. Prac.,* API (1964) 88.
19. Garland, T. M. Elliott, W. C., Jr., Dolan, Pat and Dobyns, R. P.: "Effects of Hydraulic Fracturing Upon Oil Recovery from the Strawn and Cisco Formations in North Texas", RI 5371, USBM (1957).
20. Howard, G. C.: "Evaluation of the Hydrafrac Process", unpublished report, Pan American Petroleum Corp., Tulsa, Okla. (1965).
21. Crawford, P. B., Pinson, J. M., Simmons, J. and Landrum, B. L.: "Sweep Efficiencies of Vertically Fractured Five-Spot Pattern", *Pet. Eng.* (March, 1956) B-95.
22. Crawford, P. B. and Collins, R. E.: "Estimated Effect of Vertical Fractures on Secondary Recovery", *Trans.,* AIME (1954) **201,** 192-196.
23. Landrum, B. L. and Crawford, P. B.: "Horizontal Fractures Do Affect Ultimate Recovery", *Pet. Eng.* (June, 1961) B-80.
24. Pinson, J. M., Simmons, J., Landrum, B. L. and Crawford, P. B.: "Effect of Large Elliptical Fractures on Sweep Efficiencies in Water Flooding or Fluid Injection Programs", *Prod. Monthly* (Nov., 1963) 20.
25. Morrisson, T. E. and Henderson, J. H.: "Gravity Drainage of Oil Into Large Horizontal Fractures", *Trans.,* AIME (1960) **219,** 7-15.
26. Wasson, J. A.: "The Application of Hydraulic Fracturing in the Recovery of Oil by Waterflooding: A Summary", IC 8175, USBM (1963).
27. Hartsock, J. H. and Slobod, R. L.: "The Effect of Mobility Ratio and Vertical Fractures on the Sweep Efficiency of a Five-Spot", *Prod. Monthly* (Sept., 1961) 2.
28. Guerrero, E. T.: "Fracturing Can Help Secondary Recovery", *World Oil* (July, 1958) 126.
29. Powell, J. P. and Johnston, K. H.: "Waterflood Fracturing Pays Off", *IPAA Monthly* (Oct., 1960) 22.
30. van Poollen, H. K.: "Do Fracture Fluids Damage Productivity?", *Oil and Gas J.* (May 27, 1957) 120.
31. Flickinger, D. H.: "Effect of Formation Damage During Fracturing on Well Productivity", unpublished report, Pan American Petroleum Corp., Tulsa, Okla. (June 18, 1956).
32. Crittendon, B. C.: "The Mechanics of Design and Interpretation of Hydraulic Fracture Treatments", *J. Pet. Tech.* (Oct., 1959) 21-29.

Chapter 11

Nuclear Fracturing

11.1 Introduction

The application of nuclear energy to the recovery of oil and gas has been the subject of extensive studies by every major oil company in the U.S. and by various federal agencies. Most investigators agree that the commercial application of nuclear energy for fracturing oil or gas reservoirs is imminent and that there is a need for a carefully planned test program to evaluate nuclear stimulation in tar sand, shale oil, gas and oil reservoirs. The success of the Atomic Energy Commission's (AEC) underground test shot in Nevada in 1957, the abundance of nonproductive, or low productive, petroleum reservoirs, and the possibility of increasing their ultimate recovery provide the necessary incentive.

Project Plowshare was created to explore the peaceful use of atomic bombs. This study was conducted by The U. of California Radiation Laboratory at Livermore, Calif. It was concluded that nuclear explosions can be contained in the zone of interest and are capable of rubbling a large volume of rock.

Types of Reservoirs Under Study

The most serious interest in applying nuclear stimulation to petroleum reservoirs is being directed in 1970 toward the recovery of gas from formations in New Mexico and Colorado.

The feasibility of using nuclear fracturing in the massive oil shale deposits of Colorado is also under study. Some investigators believe that oil shale can be retorted in place and the oil removed in a liquid or gaseous form from the cavity formed by a nuclear explosion, eliminating the need to mine the shale rock and process it through a retort to extract the hydrocarbons.[1] Data on nuclear-explosive fracturing at this date are inadequate and do not indicate the fracture patterns and permeabilities that can be expected under any specific condition. Fractures resulting from an underground nuclear explosion would be expected to differ from those formed by hydraulic fracturing in both the number of fractures created and the depth they penetrate the matrix about the wellbore.

Whereas a hydraulic fracturing treatment usually produces a single fracture through the wellbore, a nuclear explosion can be expected to form multiple fractures. It would be expected to form a cavity 100 to 300 ft in diameter with many fractures radiating from it and creating a large high-permeability zone. A well penetrating this zone should produce at high rates, since it would have a very large effective wellbore radius.[2]

Nuclear vs Conventional Fracturing

In any study of nuclear fracturing it must be recognized that conventional hydraulic fracturing is effective in most low-permeability reservoirs of average thickness. In the areas where hydraulic fracturing is effective, nuclear explosives cannot compete because of high costs. However, substantial volumes of hydrocarbons are in place in reservoirs that do not yield to present hydraulic fracturing techniques, and these reservoirs may therefore be potential sites for nuclear fracturing.

Geographical Areas Applicable to Nuclear Fracturing

The Bureau of Mines has made studies of various producing areas in the U.S. to determine those areas where nuclear fracturing might be applicable. The massive, sandy-shale, gas-bearing Cretaceous formations—such as the Mesaverde and Mancos—in the Rocky Mountain region offer excellent opportunities for the application of high-yield nuclear explosions. Petroleum reservoirs applicable to nuclear stimulation are in the Rocky Mountain region, West Texas and central Kansas. Fig. 11.1 shows major areas of oil and gas production in the U.S. most amenable to nuclear fracturing.

Calculations were performed by engineers of the USBM to predict increased producing rates and ultimate recoveries from various types of petroleum reservoirs under conditions assumed to result from nuclear explosions. The reservoir characteristics desired for application of nuclear explosives are (1) low productivity due to low permeability; (2) net pay thickness greater than 200 ft, preferably bounded by thick impermeable formations; and (3) low-viscosity oil under dissolved-gas drive or a natural-gas reservoir.

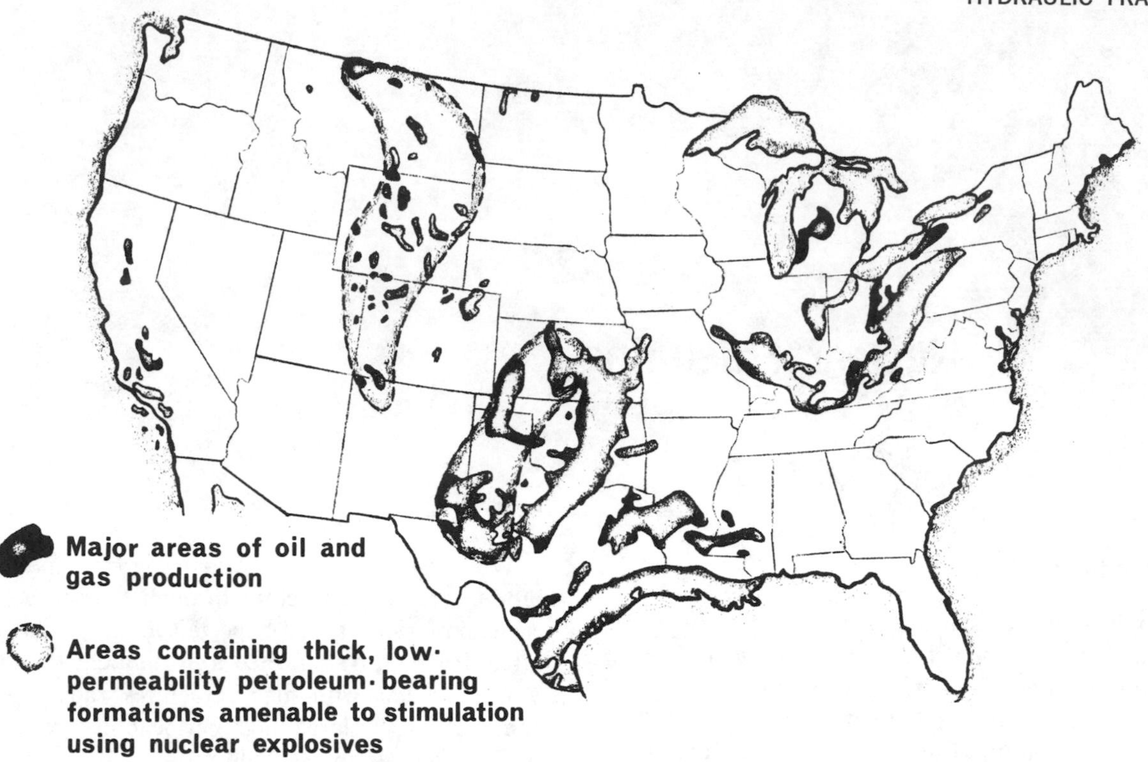

Fig. 11.1 Major areas of oil and gas production in the U. S., showing areas applicable to stimulation using nuclear explosives.[22]

11.2 Effect of Nuclear Explosion Yield on Volume of Disturbed Rock

On detonation of the nuclear device, a cavity is formed as a result of the vaporization of rock and saturating fluids. The gases contained within the cavity are initially at extremely high pressure and temperature. Compaction of the rock in the lower hemisphere, coupled with upward and lateral rock movement, results in a spheroid cavity. The gas pressure temporarily supports the overburden, thus preserving the cavity shape. Subsequent heat losses, gas leakoff through the fracture system, and vapor condensation reduce the pressure until the fractured rock above the cavity can no longer be supported. Rock collapse into the cavity forms a complex chimney-rubble zone. Most of the molten material and radioactive fission products collect in the bottom of the zones.[11]

The Gnome nuclear test, detonated at a depth of 1,184 ft in the Salado formation (Permian age) in an oil producing region southeast of Carlsbad, N.M. (Fig. 11.2), illustrates such a cavity. The void created had a total volume of about 1 million cu ft (62-ft radius sphere).[3-8] The cavity is asymmetric, probably due to departures from spherical symmetry during its growth and to changes in shape caused by implosion of the cavity walls and by uplift and dropping of the ceiling. The final cavity shown in Fig. 11.2B has an average radius of 57 ft for the lower portion of the cavity (measured from the working point to the boundary of radioactive melt); an average radius of 80 ft in the equatorial plane; and an average radius of 75 ft in the upper portion of the cavity (measured from the working point to the rock-void interface).

A 30-ft-high girdle of rock around the equatorial plane of the cavity moved radially farther from the working point than did rock on the bottom or at the top of the cavity. This bulge resulted from bedding plane weaknesses caused by horizontal clay seams.

The amount of rock melted by the explosion is estimated to be about 2,400 tons, equivalent to 800 tons/kiloton* of yield. This melt, which contains most of the

*An explosive yield of 1 kt TNT equivalent represents energy equal to 10^{12} calories.

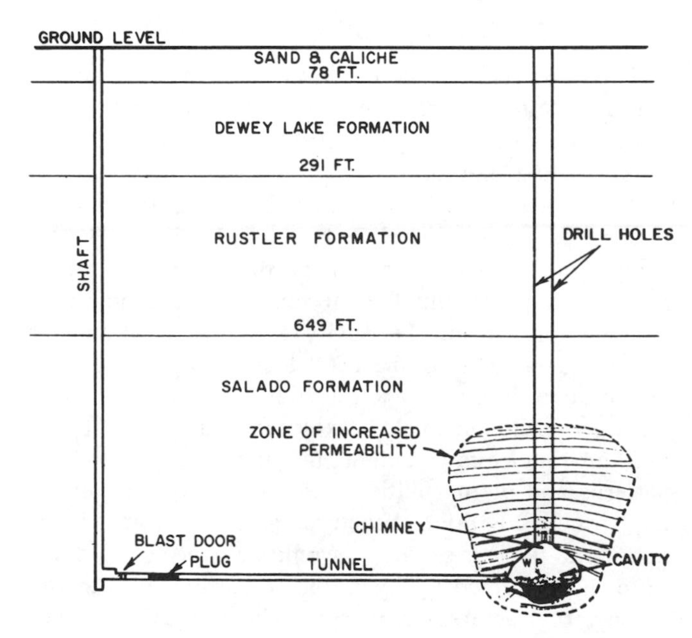

Fig. 11.2A Vertical section through Gnome environment.

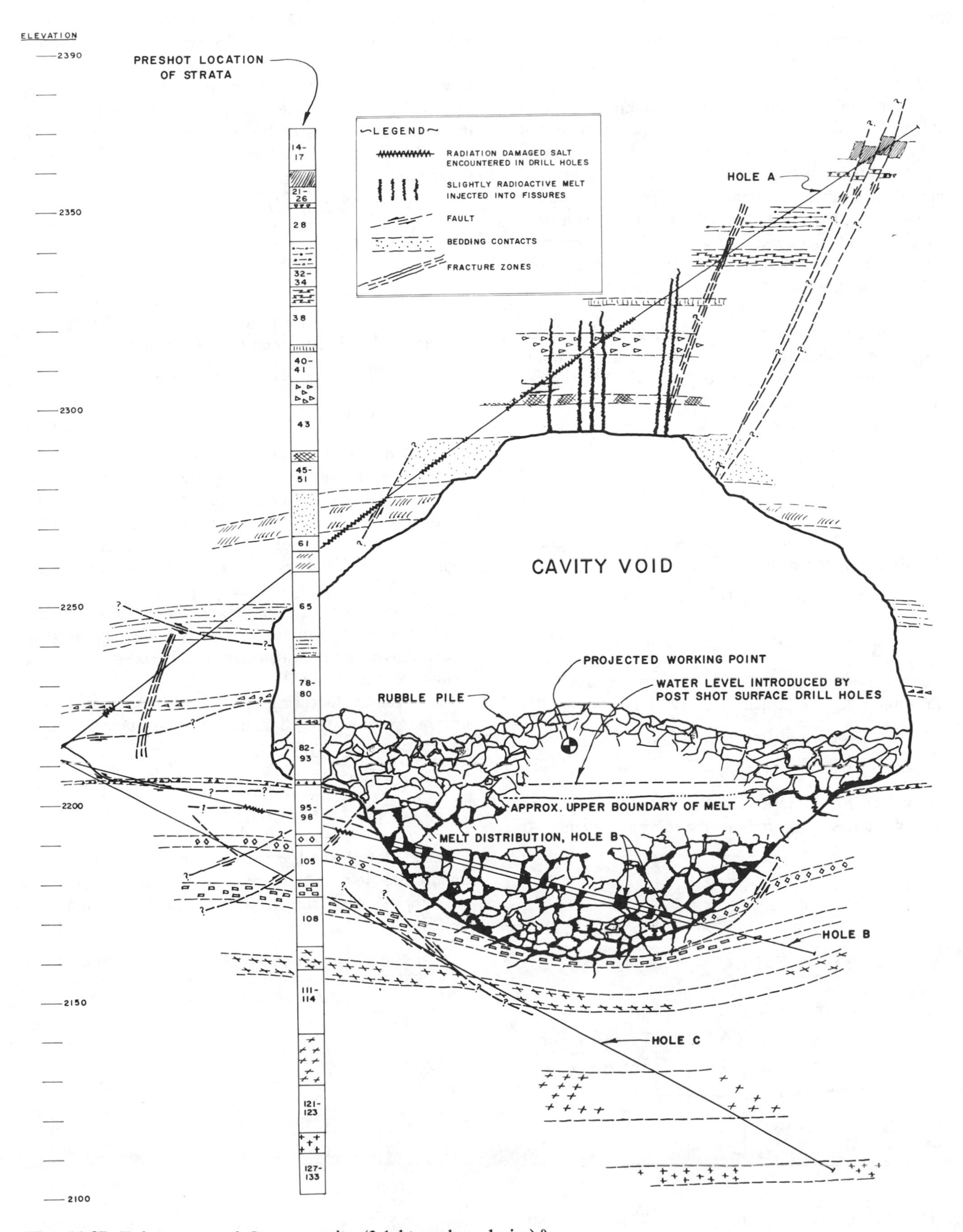

Fig. 11.2B Enlargement of Gnome cavity (3.1 kt nuclear device).[9]

radioactive debris, was intimately mixed with about 13,000 tons of salt rock that were imploded, or spalled into the cavity, probably during the first minute following the explosion. Steam pressure, produced from the 1 to 2 percent of water associated with the rock, may have been sufficient to blow off blocks of rock from the walls of the cavity.

An additional amount of rock, estimated at about 15,000 tons, collapsed from the upper hemisphere and covered the cavity base containing the radioactive melt.

Bray, Knutson, Wahl and Dew,[14] in their review of studies of a number of contained underground nuclear explosions, reported that these devices have yielded a rather consistent picture of the geometric features produced.[6,10] The major features of cavity, chimney and fractured zone are illustrated by a schematic drawing in Fig. 11.3.

Based on a large number of contained underground explosions in various rock media, the scaling equations for calculating various features of the post-shot geometry as a function of yield, depth of burial, and media are as follow.[6,9,10]

Cavity Radius

$$r_c = \frac{(F_c\ \text{kt})^{1/3}}{(\rho_{OB}\, D)^{1/4}} \quad \ldots \ldots \ldots \quad (11.1)$$

A factor of 290 may be assumed to be generally applicable to hydrocarbon reservoirs.[6]

Chimney Height

$$Z = \frac{4r_c}{3\Delta\phi} \quad \ldots \ldots \ldots \quad (11.2)$$

The net porosity increase in the rubble zone for media similar to a petroleum reservoir is about 0.25.

Radius of Permeable Zone

$$r_p = F_p\, r_c \quad \ldots \ldots \ldots \quad (11.3)$$

The values for F_p represent the most probable and maximum values observed at the Gnome event. At Gnome, the permeable zone radius was at least 150 ft, and the cavity radius was 62 ft.[9,12] Therefore the factor for extrapolating the increase in "probable" radius of permeability is 150/62 = 2.4. The "maximum" permeable-zone radius from the Gnome data is based on phenomena observed during mining operations in the new drift: damaged salt section at 215 ft and water seep at 221 ft from the shot point.[6,9,12] The factor for the maximum radius of increased permeability is 215/62 = 3.5. In granite, a factor as high as 5.3 has been reported.[11]

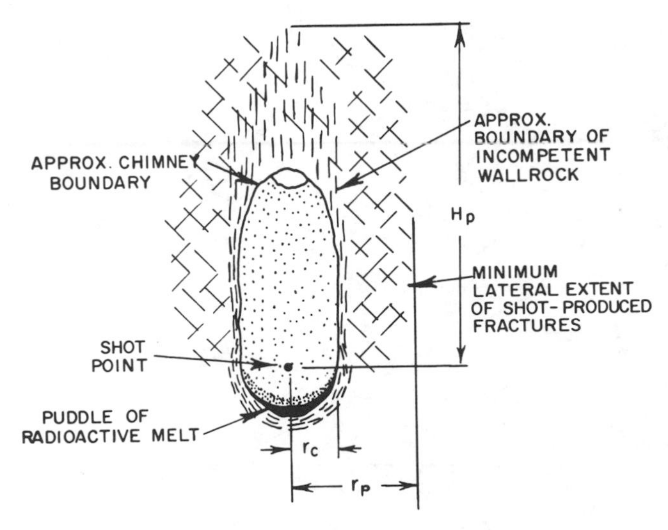

Fig. 11.3 Schematic diagram of typical post-shot environment.[10]

Height of Permeable Zone

$$Z_p = F_z\, r_c \quad \ldots \ldots \ldots \quad (11.4)$$

During post-shot drilling at Gnome, the permeable zone was observed to be 350 ft above the shot point. The value of F_z is the ratio of 350 to 62. Factors as high as 7.7 have been reported for granite.[6,10]

Scaled Depth of Burial (SDB)

$$\text{SDB} = \frac{D}{\text{kt}^{1/3}} \quad \ldots \ldots \ldots \quad (11.5)$$

Scaled depth of burial allows correlation of results from explosions having different yields. It relates these results on the basis of an equivalent 1-kt yield through the proportion $\text{SDB}/(1\ \text{kt})^{1/3} = D/(\text{kt})^{1/3}$. Empirical data from a number of reported underground explosions in competent rock material such as granite, salt and tuff indicate that a scaled depth of burial greater than 500 ft is sufficient for containment.[6] For an incompetent overburden material, such as Nevada test-site desert alluvium, the scaled depth of burial required for containment is greater.[13]

Post-nuclear fracturing productivity, assuming radial flow of gas into a well under steady-state conditions, may be expressed by[23]

$$q = \frac{10.320\ kh\ (p_d^2 - p_w^2)}{\mu \ln \dfrac{r_e}{r_w}\ T_f\ (15.025)\ z_{\bar{p}}} \quad \ldots \quad (11.6)$$

Table 11.1 shows the variation in cavity radius as a function of yield at the scaled depths of burial of 450 and 800 ft. Also shown is the estimated volume of rock, in acre-feet, in which there is expected to be a

TABLE 11.1 — ESTIMATED VOLUME OF INCREASED PERMEABILITY FOR NUCLEAR EXPLOSIONS IN RESERVOIR ROCK[9]

Yield (in kt) at Scaled Depth of Burial (ft)		Cavity Radius (ft)	Depth of Burial, D (ft)	Height of the Permeable Zone, Z_p (ft)	Volume of Gross Permeability Increase (acre-ft)
kt	SDB				
10	800	96	1720	540	2.1×10^3
	450	110	970	620	3.2×10^3
100	800	167	3720	940	1.1×10^4
	450	196	2100	1110	1.8×10^4
500	800	252	6350	1420	3.8×10^4
	450	290	3600	1640	5.9×10^4
1000	800	299	8000	1690	6.4×10^4
	450	346	4500	1950	1.0×10^5

large increase in permeability. In this extrapolation from the Gnome data it is assumed that the volume of this fractured zone increases with explosive yield and is proportional to the increase in cavity radius, r_c.

It should be emphasized that the data presented in Table 11.1 are not quantitative predictions, but are only approximations to show how these parameters are expected to vary.

From these data it appears that nuclear devices in the yield range of approximately 100 kt would be the most useful for oil reservoirs in which the producing zone thickness is about 1,000 ft at depths of 2,000 to 4,000 ft.

Using the data from Table 11.1, Coffer *et al.* calculated the amount of oil affected in the zone of altered permeability.[9] These calculations assumed: (1) a 100-kt device, (2) a reservoir thick enough (1,110 ft) to utilize the maximum extent of the permeability change, and (3) the average reservoir properties for a marginal limestone reservoir (i.e., 10 percent porosity, low permeability, 30 percent water saturation, and an oil formation volume factor of 1.2). The amount of oil in the permeable zone is [1.84×10^4 acre-ft $\times$ 10 percent (1-0.3) $\times$ 7.785 bbl/acre-ft] $\div$ 1.2 (vol factor) $- 8.3 \times 10^6$ bbl. If we assume a recovery factor of 30 percent, oil recovery is equal to (0.30) (8.3×10^6 bbl) $= 2.49 \times 10^6$ bbl.

A further review of the data in Table 11.1 shows that most of the permeability change is in a vertical direction. This is believed to be a factor of the depth of burial of the shot. Most oil and gas producing reservoirs are less than 1,000 ft thick. This factor will tend to restrict the yield of nuclear devices used for oil and gas stimulation. Effort must be directed, therefore, to focusing more of the energy in a horizontal direction; such a development would greatly expand the number of reservoirs adaptable to nuclear stimulation.

11.3 Effect of Nuclear Explosions on Rock Properties

Very important in evaluating the 1961 Gnome test were varieties of rock samples cemented in Observation Holes 3 and 15, Fig. 11.4. The rock cores were grouted into thin-walled metal containers, and the containers were then cemented into the observation hole with a cement having the same acoustical properties as the surrounding Salado formation. After detonation of the nuclear device the samples were recovered by mining and were analyzed to determine changes in physical properties (see Table 11.2).

A study of the deformation of rock samples embedded in the observation holes showed changes in physical properties greater than that observed in the surrounding Salado formation.[9] An interesting aspect of the reservoir rock reaction is the contrast between the behavior of the sandstones and that of the carbonates.

In general, the carbonates seemed to react favorably (they increased in porosity and permeability), whereas the sandstones reacted unfavorably (they showed a decrease in permeability). Part of this observed difference is due to the microfracturing that accompanies the development of the large cylindrical zone. When permeability development is associated with fractures, as it generally is in carbonates, permeability reduction noted in sandstone with intergranular permeability is compensated for by this microfracturing effect.

The permeability behavior is shown graphically in Figs. 11.5 and 11.6. The trend of the carbonate samples was toward a permeability increase that varied with peak shock pressure. The sandstones showed a rather uniform decrease in permeability.

Following the shot, all of the rock types generally decreased in compressive strength.

Thin-sections were prepared from the center of

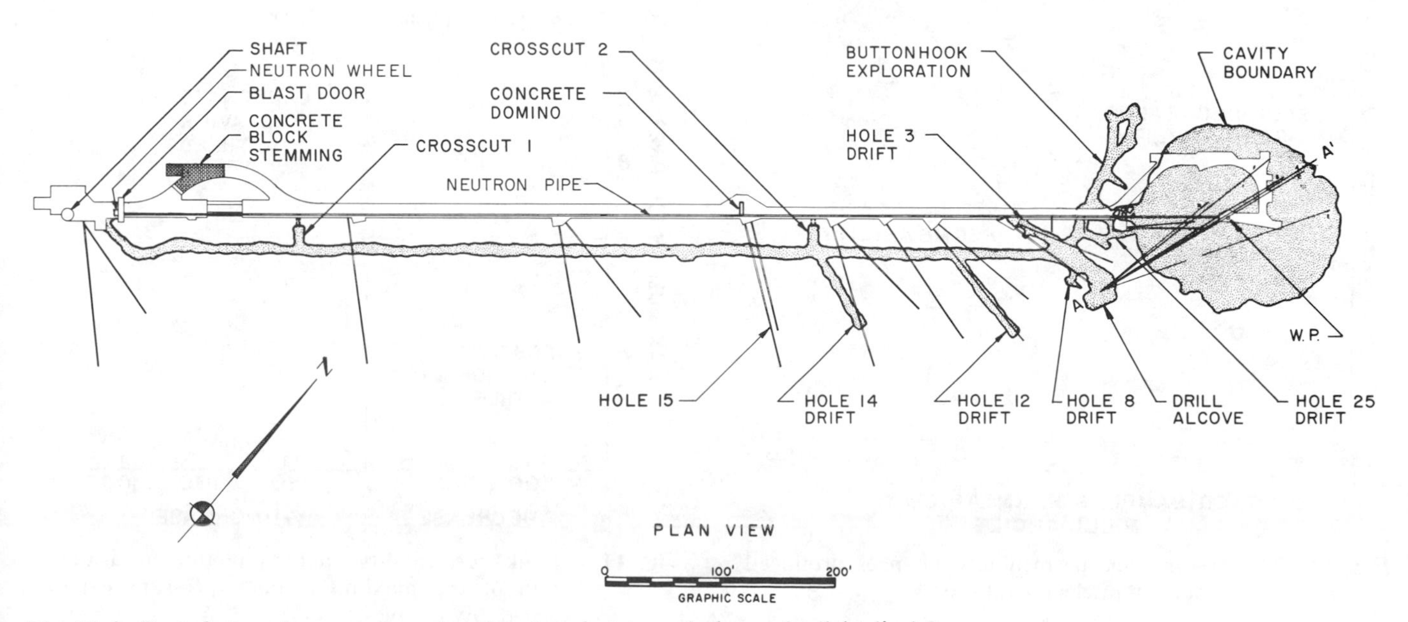

Fig. 11.4 Post-shot underground exploration (preshot tunnel shown in light line).[9]

TABLE 11.2 — RESERVOIR ROCK BEHAVIOR[9]

Sample No.		Permeability (md)		Porosity (percent)		Compressive Strength (kg/sq cm)		Change in Porosity (percent)	Change in Permeability (percent)	Change in Compressive Strength (percent)	Lithology†	Maximum Pressure (kbar)	Formation
Shot	Adjacent	Shot*	Adjacent**	Shot*	Adjacent**	Shot*	Adjacent**						
A-1s	A-1a	0.1	<0.1	4.4	3.6	850	813	+ 22	—	+ 5	LS	3.8	Amsden
A-2s	A-2a	36.0	83.0	13.8	15.4	527	714	— 10	— 57	—26	SS	3.8	Tensleep
A-3s	A-3a	27.0	10.0	18.5	16.8			+ 10	+ 170		DOL	3.7	Embar
A-4s	A-4a	13.0	26.0	22.7	19.6			+ 16	— 50		SS	3.7	Puente
B-1s	B-1s	0.2	<0.1	0.9	0.1	504	550	+900	—	— 8	LS	4.8	Madison
B-2s	B-2a	169.0	188.0	18.4	18.3	275	452	+ 1	— 10	—39	SS	4.7	Tensleep
B-3s	B-3a	0.2	26.0	10.0	12.6			— 21	— 99		DOL	4.7	Embar
B-4s	B-4a	61.0	160.0	28.6	20.2			+ 42	— 62		SS	4.7	Repetto
C-1s	C-1a	<0.1	<0.1	2.6	1.5			+ 73	—		LS	6.6	Madison
C-2s	C-2a	205.0	440.0	18.7	19.4	166	339	— 4	— 53	—51	SS	6.5	Tensleep
C-3s	C-3a	1.0	0.2	9.5	5.0	96	740	+ 90	+ 400	—87	DOL	6.5	Embar
C-4s	C-4a	16.0	150.0	21.0	24.4	44	73	— 14	— 89	—40	SS	6.5	Repetto
D-7s	D-1a	<0.1	<0.1	2.1	1.3			+ 62	—		LS	8.4	Madison
D-8s	D-2a	25.0	203.0	15.0	14.1	100	670	+ 6	— 88	—85	SS	8.4	Tensleep
D-9s	D-3a	15.0	0.6	16.8	6.5			+158	+2400		DOL	8.4	Embar
D-10s	D-4a	38.2	93.0	22.8	24.2	19	21	— 6	— 59	—10	SS	8.4	Repetto

*Rock sample subjected to nuclear detonation
**Rock sample *not* subjected to nuclear detonation
†LS=Limestone SS=Sandstone DOL=Dolomite

undeformed longitudinal slices of core (the slices were taken from the sample and then grouted into the stainless steel container). Under a petrographic microscope these thin-sections were compared with adjacent thin-sections cut from the shot samples. The sandstones showed no apparent changes, whereas the carbonates were generally microfractured.

The porosity of the shot limestone samples was usually increased, but the porosity of the shot sandstone samples showed no significant change. Fig. 11.7 is a log-log plot of the porosity of the shot samples and their adjacent control sample.

The increases in porosity and permeability and the decreases in the compressive strength of the carbonate samples are minimum values, since the samples that were visibly fractured were not used. The values reported on Table 11.2 are the results of the analysis of carefully selected samples that did not contain any visible fractures.

The values of porosity, permeability and compressive strength measured for the sandstones are more representative of the total sample behavior, since gross fracturing was not one of the phenomena observed in this lithologic type.

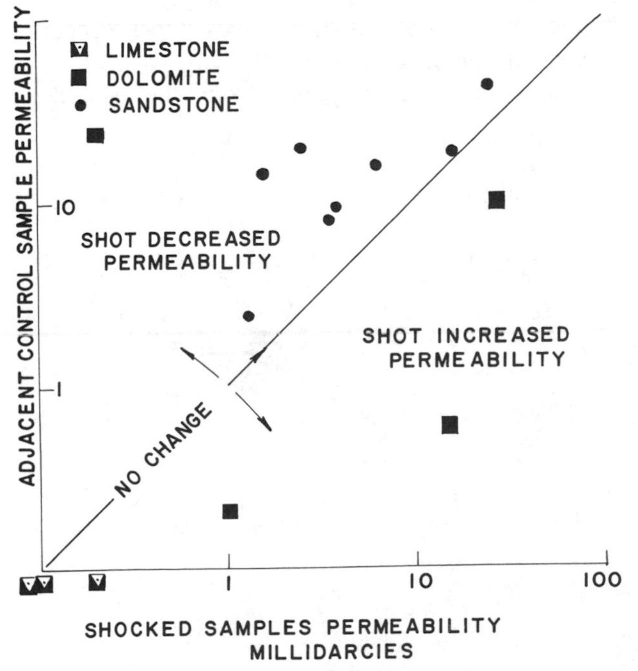

Fig. 11.5 Reservoir rock permeability changes produced by shock of nuclear explosion.[9]

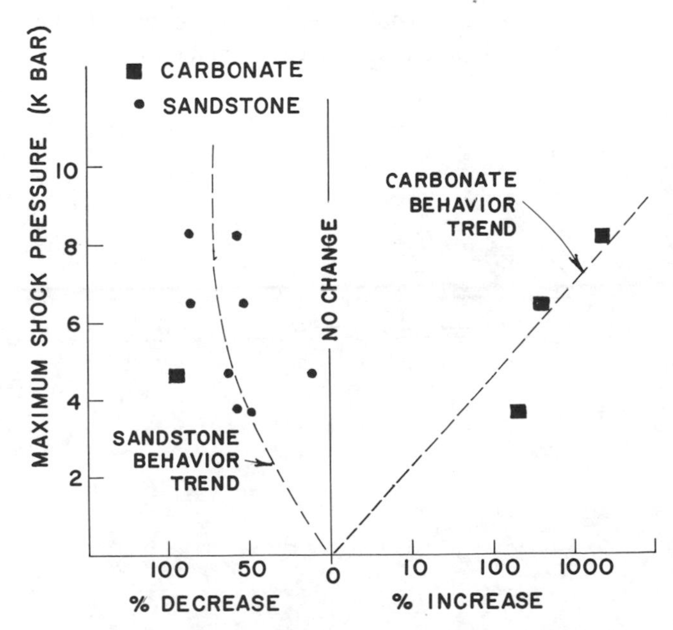

Fig. 11.6 Rock permeability changes plotted as a function of the maximum shock pressure experienced by samples.[9]

The information on the shock-produced deformation of the reservoir rock cannot be accepted without reservation because (1) the samples tested represent only a few types of reservoir rock; (2) these samples were exposed to only a portion of the energy associated with the nuclear explosion; and (3) the imperfect coupling of the samples in the hole means the pressure values reported are only "semi-quantitative". On the basis of these studies, Bray *et al.* concluded that to take maximum advantage of an explosion's uplift-rebound effect on highly lenticular, low-permeability reservoirs, the nuclear device should be located at, or somewhat below, the base of the formation.[14]

11.4 Economics of Nuclear Reservoir Stimulation

Bray, Knutson, Wahl and Dew[14] studied potential well productivity from nuclear fractured formations and showed that the increased productivity of wells tapping the post-shot cavity-chimney-fracture zone[15] should be of major importance and must be considered in any economic evaluation. Permeability increases due to fracturing beyond the chimney-cavity rubble zone have been observed in underground shots, and must be considered in evaluating the effect of nuclear explosions on well productivity.[6,11] The effect on well production of the geometry of a post-shot nuclear-fractured zone in a petroleum reservoir has been simulated on a computer by assigning concentric annular configurations to the cavity-chimney and surrounding fractured zones with the producing wellbore in the center.[14]

A computer analysis of four reservoir conditions involving the production of both gas and oil was prepared by Bray, *et al.*[14] These studies illustrate a method of

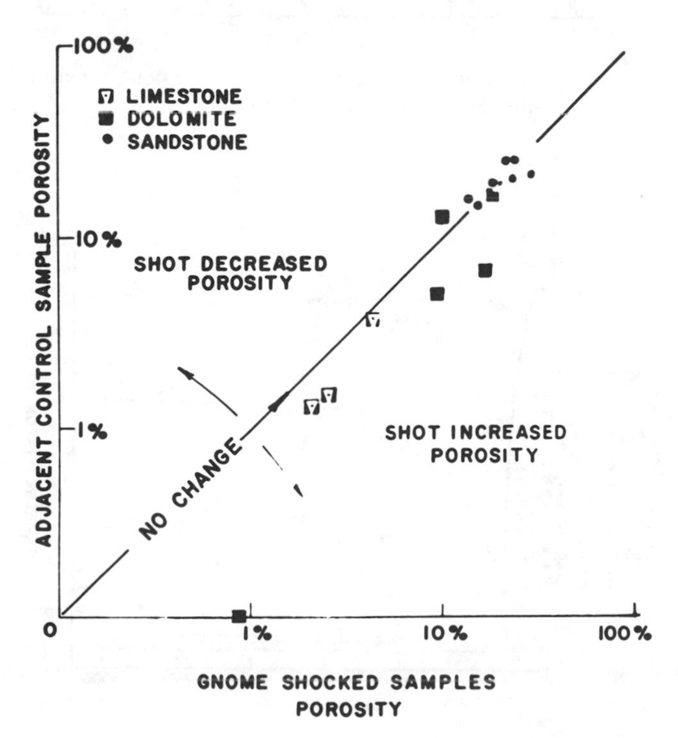

Fig. 11.7 Reservoir rock porosity changes produced by shock of nuclear explosion.[9]

economic evaluation and provide an insight into the results to be expected from nuclear fracturing of hydrocarbon-producing formations. To establish a basis for calculation, the drainage area around a wellbore was divided into 40 or 50 concentric annular cylindrical shells of specified permeability. The well production-pressure-time history was calculated for a large number of time increments. Flow capacity of the existing reservoir matrix was established from a computer fit of well test data from existing wells in the formation.

Flow capacity in the cavity-chimney area near the wellbore in the explosion-stimulated cases was conservatively assumed to be about 100 times greater than that of the unaffected formation. This ratio was decreased to unity at the unaffected formation boundary. The production scheduled called for starting at a rate consistent with an estimated salable quantity of gas from a well in the area. To provide a satisfactory delivery pressure to the gathering system, no further reduction was allowed after pressure in the wellbore reached 100 psia.

Shallow, Low-Pressure Gas Reservoir A

The formation characteristics utilized in the study of this reservoir are given in Table 11.3.

Scaled depth of burial (SDB), initial cavity radius, chimney heights, cavity volume, and volume of the cylindrical chimney-rubble zone for various sizes of nuclear devices have been calculated from the empirical explosion scaling equations (Eqs. 11.1 through 11.5). The depth of burial D of the nuclear device would be at 2,700 ft, the base of the gas-bearing formation. Two different permeability-affected zones were assumed: a fractured zone, and a maximum shot-affected zone. These relations are shown on Figs. 11.8 and 11.9. For comparison, scaled depth of burial for previously contained underground explosions in salt (Gnome), granite (Shoal, Hardhat), and ruff (Rainier, Tamalpais) are also shown.[6,10,17] Thus, devices up to 100 kt should be completely contained by the overburden and would have an SDB greater than 500.[6]

Although gas deliverability could be calculated for the yield from a number of devices, a 40-kt explosion

TABLE 11.3 — PROPERTIES FOR EXAMPLE RESERVOIR A

Depth to top of pay—hydrocarbon-bearing formation, ft	2,300
Gross thickness of producing formation, ft	400
Net effective pay thickness, ft	150
Pressure, psia	450
Temperature, °F	95
Gas specific gravity (air = 1.00)	0.647
Gas deviation (compressibility) factor	0.94
Pressure base, psia	15.025
Porosity, percent	12.7
Permeability (arithmetic average), md	5.7
Median permeability, md	1.3
Residual oil saturation, percent	5
Residual water saturation, percent	37
Irreducible water saturation, $_{IWS}$ (based on median permeability), percent	37
Bulk density, gm/cc	2.38
Hydrocarbon pore volume (cu ft/640 acres)	33.4×10^7
Initial gas in place (scf gas/640 acres)	9.96×10^9 or 9.96 Bscf/section

was chosen to illustrate the flow and economic calculations since (1) kiloton cost (nuclear) varies only slightly for the range of devices applicable for gas or oil well stimulation,[16] and (2) the scaled depth of burial of 790 ft is approximately the same as in Gnome.

A 40-kt underground nuclear explosion 2,700 ft below the surface would produce the following calculated environment in the producing formation.

1. A cavity radius of 110 ft.
2. A probable permeable zone radius of 270 ft (the maximum radius is 385 ft).
3. A probable permeable zone height of 630 ft (the maximum height is 900 ft).
4. A probable permeable zone volume of 140 MMcf, or 3,220 acre-ft (the maximum permeable zone volume could be 420 MMcf, or 9,650 acre-ft).
5. A total cavity and chimney volume of 23 MMcf or 528 acre-ft with a void volume of 5.7 MMcf.

The mathematical model was used to compute the production-pressure-time history for five cases — two unstimulated and three device-stimulated. These cases correspond to the following.

1. One conventional well per 640-acre section, with a 4¾-in. open-hole completion running through the entire producing section.
2. Four conventional wells per 640-acre section, with each well draining 160 acres.
3. One "probable" nuclear stimulated well per section, with the calculated permeable zone radius being 270 ft.
4. One "minimum" nuclear stimulated well per section. (It was assumed that no fracturing occurred beyond the initial cavity radius of 110 ft.)
5. One "maximum" nuclear stimulated well per section, with the calculated permeable zone radius being 385 ft.

Two examples of pressure-radius relationships are shown in Fig. 11.10. The gas deliverability for each of the five cases is shown in Fig. 11.11. A summary of pertinent production data is given in Table 11.4.

The AEC estimated that the cost of nuclear devices and related services ranged from $350,000 for 10 kt to $600,000 for 2 mt (megatons) (see Fig. 11.12). These charges include the nuclear materials, fabrication and assembly, and arming and firing. They do not include safety studies, site preparation, transportation, placement, or support services that might vary considerably from site to site. The diameter of the nuclear device

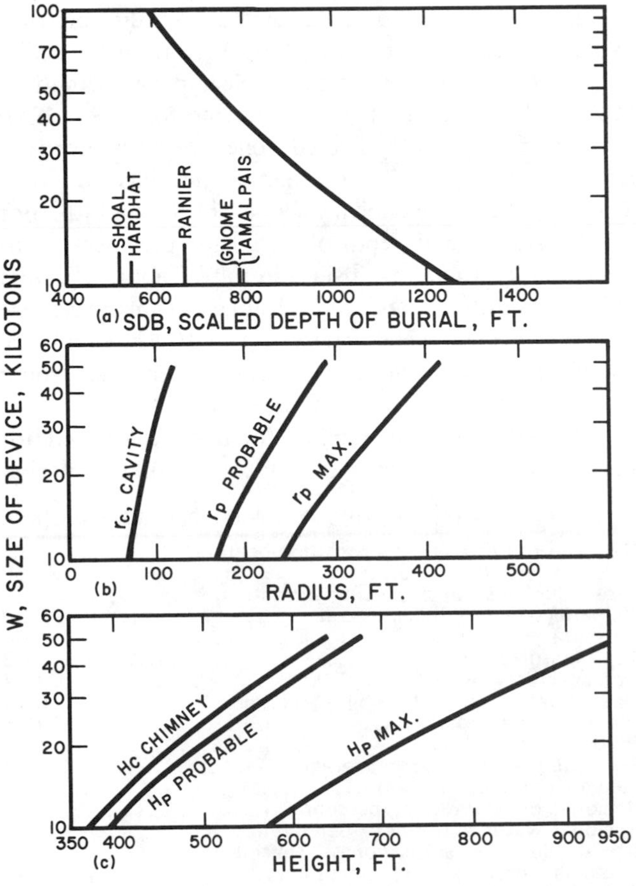

Fig. 11.8 Calculated data for nuclear explosion environment in Reservoir A at a 2,700-ft depth of burial.[14]

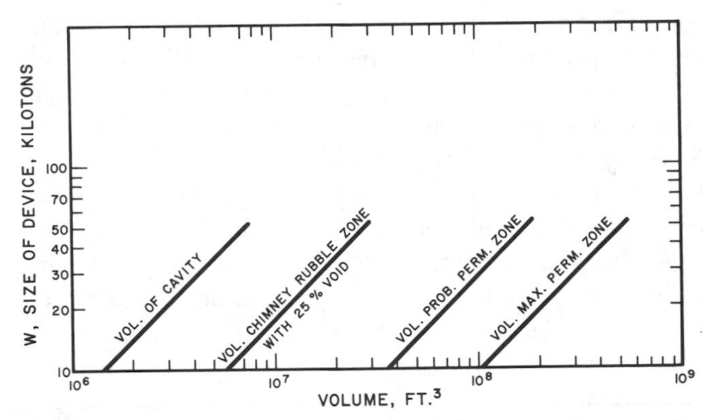

Fig. 11.9 Total volumes generated by a nuclear explosion at a 2,700-ft depth of burial.[14]

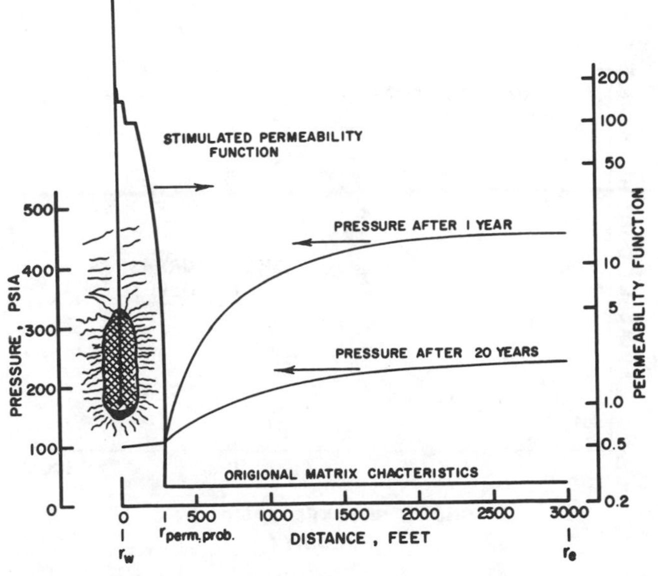

Fig. 11.10 Gas deliverabilities for conventional and nuclear stimulated wells in Reservoir A.[14]

TABLE 11.4 — EXAMPLE RESERVOIR A: PRODUCTION CHARACTERISTICS
(Based on 640 acres and 20 years' total production time, 9.96 Bscf initial gas in place.)

	Initial Rate (Mscf/D)	Time On Initial Rate (days)	Pressure at Drainage Radius After 20 Years (psia)	Production Rate at End of 20 Years (Mscf/D)	Total Production in 20 Years (Bscf)	Total Recovery (percent)
UNSTIMULATED CASES						
One Conventional well	350	4	383	176	1.54	15.4
Four Conventional wells	1400	4	257	311	4.31	43.2
STIMULATED CASES						
Probable well	2800	130	237	284	4.99	50.1
Minimum well	1000	275	286	299	3.81	38.2
Maximum well	2800	245	217	290	5.53	55.5

was estimated to be 1 ft for 10 kt, or 2 ft for 1 mt.[16]

The economic incentive for nuclear stimulation of Reservoir A is shown in Table 11.5.

Investment costs of $650,000, $400,000, $275,000 and $165,000 were evaluated for the most probable post-shot environment. In addition, the economics were calculated for the "minimum" and "maximum" post-shot permeable-zone geometry at the intermediate $400,000 investment cost. The $650,000 case includes the currently published device and service charges of $415,000 for a 40-kt device.

Table 11.5 shows that in only one case (Probable Well 4) does the cost of nuclear stimulation compete with the cost of conventional well development. Lower device and placement costs are required to make this method of stimulation economically attractive.

An economic factor that was not considered is the large wellbore storage capacity. If a nuclear-stimulated gas field were located on a gas transmission line, this large storage capacity would provide high-productivity emergency capacity in times of maximum demand.

Deep, High-Pressure Gas Reservoir B

A similar production and economic analysis was made for a much deeper, thicker gas reservoir where larger devices may be used. Nuclear device cost per unit of explosion energy decreases with increasing size, thus deeper reservoirs may provide more economic incentive.

The properties of the deep, high-pressure gas reservoir are given in Table 11.6. The reservoir is a thick sequence of interbedded sandstones and shales. Only sandstone layers having a thickness of 20 ft or more were considered. (The net sand picked on the basis of this criterion would be conservative if nuclear stimulation were used, since the large fractured area formed by the shot would interconnect most of the smaller zones, including many of limited areal extent. This factor was not considered in the calculations.)

A conventional open-hole well completion in the reservoir would cost $350,000 and result in an initial potential of 1 MMscf/D. Calculations for one conventional well per 640 acres show a total recovery of 9.8 Bscf, or approximately 5.3 percent of the gas in place in 20 years. Theoretically, six to eight times this quantity of gas can be produced in the same time period with a nuclear-stimulated well.

A 1-mt nuclear device located at a depth of 10,000 ft (base of the pay formation) would produce the following results: (1) scaled depth of burial, 1,000 ft;

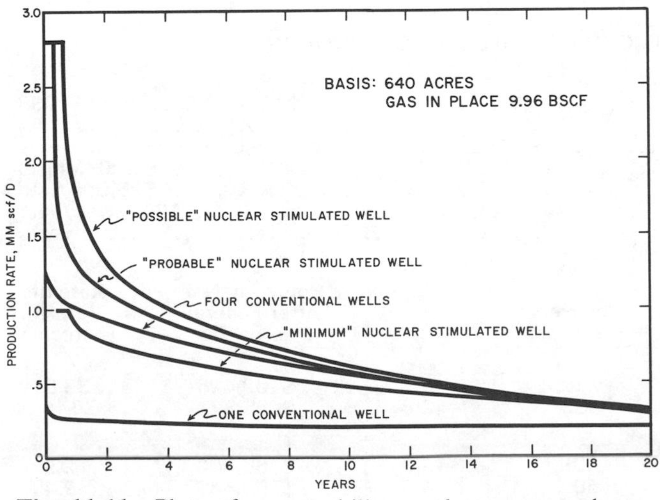

Fig. 11.11 Plot of permeability and pressure characteristics for a nuclear stimulated well in Reservoir .[14]

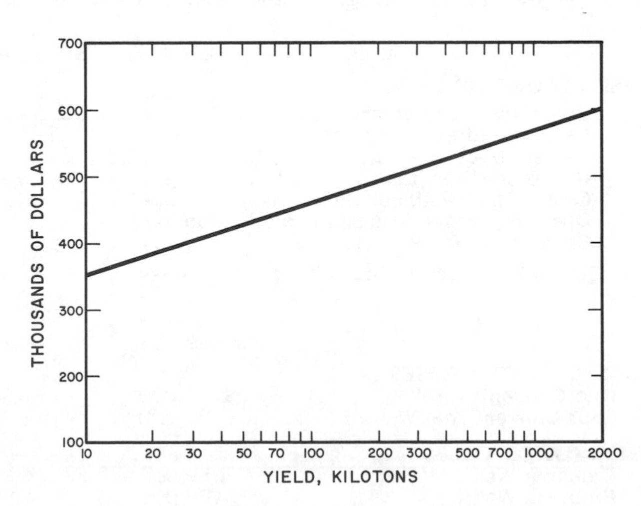

Fig. 11.12 Projected device and service charges for nuclear explosives.[14]

(2) cavity radius, 233 ft; (3) most probable permeable-zone radius, 564 ft (maximum permeable-zone radius is 810 ft); (4) most probable permeable-zone height, 1,320 ft (maximum permeable-zone height is 1,900 ft).

Economic calculations were made using the production and economic data given in Table 11.7. Total investment costs of $1,250,000 and $1,750,000 were used to allow for possible variations of emplacement hole costs and to show the variation of economics with investment cost. In both cases the charge for the nuclear device was $570,000, and the post-shot production well cost was $250,000. The results given in Table 11.7 show sufficient improvement in economics to warrant serious planning for the use of nuclear stimulation procedures.

Deep Oil Reservoir C

The mathematical model used for the oil reservoir cases involves radial, unsteady-state, two-phase flow in porous media. The effects of capillary pressure and gravitational forces are not considered. In the model, distance to the drainage radius is divided into 30 cells that vary logarithmically in size, the smallest at the wellbore and the largest at the drainage radius. This allows the model to simulate radial flow, with permeability, porosity and saturation varying with distance from the wellbore. Saturations, pressures and flow rates are calculated for each cell as a function of time (according to the fluid characteristics), permeability, and relative permeability characteristics prescribed. The calculations continue until the producing rate declines to some preselected limit.

The reservoir selected for study was a thick, low-permeability oil zone. Both prorated and nonprorated production schedules were considered. The reservoir characteristics are summarized in Table 11.8.

A permeable water sand is 800 ft above the base of the producing formation. The top of the fractured zone was planned to be 200 ft below the base of the water sand to prevent fracturing into the sand. This restricted the nuclear device size to 80 kt or less. The post-shot geometry for an 80-kt device is: (1) scaled depth of burial, 1,910 ft; (2) cavity radius, 106 ft; (3) probable permeable-zone radius, 256 ft; (4) probable permeable-zone height, 600 ft; (5) probable permeable-zone volume, 123 $\times$ 10^6 cu ft; (6) probable chimney and cavity volume, 20.2 $\times$ 10^6 cu ft; and (7) probable chimney void volume, 5 $\times$ 10^6 cu ft.

The well spacing was assumed to be 80 acres for both the nuclear stimulated and the nonstimulated cases. Initially, the producing rate was set at an allowable of 55 BOPD. When this rate could no longer be maintained, the production was assumed to be the maximum obtainable with 150 psi backpressure or a 2,000 cu ft/bbl penalty GOR.

The economic feasibility of nuclear stimulation was evaluated for a variety of devices and a range of emplacement costs. The rate of return and payout period were determined for each case and are summarized in Table 11.9.

In considering the rate of return, Fig. 11.13 shows that to make the nuclear stimulated well as attractive as the conventional prorated well, the cost of the nuclear stimulation device and service charge would have to be limited to approximately $90,000, or $250,000 for nuclear device and service charge for nonprorated wells.

TABLE 11.6 — PROPERTIES FOR EXAMPLE RESERVOIR B

Depth to top of pay, hydrocarbon-bearing formation, ft	8,000
Gross thickness of producing formation, ft	2,000
Net effective pay thickness, ft	500
Pressure, psig	5400
Temperature, °F	170
Gas specific gravity (air = 1.00)	0.605
Gas deviation (compressibility) factor	1.025
Pressure base, psia	15.025
Porosity, sand lenses, percent	8
Permeability, md	0.1
Water saturation, percent	45
Bulk density (assumed), gm/cc	2.4
Hydrocarbon pore volume (cu ft/section)	6.14 $\times$ 10^6
Initial gas in place (scf gas/640 acres)	186.8 $\times$ 10^9 or 186.8 Bscf/section

TABLE 11.5 — EXAMPLE RESERVOIR A: ECONOMIC CALCULATION SUMMARY

FIXED ECONOMIC DATA:

Company net interest	0.81933
Federal and state income tax, percent	54
Gas contract	$0.145/Mscf with 1 cent escalation each 5 years
Gas production tax	$3.00/MMscf
Cost of conventional well	$25,000/well
Operating costs (excluding production tax)	$900/well/year
Calculation period	20 years

DEVELOPMENT ECONOMICS (Basis—640 Acres)

	Total Investment	Gas Produced (Bscf)	Recovery (percent)	Payout (years)	Company Net After Federal Income Tax	Rate of Return (percent)
UNSTIMULATED CASES						
One Conventional Well	$ 25,000	1.54	1.54	2.7	$102,425	33.8
Four Conventional Wells	$100,000	4.31	43.2	2.8	$255,871	29.4
STIMULATED CASES						
Probable Well 1	$650,000	4.99	50.1	13.0	$ 85,500	2.9
Probable Well 2	$400,000	4.99	50.1	6.6	$193,689	9.4
Probable Well 3	$275,000	4.99	50.1	4.2	$252,652	16.2
Probable Well 4	$165,000	4.99	50.1	2.4	$295,418	28.6
Minimum Well	$400,000	3.81	38.2	10.5	$112,775	5.2
Maximum Well	$400,000	5.53	55.5	5.3	$229,629	11.7

TABLE 11.7 — EXAMPLE RESERVOIR B: ECONOMIC CALCULATION SUMMARY

FIXED ECONOMIC DATA:

Company net interest	0.875
Federal and state income taxes, percent	54
Gas contract	$0.145 Mscf with 1 cent escalation each 5 years
Gas production tax	$4.00/MMscf
Conventional production well cost	$350,000
Operating cost (excluding production tax)	$2,000/year
Calculation period	20 years

DEVELOPMENT ECONOMICS (Basis—640 Acres)

	Gas Produced (Bscf)	Recovery (percent)	Investment Cost (1,000's)	Payout (years)	Company Net After Federal Income Tax (1,000's)	Rate of Return (percent)
One Conventional well	9.856	5.3	$ 350	4.3	$ 658	19.4
Probable shot well	59.55	31.9	$1,250	1.3	$4,284	50.6
Probable shot well	59.55	31.9	$1,750	1.8	$4,054	36.1
Maximum shot well	72.34	38.8	$1,250	1.3	$5,348	56.6
Minimum shot well (No fracturing beyond cavity)	35.28	18.9	$1,250	2.7	$2,323	24.4

TABLE 11.8 — PROPERTIES FOR EXAMPLE RESERVOIR C

Spacing, acres	80
Depth to top of pay, ft	7,800
Gross thickness of producing formation, ft	400
Net thickness of pay, ft	200
Initial pressure, psi	3,800
Reservoir temperature, °F	140
Bubble point, psi	1,500
Solution GOR, scf/bbl	500
Gas viscosity at bubble point, cp	0.0148
Oil viscosity at bubble point, cp	0.88
Gas formation volume factor at bubble point, cu ft/scf	0.0095
Oil formation volume factor at bubble point, bbl/STB	1.34
Porosity, percent	10
Median oil permeability at IWS, md	0.1
Oil saturation, original, percent	55
Water saturation, original, percent	45
Bulk density, gm/cc	2.52
Original oil in place	5.2×10^6 STB/80 acres
Original gas in place	2.6 Bscf/80 acres

TABLE 11.9 — EXAMPLE RESERVOIR C: ECONOMIC CALCULATION SUMMARY

FIXED ECONOMIC DATA:

Company gross interest, percent	100.00
Company net interest, percent	87.5
Federal and state income tax, percent	54
Oil sale price per barrel	$2.75
Gas sale price per Mcf	$0.15
Operating cost per well per year	$3,415
Cost of conventional well	$110,000
Calculation period, years	40

ECONOMIC EVALUATION:

	Total Invest-ment	Payout Period (years)	Company Net After Federal Income Tax	Rate of Return (percent)
PRORATED CASES				
Conventional Well	$110,000	3.7	$224,000	17.8
Stimulated Wells	$110,000	2.8	$767,000	33.0
	$480,000	8.0	$620,000	10.7
	$580,000	9.6	$570,000	8.5
	$680,000	11.2	$520,000	6.9
NONPRORATED CASES				
Conventional Well	$110,000	3.5	$225,000	18.9
Stimulated Wells	$275,000	0.9	$842,000	56.3
	$480,000	3.1	$708,000	21.1
	$580,000	4.2	$658,000	15.9
	$680,000	5.3	$608,000	12.4

Deep Oil Reservoir D

The final case considered by Bray *et al.* involves a well characterized by two thin, low-productivity formations separated by a shale barrier but close enough to be stimulated by one large nuclear device. The formation characteristics and economic factors are shown in Table 11.10.

Production assumptions were: (1) conventional well, maximum recovery using dual completion; (2) minimum recovery from both zones with nuclear stimulation; (3) maximum recovery from both zones with nuclear stimulation; and (4) maximum recovery from Zone 1 only with nuclear stimulation.

The capacity of the nuclear devices was 40 kt for the single zone and 100 kt for both zones. Several different device and service charges were considered for each case.

The economic feasibility of this study is presented graphically as Fig. 11.14. This figure shows that when maximum recovery from both zones is achieved with one device, rates of return that are more attractive than those for conventional wells can be effected, with device

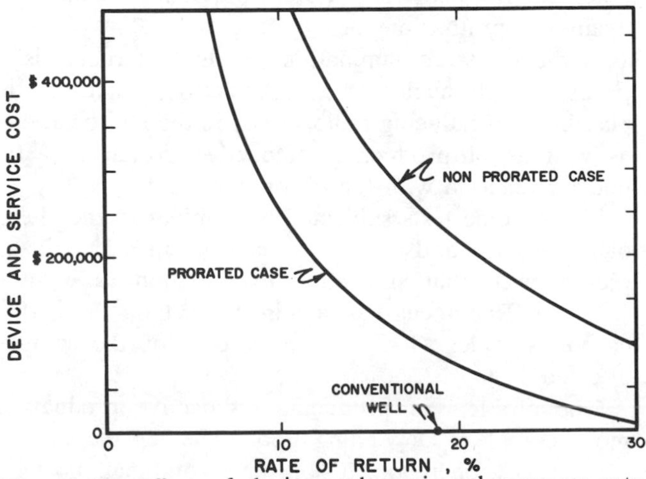

Fig. 11.13 Effect of device and service charges on rate of return, Reservoir C.[14]

TABLE 11.10 — EXAMPLE RESERVOIR D

RESERVOIR SUMMARY:

Zone 1 Max. — 80,000 STB	Zone 2 Max. — 40,000 STB
Zone 1 Min. — 50,000 STB	Zone 2 Min. — 25,000 STB

ECONOMIC SUMMARY:

Acres per well, conventional	80
Acres per well, nuclear stimulated	320
Lifting cost per well per year	$7,000
Oil price per barrel	$2.25 to $2.55
Gas price per Mscf	$0.13 to $0.15
Cost per well, conventional	$95,000
Cost per well, emplacement and reconditioning of well for production	$190,000

and service costs of $470,000 or less. If nuclear stimulation results in minimum recovery from both zones, or maximum recovery form only Zone 1, economically attractive comparisons occur at device and service costs of $220,000 or less.

Summary of Reservoir Case Studies

Reservoir Case A showed that a shallow, low-productivity gas reservoir may not be a good prospect for nuclear stimulation when compared with multiple, conventional well development.

Reservoir Case B showed that a deep, thick, low-productivity gas reservoir is a good prospect for nuclear stimulation as opposed to a conventional completion.

Oil Reservoir Case C showed that device and service charges must be reduced from present published values to compete economically with the cost of conventional well development. Also, nuclear stimulation will be more attractive in nonprorated than in prorated areas.

Oil Reservoir Case D indicates that multiple, thin, low-productivity reservoirs are amenable to nuclear stimulation if the zones can be stimulated by a single large device.

11.5 Post-Nuclear-Fracturing Operating Safety

Cauthen reports only minor damage to surface facilities as a result of nuclear fracturing.[18] Buried pipelines have presented no problem.[14,18] Cased holes in excess of (600) $\times$ (kilotons in the nuclear device)$^{1/3}$ ft from point of detonation have not been damaged *if* both the hole and the explosion are in alluvium. No data are available on other media.[14]

Radioactive contamination of hydrocarbons is a problem with nuclear well stimulation. There is the possibility of inducing radioactivity in the hydrocarbons, as well as of producing radioactive products of the nuclear reaction with the oil or gas.

The Gnome tests subjected oil samples to the effects of gamma ray and neutron bombardment.[9] These samples showed that subjection to radiation as high as 7×10^5 Roentgens had a minor effect on the hydrocarbon samples (less than 5 percent of the samples affected).

The problem of producing radioactive products is more complex. Depending upon whether the nuclear event is one of fission, fusion, or a combination of the two, the amounts of gaseous and oil-soluble radioactive products can be predicted.[19,20] Probably, krypton-85 and tritium will be the problem nuclides in the gas and oil.

Other particulate material could be dissolved or physically entrained in produced liquids. Water produced with the hydrocarbons may leach additional radioactive material from the fused zone. These problems will have to be dealt with through the choice of producing techniques and the selection of reservoirs to be stimulated with nuclear explosions.

11.6 Exploitation of Oil Shale with Nuclear Energy

It is beyond the scope of this book to discuss in detail the use of nuclear energy to recover oil from the massive shale oil deposits of the Rocky Mountain region of the U.S.

As early as Jan., 1959, the AEC, the USBM and the oil industry were seriously discussing the feasibility of a series of tests to evaluate nuclear energy as the rock crushing or rubbling force in the shale oil extraction process.

Hydraulic fracturing cannot compete with nuclear energy as a means of rubbling rock; however, where in-situ thermal recovery processes are planned that are to be conducted through a fracture system, hydraulic fracturing offers advantages over the mass rubbling process.

Considerable difficulty has been experienced in fracturing the oil shale. The keys to hydraulic fracturing in a matrix of this type are (1) isolation of the zone with good primary cementing of the well casing, and (2) the use of deeply cut, sand-abraded notches to weaken the formation so that fractures will be initiated in a direction designed to effect maximum oil recovery.

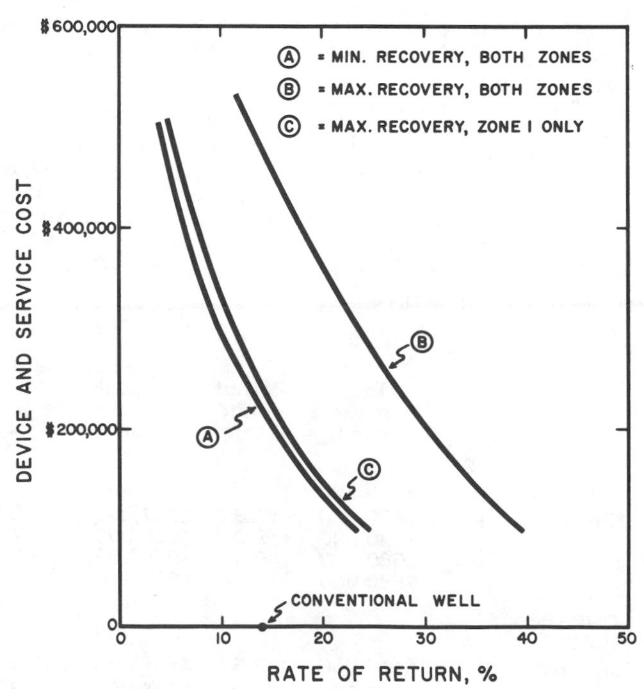

Fig. 11.14 Effect of device and service charges on rate of return, Reservoir D.[14]

11.7 Project Gasbuggy

Project Gasbuggy, a nuclear energy stimulation test, was conducted in the Pictured Cliffs formation of New Mexico on Dec. 10, 1967.

This project was by far the most carefully planned test ever proposed for the evaluation of nuclear energy as an oil and gas formation stimulation device. The broad outline of the project and the predicted and observed results are given here to illustrate the thinking of many engineers on the potential of this stimulation process.

Location of Test Site

The test location is in the SW/4 of Sec 36, Twp 29N, Rge 4W, N.M.P.M., Rio Arriba County, N.M., on the eastern side of the San Juan basin, a structural feature of the Colorado Plateau Province in northwestern New Mexico and southwestern Colorado.

Description of Test Formation

There are eight gas producing formations of Cretaceous age in the central basin. These formations, in descending order, are as follows: Kirtland, Fruitland, Pictured Cliffs, Lewis, Mesaverde, Mancos, Graneros, and Dakota. However, the significant production is obtained from the Pictured Cliffs, Mesaverde, and Dakota formations. The sequence and thickness of these formations at the test site are shown in Fig. 11.15.

The Pictured Cliffs formation is composed of interbedded sand and shale deposits, and at the test site it is a low-productivity reservoir due to the generally low permeability of the sand deposits except where networks of natural fractures exist. The reservoir is contained by the overlying Fruitland formation, 100 ft thick, and the underlying Lewis formation, more than 1,500 ft thick. The characteristics of this formation are given in Table 11.11.

Well spacing for the Pictured Cliffs reservoir is 160 acres; therefore the reservoir would contain approximately 5,280 MMcf of in-place gas.[23,24] However, performance and decline curves indicate only 13 percent of the in-place gas will ultimately be recovered, 10 percent in a 20-year period, from a well at this location completed with present stimulation methods based on gross gas in place. Data from producing wells in the area indicate that the perforating and fracturing method now used results in an initial stabilized deliverability rate of 275 Mcf/D and average production of 74 Mcf/D calculated over a 20-year period.

Field Test Procedure

There were certain experimental objectives of the Gasbuggy project:

1. To determine the change in productivity of existing wells within effective range of the nuclear detonation.
2. To determine the productivity of the post-shot well drilled into the chimney, as compared with the productivity of the unstimulated formation.
3. To determine the increase in producible reserves.
4. To determine the extent of radioactive contamination of produced gas.
5. To determine the extent of mixing of formation gas with contaminated chimney gas.

TABLE 11.11 — PICTURED CLIFFS FORMATION

Depth to top of formation, ft	3,850
Gross thickness of formation, ft	300
Net thickness of formation, ft	190
Average permeability, md	0.14
Average porosity, percent	11
Average gas saturation, percent of PV	41
Original gas in place, cu ft/acre	33×10^6

Following is the experimental procedure.

1. *Pre-shot test wells* were drilled to confirm by intensive coring and analysis the absence of mobile water in the Ojo Alamo and Fruitland formations, and the reservoir conditions and flow behavior of the Pictured Cliffs formation.
2. *The emplacement hole* was drilled to accommodate the nuclear device and furnish data as to depth of casing collapse after detonation.
3. *The post-shot re-entry well* was drilled into the rubble or chimney zone overlying the point of detonation to obtain production from the formation after nuclear stimulation.
4. *Additional post-shot wells* were drilled to evaluate the effectiveness of nuclear stimulation.

Test Results (Gasbuggy)

The maximum yield device that could be used for stimulation of the Pictured Cliffs formation, assuming the presence of the Ojo Alamo as an aquifer approximately 600 ft above the shot level, is 30 kt. The use of a 30-kt device detonated 40 ft below the Pictured Cliffs formation was estimated to create a lateral fracture radius of 280 ft (conservative estimate) to 600 ft (optimistic estimate), and recovery for a 20-year period would range from approximately 3,625 to 8,734 MMcf, depending upon well spacing and fracture radius.

Tests[25] conducted after the detonation of the 26-kt device in the Gasbuggy Test Well GB-1 on Dec. 10, 1967, showed a spherical cavity 166 ft in diameter and chimney dimensions of 160 ft in diameter and 133 ft in height. The radius of the fractured zone around the well was approximately 400 ft and the vertical extent of this zone was 430 ft.

Subsurface ground velocities measured in Instrument Hole GB-D located 1,470 ft from the emplacement hole range from about 1.15 to 1.6 m/sec depending on the gauge location. These values are somewhat but not surprisingly higher than expected. Peak surface velocities range from 1.6 m/sec at the surface above the explosion to about 0.4 m/sec 8,400 ft away. These data, as well as preliminary values of ground motion out to about 60 miles, are in very good agreement with preshot expectations. No damage was sustained by any of the conventional gas wells, which were as close as

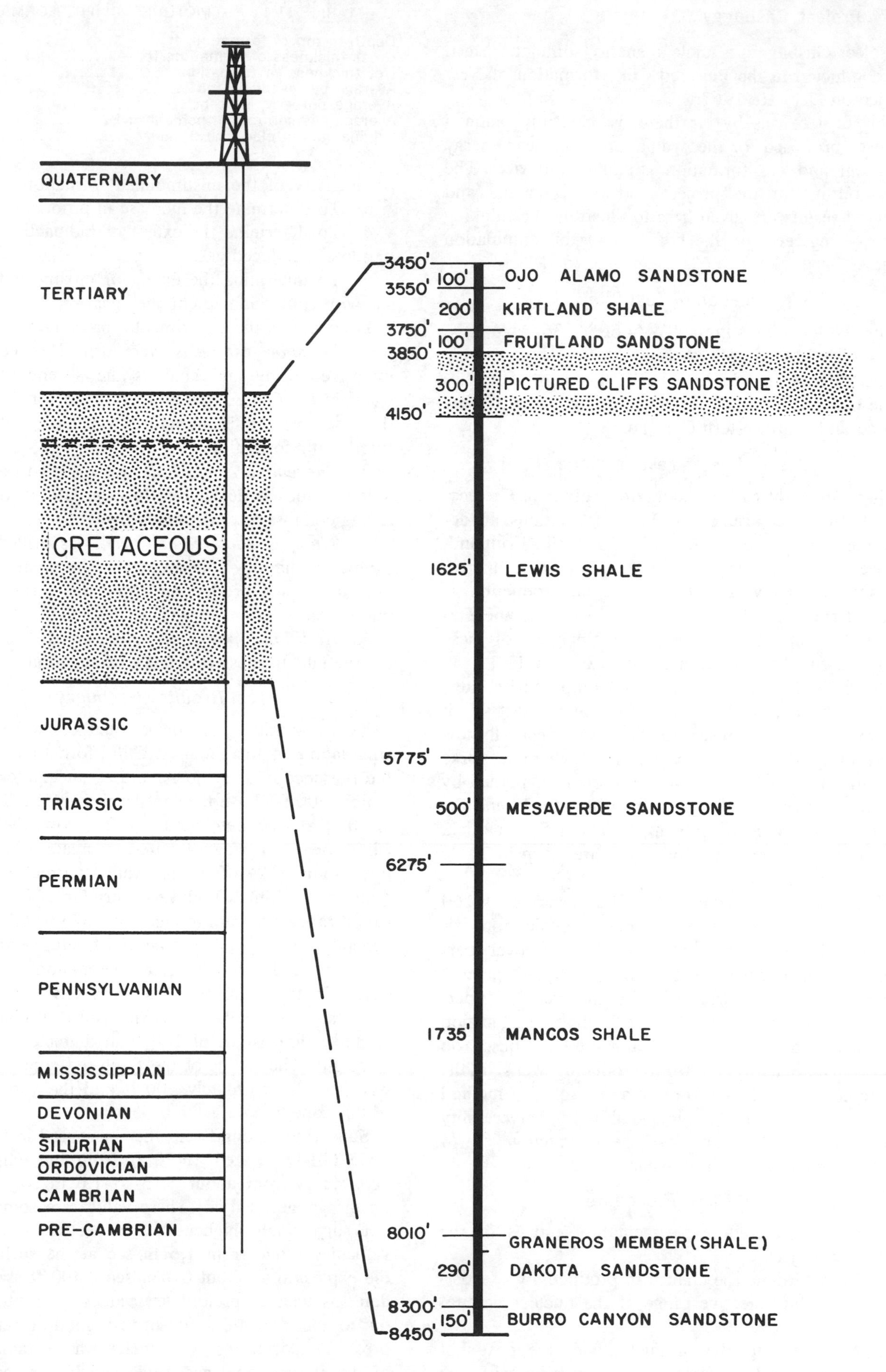

Fig. 11.15 Generalized geologic time column of the San Juan basin.[23]

2,600 ft to the explosion, and no structural damage has been reported from the surrounding area.

Tables 11.12 and 11.13 contain estimates, for a 20-year period, of the results obtained from a 30-kt nuclear device and from conventional stimulation using well spacings of 160, 320, and 640 acres.

Post-treatment production tests after firing the 26-kt device were started on June 28, 1968. These tests consisted of producing at a rate of 5 MMcf/D for 6 days, shutting the well in 1 day for bottom-hole pressure and temperature measurements, and resuming the 5 MMcf/D flow for 5 days. Again shut in briefly for measurement, the well was re-opened at a rate of 750 Mcf/D for 4 days. Approximately 57 MMcf of gas was produced during the test.

At the beginning of the test, surface wellhead pressure was 954 psia and bottom-hole pressure, measured at 3,790 ft, was about 1,100 psia. During the brief shut-in period prior to 750 Mcf/D flow, bottom-hole pressure was about 800 psia and bottom-hole temperature was 247°F. Flowing wellhead temperature during the test rose from 68° initially to 154° during production at 5 MMcf/D and declined to 82° toward the end of production at 750 Mcf/D.

Field analysis of gas samples collected during the initial production tests shows that the 36 percent carbon dioxide content found initially decreased continuously and that hydrocarbons increased continuously, as expected.

Measurements for radioactive constituents were made throughout the flow test. Both krypton-85 and tritium were detected, but radioactivity was below health hazard levels. Later production tests during mid-1969 showed that if the well continues to produce at its post-treatment rate—and assuming gas in place at 4.5 Bcf for 160 acres—ultimate recovery will be about 22 percent.

The major benefit from nuclear energy stimulation is an increase in the recoverable reserves that would be attributable to each tract so stimulated. Of additional benefit is the storage capacity of a large effective wellbore. Volumes of approximately 100 to 300 MMcf of gas could be produced upon demand at about 10 MMcf/D for intervals of 10 to 30 days, with subsequent deliverability probably in excess of 1 MMcf/D.

A substantial amount of acreage in the San Juan basin contains greater amounts of in-place gas than does the Pictured Cliffs formation. One of the better areas examined covers approximately 23,000 acres and contains 44 MMcf/acre, which would result in a significant increase above the recovery figures shown on Tables 11.12 and 11.13. In such an area deliverability levels would also be substantially higher, and nuclear energy stimulation would be much more economical.

Project Cost (Gasbuggy)

It is estimated that the total cost of the experiment was $3 million, exclusive of the cost of the explosive (see Table 11.14).

The latest information made available by the AEC projects the range of charges for nuclear explosives from $350,000 for yields of 10 kt, to $600,000 for yields of 2 mt. These charges are projected assuming quantity of production, and include arming and firing, but not safety studies, site preparation, transportation, emplacement, or support. However, significant reductions in cost might be achieved through improvements in the technology of explosive design and production techniques.

Reducing the diameter of the explosive would result in considerable savings because the cost of an emplacement hole at any given depth increases exponentially with hole size. An explosive that could be emplaced in 7-in. OD casing would reduce the cost of the emplacement hole to that of a conventionally drilled well and might entirely eliminate the emplacement hole cost where such wells already exist.

Another desired development would be a means of channeling the force of the nuclear explosion laterally

TABLE 11.12 — PROJECT GASBUGGY ESTIMATED RECOVERY

(30 kt, Pictured Cliffs Formation)

Gas In Place, MMcf (well spacing, acres)	MMcf (Percent In-Place Gas) Conventional Stimulation	Nuclear Stimulation Conservative	Nuclear Stimulation Optimistic
5280 (160)	537 (10)	3625 (69)	3827 (73)
10560 (320)		5544 (53)	6564 (62)
21120 (640)		6808 (32)	8734 (41)

TABLE 11.13 — PROJECT GASBUGGY ESTIMATED DELIVERABILITY

(30 kt, Pictured Cliffs Formation)

CONVENTIONAL STIMULATION

Well Spacing (acres)	Initial Stabilized Deliverability (Mcf/D)	Average Production (Mcf/D)
160	275	74

NUCLEAR STIMULATION

Well Spacing (acres)	Initial Stabilized Deliverability (Mcf/D)		Average Production (Mcf/D)	
	Conservative	Optimistic	Conservative	Optimistic
160	1918	3539	497	524
320	1588	2551	760	899
640	1358	2000	933	1196

TABLE 11.14—COST ESTIMATE FOR PROJECT GASBUGGY*

Item	Cost
Two preshot test wells	$ 140,000
Emplacement hole	200,000
Miscellaneous construction	265,000
Post-shot re-entry well	150,000
Two post-shot test wells	270,000
Engineering and inspection	200,000
Well testing	50,000
Safety	425,000
Support	910,000
Contingency	370,000
Total	$2,980,000

*Does not include cost of nuclear device

into just the preselected productive zone to obtain greater concentration of its effect. This might be accomplished by firing arrays of shots at the same depth simultaneously or sequentially.

If safety studies are made on a field-wide basis, prorated cost to individual wells may be insignificant. Similarly, the cost per well for any necessary decontamination or dilution of the produced gas should be low. As a preliminary estimate, costs of drilling and completion of a post-shot re-entry well should be no greater than twice that of a conventionally completed well.

11.8 Project Rulison[26-28]

The second nuclear blast designed to stimulate natural gas production from the Mesaverde formation was conducted in Garfield County, Colo., about 40 miles northeast of Grand Junction, on Sept. 10, 1969.

The shot point in the Rulison test was at 8,427 ft depth and was a fission-type nuclear device nominally equivalent in power to 40,000 tons of TNT. The rubble-filled, roughly cylindrical-shaped chimney that should result from the explosion will be an estimated 216 ft in diameter × 450 ft high. Diameter of the fractured rock zone around the chimney is expected to reach 580 ft.

The Rulison device (which was about 1.5 times as powerful as the Gasbuggy explosive) weighed only 1,500 lb, was run on wireline, and did not require an elaborate environmental control capsule. In addition, drill-hole and casing sizes were standard, whereas Gasbuggy required special drilling and completion techniques.

The nuclear explosive was installed in Well R-E in mid-August. The device was lowered in the well on ¾-in. wireline suspended from a specially constructed wireline winch. A switch on the bottom of the device was activated to confirm proper positioning when the explosive was seated on the cement plug.

After the explosive was seated in Well R-E, the casing was filled to within 200 ft of the surface (stemmed) with alternating layers of sand and gravel. Test well R-EX, 300 ft from the emplacement well (Well R-E), was stemmed with cement. Well R-E was sounded with wireline during stemming to determine fillup. Wellhead assemblies were tested to 3,000 psi. The explosive was detonated and monitored from a control point only 2.6 miles from Well R-E, which was also surface ground zero.

In fission-type explosions, several radioactive contaminants are produced. The most common are xenon-133, iodine-131, krypton-85 and tritium. In a fusion or so-called hydrogen-type explosion, as was used in the Gasbuggy experiment, varying amounts of the same contaminants are formed, except that tritium concentration is much higher. Of all contaminants, tritium causes the most concern.

Because of the 6-month waiting period before the chimney can be entered, results of the nuclear shot will not be known until about April, 1970. If possible, Well R-E will be cleaned out and used for test purposes.

Alternative plans call for re-entry of R-EX if mechanical problems block the wellbore in R-E. A similar mechanical problem at shallow depth in R-EX might result in starting a completely new well.

The re-entry well will be extensively tested to determine chimney and fracture zone geometry and flow capacity resulting from the stimulation treatment. Test results will be used to modify the preshot reservoir model so that better deliverability predictions for nuclear stimulation can be made. Also, it will then be possible to predict nuclear stimulation potential of the Rulison area.

During the test period the gas also will be carefully checked for radiation hazard and possible "clean up" techniques, if required, will be evaluated.

11.9 Summary

The effect of a shock wave from the Gnome nuclear explosion on reservoir rock was a function of the brittleness of the rock. Limestones and dolomites, for example, were extensively fractured and showed large increases in permeability and porosity. Sandstones were reduced in permeability while their porosity was essentially unchanged. These results indicate that the use of nuclear explosives to fracture low-productivity gas or oil reservoirs is technically feasible. The shock wave from such an explosion in a reservoir would create a cavity, chimney, and a broken or fractured zone that would increase deliverabilities of gas or oil to the wellbore. Stabilized post-shot producing rates would be controlled by the size of the broken-rock zone and the permeability of the matrix, because the flow restriction would be at the junction of the broken rock and the unchanged zones.

One primary criterion for the application of nuclear exposives is that the formation be sufficiently thick and deeply buried to confine the fractured zone to the hydrocarbon-bearing formation, or at least to adjacent impermeable formations.

The first test of nuclear stimulation, Project Gasbuggy (26-kt device), was in a low-productivity gas zone. Here, the possible contamination by leaching of particulate matter from the fused zone would be eliminated, so that the only contaminant would be radioactive gases. Verification of the calculations in such a gas reservoir should provide the base for applications in oil reservoirs. The second nuclear stimulation utilizing a 40-kt device, Project Rulison, will provide additional data for evaluating this type of well stimulation.

In the final analysis, economics will govern the application of this new stimulation technique. The published price of $350,000 for a small device and $600,000 for a large one restricts the use of this energy source. Reduction of the cost would accelerate the development of nuclear stimulation techniques. Successful utilization of nuclear explosions would materially increase the world hydrocarbon reserves.

References

1. "Oil's Interest in A-Bomb Grows, But . . .", *Pet. Week* (Sept. 26, 1958) 13.

2. Atkinson, C. H. and Lekas, M. A.: "Atomic-Age Fracturing May Soon Open Up Stubborn Reservoirs", *Oil and Gas J.* (Dec. 2, 1963) **LXI,** No. 48, 154-156.

3. Rawson, D. E.: "Review and Summary of Some Project Gnome Results", *Trans.,* AGU (March, 1963).

4. Nathans, M. W. and Holzer, A.: "Shock-Induced Transitions—Sample History and Sample Recovery", Report PNE-112F, Part 1, AEC-UCRL (June, 1963).

5. Rawson, D. E., Boardman, C. R. and Jaffe, N.: "The Environment Created by a Nuclear Detonation in Salt", Report PNE-107F, AEC (to be published).

6. Boardman, C. R., Rabb, D. D. and McArthur, R. D.: Report UCRL 7350, Lawrence Radiation Laboratory, Livermore, Calif. (May, 1963); "Characteristic Effects of Contained Nuclear Explosives for Mining Application", Rev. I, UCRL 7350 (Sept., 1963).

7. Flangas, W. G. and Rabb, D. D.: Report UCRL 6636, Lawrence Radiation Laboratory, Livermore, Calif. (Sept., 1961).

8. Rabb, D. D.: "Large Borehole Drilling Cost", Lawrence Radiation Laboratory, Livermore, Calif. (to be published).

9. Coffer, H. F., Bray, B. G., Knutson, C. F. and Rawson, D. E.: "Effects of Nuclear Explosions on Oil Reservoir Stimulation", *J. Pet. Tech.* (May, 1964) 473-480.

10. Boardman, C. R., Rabb, D. D. and McArthur, R. D.: "Contained Nuclear Detonations in Four Media—Geologic Factors in Cavity and Chimney Formation", *Engineering with Nuclear Explosives,* USAEC TID 7695 (1964) 109-126.

11. Atkinson, C. H.: "Subsurface Fracturing from Shoal Nuclear Detonation", Report PNE 3001, USAEC-USBM (Nov., 1964).

12. Rawson, D. E.: "Review and Summary of Some Project Gnome Results", Report UCRL 7166, Lawrence Radiation Laboratory, Livermore, Calif. (Dec., 1962).

3. Nordyke, M. D.: "Cratering Experience with Chemical and Nuclear Explosives", *Engineering with Nuclear Explosives,* USAEC TID 7695 (1964) 51-73.

4. Bray, B. G., Knutson, C. F., Wahl, H. A. and Dew, J. N.: "Economics of Contained Nuclear Explosions Applied to Petroleum Reservoir Stimulation", *J. Pet. Tech.* (Oct., 1965) 1145-1152.

15. Coffer, H. F., Bray, B. G. and Knutson, C. F.: "Applications of Nuclear Explosives to Increase Effective Well Diameters", *Engineering with Nuclear Explosives,* USAEC TID 7695 (1964) 269-288.

16. Frank, W. J.: "Characteristics of Nuclear Explosives", *Engineering with Nuclear Explosives,* USAEC TID 7695 (1964) 7-10; also Report UCRL 7870 (1964).

17. Johnson, G. W. and Violet, C. E.: "Phenomenology of Contained Nuclear Explosions", Report UCRL 5124, Rev. I, Lawrence Radiation Laboratory, Livermore, Calif. (Dec., 1958).

18. Cauthen, L. J., Jr.: "Survey of Shock Damage to Surface Facilities and Drilled Holes Resulting from Underground Nuclear Detonations", Report UCRL 7694, Lawrence Radiation Laboratory, Livermore, Calif. (July, 1964).

19. Batzel, R. E.: "Radioactivity Associated with Underground Nuclear Explosions", Report UCRL 5623, Lawrence Radiation Laboratory, Livermore, Calif. (June, 1959).

20. Stead, F. W.: "Distribution in Groundwater of Radionuclides from Underground Nuclear Explosions", *Engineering with Nuclear Explosions,* USAEC TID 7695 (1964) 127-138.

21. "Can Underground Blast Make Shale Oil?", *Oil and Gas J.* (Jan. 12, 1959) **LVII**, No. 2, 58.

22. Coffer, H. F. and Aronson, H. H.: "Commercial Applications of Nuclear Explosives", Paper 851-40-E, presented at Mid-Continent District, Div. of Production Meeting, API, Tulsa, Okla., March 30-April 1, 1966.

23. "Project Gasbuggy—Feasibility Study", El Paso Natural Gas Co., USAEC and USBM (May 14, 1965).

24. Gevertz, Harry: "Project Gasbuggy—Fracturing with Nuclear Explosive", *Bull.,* AAPG (1968).

25. Holzer, Fred: "Gasbuggy—Preliminary Post-shot Summary Report", Report PNE 1003, Lawrence Radiation Laboratory, Livermore, Calif.

26. "Project Rulison—New Try at Nuclear Stimulation", *World Oil* (Sept., 1969) 67-71.

27. "Rulison Goes off Without Hitch after Needless Delay", *Oil and Gas J.* (Sept. 15, 1969) 40.

28. "Rulison Nuclear Device Is Finally Detonated", *World Oil* (Oct., 1969).

Nomenclature

Chapter	Symbol	Definition
9	a	= interest rate, decimal fraction
3	A	= drainage area
4, 9	A	= area
2	A, A_1, A_2	= constants in the variable cohesive-strength theory
4	A_C	= cross-sectional area of core through which filtrate tests are made
4	A_{ff}	= area of one fracture face
6	A_p	= wetted area of the proppant
8	A_p	= area of pump plunger
6	A_s	= wetted surface
6	A_w	= total wetted area per square inch
2	b	= zone of plasticity
7	bpm	= barrels per minute (bbl/min)
3	B	= oil formation volume factor
6	B	= constant, characteristic of the formation
9	B	= variable correction factor
2	B, B_1, B_2	= constants in the variable cohesive-strength theory
	BOPD	= barrels oil per day
2	c	= one-half major axis of ellipse
3, 10	c	= compressibility psi^{-1}
4	c	= isothermal coefficient of compressibility of the fluids
4	C	= constant, "fracturing fluid coefficient", a measure of the flow resistance of the fluid
8	C	= conversion constant
9	C	= constant which is characterized by the fluid
4	C_D	= particle drag coefficient
4	C_i	= injection concentration of propping agents
6	C_K	= Kozeny-Carman constant
6	C_o	= shape factor for flow in a particular cross-section, a value of about 2.5 for a bed of spheres
2	C_{oL}	= atmospheric compressive strength of the randomly oriented long-crack material
2	C_{oLc}	= atmospheric compressive strength for the most critical orientation of α
2	C_{os}	= atmospheric compressive strength of the randomly oriented short-crack material
4	C_{pa}	= propping agent concentration within fracture
3	CR	= condition ratio
6	C_S	= sand concentration
4	C_t	= combined fracturing fluid coefficient (viscosity, fluid loss, and compressibility)
6	C_v	= volumetric solids concentration
4	C_I	= fracturing fluid coefficient (viscosity and relative permeability)
4	C_{II}	= fracturing fluid coefficient (reservoir fluid viscosity compressibility)
4	C_{III}	= fracturing fluid coefficient (wall building)
4	d_D	= particle shape factor, d_{max}/d_{min}
6	d_i	= diameter of the impression made during embedment of the proppant into the fracture face
4, 6	d_{pa}	= diameter of propping agent particles
6	D	= depth of proppant embedment

Chapter	Symbol	Definition
11	D	= depth of burial
4, 9	e	= transcendental number, 2.718
8	ehp	= engine or brake horsepower
4	erfc(x)	= complementary error function of x
8	erpm	= engine revolutions per minute
2, 4	E	= Young's modulus of elasticity
2	f_L	= coefficient of friction, long crack
2	f_s	= coefficient of friction, short crack
11	F_c	= a factor depending upon rock and fluid content relating cavity radius to other quantities (ranges from 261 to 343)
9	F_k	= k_f/k_e
11	F_p	= factor relating radius of cavity and permeable zone (ranges from 2.4 to 3.5)
11	F_z	= a factor relating height of permeable zone to cavity radius (approximately 5.7)
2, 3, 9, 10, 11	h	= formation thickness
4	h_f	= height of vertical fracture
6, 10	h_n	= net pay thickness
6	h_o	= section open above settled sand
6	h_s	= height of settled sand bed
4	$h_t/2$	= one-half the formation thickness
7, 8	hhp	= hydraulic horsepower
4, 6, 7	i	= injection flow rate
	i_{bpm}	= pump rate barrels per minute
4	i_L	= volume rate of fluid loss to formation
9	I_c	= cumulative interest on operating expense
3, 10	J	= *generalized* productivity index (PI) in B/D/psi drop in pressure
6	J_f	= productivity after fracturing
6, 10	J_o	= productivity before fracturing
3, 4, 6	k	= permeability of formation to oil
10	k	= permeability of the unfractured formation
11	k	= permeability of the formation, md

Chapter	Symbol	Definition
4	k'	= a measure of the flow properties of a non-Newtonian fluid
9, 10	k_{avg}	= average permeability of formation after fracturing
3	k_{BU}	= permeability as determined from pressure buildup test
6, 9, 10	k_e	= original horizontal permeability of formation
6, 9, 10	k_f	= fracture permeability
9, 10	$k_f w$	= flow capacity of fracture
6	$k_f W_f$	= fracture capacity
10	k_h	= horizontal permeability
10	$(kh)_f$	= k_n of fracture
10	$(kh)_o$	= k_n of unfractured formation
10	k_n	=
10	k_o	= effective permeability to oil
3	k_{PI}	= permeability as determined from productivity-index test
11	kt	= explosive yield of 1 kt of TNT equivalent, represents energy equal to 10^{12} calories
6	kW_f	= millidarcy-feet
9	k_w/k_o	= relative permeability (water to oil)
10	k_1	= effective horizontal permeability, or permeability of damaged formation
10	k_2	= original horizontal permeability, or permeability of undamaged formation
2	K	= constant, as high as 3
4, 10	L	= length of vertical fracture measured from wellbore
6	L	= length in bed
9	L	= operating or lifting cost, dollars per day
6	L_B	= length of bed or porous medium
6	L_e	= length of path taken by fluid in traversing L_B
8	L_s	= length of stroke per crank revolution
4	L_{sp}	= spurt distance
3	m	= slope of build up curve, psi/cycle
4	m	= slope of the line of filtrate volume vs time
6	m	= constant, characteristic of the formation

Chapter	Symbol	Definition
9	m	= slope of production decline curve
9	$\|m\|$	= absolute value of the slope m
11	mt	= megaton = 1,000 kt
11	M	= 1,000
8	ME	= mechanical efficiency
2	n	= exponent in matrix cohesive equation (Eq. 2.38)
4	n	= integers
4	n'	= a measure of the flow properties of a non-Newtonian fluid
6	n_{pa}	= number of propping agent particles per square inch of fracture surface (one face)
9	N_{nv}	= net value of ultimate recoverable oil
8	N_p	= number of plungers
9	N_p	= cumulative production barrels
4	N_{Re}	= fluid Reynolds number
4, 8	p	= pressure, psi
3	$\bar{p}$	= average pressure
10	$\bar{p}$	= average pressure within the drainage area
7	p_{bf}	= bottom-hole fracture pressure
11	p_d	= formation pressure at r_d
2	p_e	= external pressure
9	p_e	= static bottom-hole pressure
7	p_F	= casing or tubing friction loss
7	p_h	= hydrostatic pressure
2, 3	p_i	= internal pressure, or pressure in well after infinite closed-in time in an infinite reservoir
7	p_{ISI}	= instantaneous shut-in pressure
2	p_{OB}	= weight of the overlying mass per unit area
6	p_{OBe}	= effective overburden pressure
7	p_{pF}	= perforation friction
7	p_{ti}	= surface injection pressure
11	p_w	= flowing pressure at the wellbore
3	p_{wf}	= pressure just prior to buildup test
10	p_{wf}	= bottom-hole flowing pressure
3	p_{ws}	= bottom-hole pressure at time t_n, or static reservoir pressure, or pressure in well during buildup
3	p_1	= pressure 1 hour after closing in
8	prpm	= pump revolutions per minute
8	PA	= plunger area
8	PL	= plunger load
8	PR	= pump reduction ratio
8	P-V	= pressure-volumetric rate data
9	PW	= present worth of future production
3	q	= oil-production rate at surface conditions
8	q	= volumetric pump rate
9	q	= production rate at any particular cumulative production N_P
10	q	= stabilized producing rate
11	q	= rate of gas flow, cu ft/D at 15.025 psia and 60°F
9	q_a	= production rate at abandonment
10	q_d	= damaged well productivity
10	q_f	= well productivity after fracturing
9	q_i	= constant injection rate
3	$\bar{q}_n$	= average flow rate between = $\triangle t_{ws,n}$ and $\triangle t_{ws,n-1}$
9, 10	q_o	= production rate before treatment
10	q_u	= undamaged well productivity
4	Q_f	= volume of fracture
10	r	= well spacing
6	r_B	= distance sand bed extends from wellbore
11	r_c	= radius of cavity, ft
2	r_e	= external radius
3, 6, 9, 10, 11	r_e	= radius of drainage
2	r_{ee}	= radius of the end of the ellipse
4, 9, 10	r_f	= fracture radius
2	r_h	= radius of the hole
2	r_i	= radius of hole or internal radius
11	r_p	= radius of the permeable zone
2	r_r	= distance from center of hole or radius to any point to be considered

Chapter		
10	r_v	= vertical fracture penetration
3, 6, 9, 10	r_w	= wellbore radius
4	r_x	= fracture radius at intermediate radii x
3	s	= dimensionless skin effect
7	scf	= standard cubic feet
6	S	= area of particle surface per unit volume of packed space
6	S_p	= wetted surface area, per unit volume of the propping agent
11	SDB	= scaled depth of burial
8	SOR	= sand-oil ratio
3, 4, 6, 9	t	= time, total pumping time, flow time
9	t_f	= injection time
9	$\bar{t}_{im}$	= mean time within interval Δt_i
10	t_s	= time to approximately steady state
9	T	= total producing life of well, days
11	T_f	= formation temperature, °F
2	T_{ma}	= uniaxial tensile strength or cohesive strength of matrix
2	T'_{ma}	= uniaxial tensile strength or cohesive strength of matrix in the plane of weakness
2	T_o	= uniaxial tensile strength
8	TR	= transmission reduction ratio
4	v	= velocity of flow perpendicular to the fracture face
4, 6	v_b	= velocity of slurry advance
4	v_f	= fluid velocity
6	v_{fH}	= horizontal fluid velocity
4	$(v_f)_{max}$	= maximum fluid velocity from cross-sectional velocity profile
6	v_g	= sand settling rate
4	v_L	= leakoff rate divided by filter area (leakoff velocity)
6	v_p	= volume of the proppant per unit area (one face) of the fracture
4, 6	v_{pa}	= rate of proppant advance; propping agent velocity
4	$(v_{pa}/v_f)_w$	= ratio of particle to fluid velocity at the wellbore
4	V	= filtrate volume
4	V_C	= cumulative fluid loss per unit area

Chapter		
4	V_{sp}	= spurt loss volume
8	VE	= volumetric efficiency
8	VPR	= volume per pump revolution
4, 9	W	= fracture clearance; or fracture width during injection
4	W'	= increased fracture clearance
4, 6, 9, 10	W_f	= fracture width (propped)
4	$\overline{W}_F$	= average fracture width
4	W_{fx}	= fracture width at radius x
4	x	$= 2C\sqrt{\pi t/W}$
9	X	$= 2C\sqrt{\pi t_f/W}$
10	X	$= \dfrac{\text{depth of damaged formation}}{\text{total formation thickness}}$
4	y	$= (2V_L) \div w + 2w_s$
3	Y	= function of the drainage boundary of a well
9	Y	$= 4\pi C^2 A/q_i W$
4	z	= direction normal to the fracture face
11	z	= chimney height
11	$z_{\bar{p}}$	= gas deviation factor
11	Z_p	= height of permeable zone
2	α	= angle between bedding plane and direction of least principal stress
4	β	$= \dfrac{\pi^2\alpha}{4L^2}$ where $\alpha = \dfrac{k}{\phi\mu c_f}$
8	γ	= specific gravity
9, 10	Δp	= producing pressure differential
3	Δp(1 cycle)	= slope buildup curve
3	Δp_n	= buildup pressure $= (p_{ws} - p_{wf})_n$
3	$\Delta p'_n$	= corrected buildup pressure
3	Δp_s	= pressure drop in the skin
3	Δt	= closed-in time of well
9	Δt_i	= time increment
3	$\Delta t_{ws,n}$	= buildup time, hours
9	ΔV_i	= an increment of production during time interval Δt_i
2	$\Delta\sigma_v$	= change in vertical stress imposed by σ_i
11	$\Delta\phi$	= fractional net increase of porosity in rubble zone (ranges from 0.2 to 0.3)
2	ε	= deformation
2	ε_H	= horizontal strain
2	θ	= angle of plane on which principal stress acts
4	θ	= angle of inclination of fracture
4	λ	= time required for the fluid to reach a given point

Chapter		
3, 4	μ	= oil viscosity at reservoir conditions
9, 10	μ	= viscosity of oil
11	μ	= viscosity of gas, cp
6	μ_o	= oil viscosity
2	ν	= Poisson's ratio
2	ν_L	= Poisson's ratio for randomly oriented long crack
2	ν_s	= Poisson's ratio for randomly oriented short crack
2	ξ	= orientation of α that has a minimum value of τ_o
4, 6	π	= 3.1416
4, 8	ρ_f	= fluid density
8	ρ_m	= density of mix
11	ρ_{OB}	= overburden density, gm/cc
4	ρ_{pa}	= propping agent density
2	σ	= normal stress
2	$\bar{\sigma}$	= average stress
2	σ_c	= a constant stress
2	σ_e	= tensile stress at the elastic limit
2	σ_H	= horizontal stress
2	σ_i	= initial stress perpendicular to axis of hole
2	σ_{ie}	= external initial stress
2	σ_{ii}	= internal initial stress
2	σ_{iv}	= vertical initial stress
2	σ_o	= yield stress in pure tension

Chapter		
2	σ_{oct}	= octahedral shear stress
2	σ_r	= radial stress
2	σ_t	= tangential stress
2	σ_v	= vertical stress
2	σ_x	= horizontal stress in x direction
2	σ_y	= normal stress in the Mohr circle or horizontal stress in the y direction
2	σ_1	= principal stress or confining stress
2	σ_2	= principal stress
2	σ_3	= least principal stress
2	τ	= shear stress
4	τ	= fluid yield shear stress
2	τ_{ma}	= shear stress matrix material
2	τ_{oct}	= octahedral shear stress
3, 4, 6, 10	ϕ	= porosity
6	ϕ_p	= porosity of a partial monolayer
2	Φ	= angle of failure of the rock
2	$\tan \Phi$	= coefficient of friction

Subscripts:

L	= long
max	= maximum
min	= minimum
oct	= octahedral
s	= short

Bibliography

Alderman, E. N. and Woodard, G. W.: "Prevention of Secondary Deposition from Waterflood Brines", *Drill. and Prod. Prac.,* API (1957) 98.

Anderson, Terry O. and Stahl, Edwin J.: "A Study of Induced Fracturing Using an Instrumental Approach", *J. Pet. Tech.* (Feb., 1967) 261-267.

Arps, J. J.: "Estimation of Primary Oil Reserves", *Trans.,* AIME (1956) **207,** 182-191.

Atkinson, C. H.: "Subsurface Fracturing from Shoal Nuclear Detonation", Report PNE 3001, USAEC-USBM (Nov., 1964).

Atkinson, C. H. and Lekas, M. A.: "Atomic-Age Fracturing May Soon Open Up Stubborn Reservoirs", *Oil and Gas J.* (Dec. 2, 1963) **LXI**, No. 48, 154-156.

Batzel, R. E.: "Radioactivity Associated with Underground Nuclear Explosions", Report UCRL 5623, Lawrence Radiation Laboratory, Livermore, Calif. (June, 1959).

Bearden, W. G.: "Technical Information on the Hydrafrac Service", unpublished report, Pan American Petroleum Corp., Tulsa, Okla. (Jan., 1953).

Bernard, G. G., Holm, L. W. and Jacobs, W. L.: "Effect of Foam on Trapped Gas Saturation and on Permeability of Porous Media to Water", *Soc. Pet. Eng. J.* (Dec., 1965) 295-300.

Black, Harold N. and Hower, Wayne E.: "Advantageous Use of Potassium Chloride Water for Fracturing Water Sensitive Formations", paper 851-39-F presented at Mid-Continent District Meeting, API Div. of Production, Wichita, Kans., March 31-April 2, 1965.

Blaster's Handbook (du Pont), Sesquicentennial ed., 49 and 443.

Boardman, C. R., Rabb, D. D. and McArthur, R. D.: "Contained Nuclear Detonations in Four Media—Geologic Factors in Cavity and Chimney Formation", *Engineering with Nuclear Explosives,* USAEC TID 7695 (1964) 109-126.

Boardman, C. R., Rabb, D. D. and McArthur, R. D.: Report UCRL 7350, Lawrence Radiation Laboratory, Livermore, Calif. (May, 1963); "Characteristic Effects of Contained Nuclear Explosives for Mining Application", Rev. I, UCRL 7350 (Sept., 1963).

Boriskie, R. J.: "An Investigation of Productivity Increases from Hydraulic Fracturing Treatments", MS thesis, Texas A&M U., College Station (Aug., 1963).

Bray, B. G., Knutson, C. F., Wahl, H. A. and Dew, J. N.: "Economics of Contained Nuclear Explosions Applied to Petroleum Reservoir Stimulation", *J. Pet. Tech.* (Oct., 1965) 1145-1152.

Brown, J. L. and Landers, Mary M.: U. S. Patent No. 2,779,735 (April 18, 1956).

Brown, R. W., Neill, G. H. and Loper, R. G.: "Factors Influencing Optimum Ball Sealer Performance", *J. Pet. Tech.* (April, 1963) 450-454.

Campbell, J. B.: "The Effect of Fracturing on Ultimate Recovery", paper 851-31-L presented at the Mid-Continent District Meeting, API Div. of Production, Tulsa, Okla., April 10-12, 1957.

"Can Underground Blast Make Shale Oil?", *Oil and Gas J.* (Jan. 12, 1959) **LVII**, No. 2, 58.

Card, David C., Jr.: "Review of Fracturing Theories", UCRL 13040, Colorado School of Mines Research Foundation, Inc., Golden, Colo. (April 16, 1962) 14-20.

Carman, P. C.: *Trans.,* Inst. Chem. Eng., London (1937) **15,** 150.

Cauthen, L. J., Jr.: "Survey of Shock Damage to Surface Facilities and Drilled Holes Resulting from Underground Nuclear Detonations", Report UCRL 7694, Lawrence Radiation Laboratory, Livermore, Calif. (July, 1964).

Churchill, R. V.: *Modern Operational Mathematics in Engineering,* McGraw-Hill Book Co., Inc., New York (1944) 107, 111.

Clark, J. B.: "A Hydraulic Process for Increasing the Productivity of Wells", *Trans.,* AIME (1949) **186**, 1-8.

Clark, J. B., Fast, C. R. and Howard, G. C.: "A Multiple-Fracturing Process for Increasing Productivity of Wells", *Drill. and Prod. Prac.,* API (1952) 104.

Clark, R. C., Freedman, H. G., Bolstead, J. H. and Coffer, H. F.: "Application of Hydraulic Fracturing to the Stimulation of Oil and Gas Production", *Drill. and Prod. Prac.,* API (1953) 113.

Coburn, R. W.: "Custom Designed Well Stimulation", *Oil and Gas J.* (March 4, 1957) 102.

Coffer, H. F., Bray, B. G., Knutson, C. F. and Rawson, D. E.: "Effects of Nuclear Explosions on Oil Reservoir Stimulation", *J. Pet. Tech.* (May, 1964) 473-480.

Coffer, H. F., Bray, B. G. and Knutson, C. F.: "Applications of Nuclear Explosives", paper 851-40-E presented Diameters", *Engineering with Nuclear Explosives,* USAEC TID 7695 (1964) 269-288.

Coffer, H. F. and Aronson, H. H.: "Commercial Applications of Nuclear Explosives", paper 851-40-E presented at Mid-Continent District Meeting, API Div. of Production, Tulsa, Okla. March 30-April 1, 1966.

Cottrell, A. H.: *Theoretical Aspects of Fracture in Fracture:* Averbach, B. L. *et al.,* Eds., Technology Press and John Wiley and Sons, Inc., New York (1959).

Coulter, A. W. and David, H. E.: "Fracturing Fluids and Additives", unpublished report, Dowell Div. of Dow Chemical Co. (Jan., 1967).

Craft, B. C., Holden, W. R. and Graves, E. D., Jr.: *Well Design: Drilling and Production,* Prentice-Hall, Inc., Englewood, Cliffs, N. J. (1962) 494.

Crawford, P. B. and Collins, R. E.: "Estimated Effect of Vertical Fractures on Secondary Recovery", *Trans.,* AIME (1954) **201,** 192-196.

Crawford, P. B. and Landrum, B. L.: "Effect of Unsymmetrical Vertical Fractures on Production Capacity", *Trans.,* AIME (1955) **204**, 251-254.

Crawford, P. B. and Landrum, B. L.: "Estimated Effect of Horizontal Fractures on Production Capacity", paper 414-G presented at the Annual Fall Meeting of the Petroleum Branch of AIME, San Antonio, Tex., Oct. 17-20, 1954.

Crawford, P. B., Pinson, J. M., Simmons, J. and Landrum, B. L.: "Sweep Efficiencies of Vertically Fractured Five-Spot Pattern", *Pet. Eng.* (March, 1956) B-95.

Crittendon, B. C.: "The Mechanics of Design and Interpretation of Hydraulic Fracture Treatments", *J. Pet. Tech.* (Oct., 1959) 21-29.

Darcy, H.: *Memoires a l'Academic des Sciences de l'Institute imperial de France* (1858) **15,** 141.

Darin, S. R. and Huitt, J. L.: "Effect of a Partial Monolayer of Propping Agent on Fracture Flow Capacity", *Trans.,* AIME (1960) **219,** 31-37.

Dickey, P. A. and Andersen, K. H.: "Behavior of Water Input Wells — Part 4", *Oil Weekly* (Dec. 10, 1945).

Durand, R.: "Basic Relationships of the Transportation of Solids in Pipes—Experiment Research", *Proc.,* Minnesota Hydraulic Convention, Part I (1953).

Dyes, A. B., Kemp, C. E. and Caudle, B. H.: "Effect of Fractures on Sweep-Out Pattern", *Trans.,* AIME (1958) **213,** 245-249.

Essary, Roy L.: "Fracture Treatments in S.E. New Mexico", *World Oil* (March, 1962) 99.

Farris, R. F.: unpublished report, Pan American Petroleum Corp., Tulsa, Okla. (1946).

Fast, C. R., Flickinger, D. H. and Howard, G. C.: "Effect of Fracture-Formation Flow Capacity Contrast on Well Productivity", *Drill. and Prod. Prac.,* API (1961) 145.

Fast, C. R., Nabors, F. L. and Mase, G. D.: "Propping-Agent Spacer Effective in Gas-Well Fracturing", *Oil and Gas J.* (Sept. 27, 1965) 97.

Fenner, R.: "Untersuchung zur Erkenntnis des Gebirgsdrucks: Glueckauf" (Aug.-Sept., 1938).

Flangas, W. G. and Rabb, D. D.: Report UCRL 6636, Lawrence Radiation Laboratory, Livermore, Calif. (Sept., 1961).

Flickinger, D. H.: "Effect of Formation Damage During Fracturing on Well Productivity", unpublished report, Pan American Petroleum Corp., Tulsa, Okla. (June 18, 1956).

Flickinger, D. H. and Fast, C. R.: "The Engineering Design of Well Stimulation Treatments", paper presented at Annual Meeting of Petroleum Branch of the CIM, Edmonton, Alta., April 1-3, 1963.

Foshee, W. C. and Hurst, R. E.: "Improvement of Well Stimulation Fluids by Including a Gas Phase", *J. Pet. Tech.* (July, 1965) 768-772.

Frac Guide Data Book, Dowell Div. of Dow Chemical Co., Tulsa, Okla. (1965).

"Fracturing: Big and Getting Bigger", *Petroleum Week* (May 13, 1955) 18.

Frank, W. J.: "Characteristics of Nuclear Explosives", *Engineering with Nuclear Explosives,* USAEC TID 7675 (1964) 7-10; also Report UCRL 7870 (1964).

Fraser, C. D. and Pettitt, B. E.: "Results of a Field Test to Determine the Type and Orientation of a Hydraulically Induced Formation Fracture", *J. Pet. Tech.* (May, 1962) 463-466.

Garland, T. M., Elliott, W. C., Jr., Dolan, Pat and Dobyns, R. P.: "Effects of Hydraulic Fracturing Upon Oil Recovery from the Strawn and Cisco Formations in North Texas", RI 5371, USBM (1957).

Gevertz, Harry: "Project Gasbuggy—Fracturing with Nuclear Explosive", *Bull.,* AAPG (1968).

Ghauri, W. K.: "Results of Well Stimulation by Hydraulic Fracturing and High Rate Oil Backflush", *J. Pet. Tech.* (June, 1960) 19-27.

Gilbert, Bruce: "The F I Process . . . Theory and Practice", paper 1021-G presented at Fourth Annual Joint Meeting, Rocky Mountain Petroleum Sections of AIME, Denver, Colo., March 3-4, 1958.

Gilman, J. J.: "Fracture", *Proc.* Intl. Conference, Swampscott, Mass., M.I.T. Press, Cambridge, Mass. (1959) **II,** 193.

Gilman, J. J.: "Fracture in Solids", *Scientific American* (Feb., 1960) **CCII,** 94.

Gladfelter, R. E., Tracy, G. W. and Wilsey, L. E.: "Selecting Wells Which Will Respond to Production-Stimulation Treatment", *Drill. and Prod. Prac.,* API (1955) 117.

Grant, B. F., Duvall, W. I., Obert, L., Rough, R. L. and Atchison, T. C.: "Research on Shooting Oil and Gas Wells", *Drill. and Prod. Prac.,* API (1950) 303.

Gras, E. H.: "Hydraulic Fracturing Mechanical Equipment", unpublished report, Halliburton Co. (Oct., 1966).

Gray, D. H. and Rex, R. W.: "Formation Damage in Sandstones Caused by Clay Dispersion and Migration", *Proc.,* Fourteenth Annual National Conference on Clays and Clay Minerals (1965) **14**.

Grebe, J. J.: "Tools and Aims of Research", *Chem. and Eng. News* (Dec. 10, 1943) **21,** No. 23, 2004.

Grebe, J. J. and Stosser, S. M.: "Increasing Crude Production 20,000,000 Barrels from Established Fields", *World Petroleum* (Aug., 1935) **6,** No. 8, 473.

Griffith, A. A.: "The Phenomena of Rupture and Flow in Solids", *Philos. Trans.,* Royal Soc. (1920) **221A**, 163.

Guerrero, E. T.: "Fracturing Can Help Secondary Recovery", *World Oil* (July, 1958) 126.

Guest, J. J.: *Phil. Mag.* (1900) **50**, 69.

Hall, C. D., Jr., and Dollarhide, F. E.: "Effects of Fracturing Fluid Velocity on Fluid-Loss Agent Performance", *J. Pet. Tech.* (May, 1964) 555-560.

Hartsock, J. H. and Slobod, R. L.: "The Effect of Mobility Ratio and Vertical Fractures on the Sweep Efficiency of a Five-Spot", *Prod. Monthly* (Sept., 1961) 2.

Hassebroek, W. E. and Waters, A. B.: "Advancements Through 15 Years of Fracturing", *J. Pet. Tech.* (July, 1964) 760-764.

Hawsey, Jerry D., Whitesell, L. B. and Kepley, N. A.: "Injection of a Bactericide-Surfactant During Hydraulic Fracturing — A New Method of Corrosion Control", paper SPE 978 presented at SPE 39th Annual Fall Meeting, Houston, Tex., Oct. 11-14, 1964.

Hendrickson, A. R., Nesbitt, E. E. and Oaks, B. D.: "Soap-Oil Systems for Formation Fracturing", *Pet. Eng.* (May, 1957) B-58.

Hendrickson, A. R. and Wieland, D. R.: "Personal Interview with R. C. Phillips, Union Oil Co. of California", Dowell Div. of Dow Chemical Co., Tulsa, Okla. (July 5 and 14, 1966).

Hewitt, Charles H.: "Analytical Techniques for Recognizing Water-Sensitive Reservoir Rocks", *J. Pet. Tech.* (Aug., 1963) 813-818.

Holzer, Fred: "Gasbuggy—Preliminary Post-Shot Summary Report", Report PNE 1003, Lawrence Radiation Laboratory, Livermore, Calif.

Horner, D. R.: "Pressure Buildup in Wells", *Proc.*, Third World Pet. Cong., E. J. Brill, Leiden (1951) **II.**

Howard, G. C.: "Evaluation of the Hydrafrac Process", unpublished report, Pan American Petroleum Corp., Tulsa, Okla. (1965).

Howard, G. C.: "Special Fracturing Test, Pine Island Field, La.", unpublished report, Pan American Petroleum Corp., Tulsa, Okla. (1954).

Howard, G. C. and Fast, C. R.: "Factors Controlling Fracture Extension", paper presented at the Spring Meeting of the Petroleum and Natural Gas Div. of CIM, Edmonton, Alta., May, 1957.

Howard, G. C. and Fast, C. R.: "Optimum Fluid Characteristics for Fracture Extension", *Drill. and Prod. Prac.*, API (1957) 261.

Howard, G. C. and Fast, C. R.: "Squeeze Cementing Operations", *Trans.*, AIME (1950) **189**, 53-64.

Howard, G. C., Flickinger, D. H., Fast, C. R. and Evans, R. B.: "Deriving Maximum Profit from Hydraulic Fracturing", *Drill. and Prod. Prac.*, API (1958) 91.

Howard, G. C. and Scott, P. P., Jr.: "An Analysis and the Control of Lost Circulation", *Trans.*, AIME (1951) **192,** 171-182.

Hubbert, M. K. and Willis, D. G.: "Mechanics of Hydraulic Fracturing", *Trans.*, AIME (1957) **210,** 153-166.

Huitt, J. L. and McGlothin, B. B., Jr.: "The Propping of Fractures in Formations Susceptible to Propping-Sand Embedment", *Drill. and Prod. Prac.*, API (1958) 115.

Huitt, J. L., McGlothin, B. B., Jr., and McDonald, J. F.: "The Propping of Fractures in Formations in Which Propping Sand Crushes", *Drill. and Prod. Prac.*, API (1959) 120.

Huitt, J. L., Pekarek, J. L., Swift, V. N. and Strider, H. L.: "Mechanical Tool for Preparing a Well Bore for Hydraulic Fracturing", *Drill. and Prod. Prac.*, API (1960) 129.

Hurst, W.: "Establishment of the Skin Effect and Its Impediment to Fluid-Flow Into a Well Bore", *Pet. Eng.* (Oct., 1953) B-6.

Jacob, C. E.: *Trans.*, AGU, Part II (1940).

Jaeger, J. C.: *Elasticity, Fracture and Flow,* 2nd ed., John Wiley and Sons, Inc., New York (1962) 208.

Jaeger, J. C.: "Shear Failure of Anisotropic Rocks", *Geologic Magazine* (1960) **XCVII**, 65-72.

Jennings, E. R. and Vincent, R. P.: "A Glass Fabric-Plastic Liner Casing Repair Method", *Drill. and Prod. Prac.*, API (1959) 67.

Johnson, G. W. and Violet, C. E.: "Phenomenology of Contained Nuclear Explosions", Report UCRL 5124 Rev. I, Lawrence Radiation Laboratory, Livermore, Calif. (Dec., 1958).

Jones, Frank O.: "Influence of Chemical Composition of Water on Clay Blocking", *J. Pet. Tech.* (April, 1964) 441-446.

Jones, Frank O. and Fast, C. R.: U. S. Patent No. 3,179,173 (April 20, 1965).

Kastrop, J. E.: "Newest Aid to Multi-Stage Fracturing", *Pet. Eng.* (Dec., 1956) B-40.

Kaufman, M. J.: "Well Stimulation by Fracturing", *Pet. Eng.* (Sept., 1956) B-53.

Kern, L. R., Perkins, T. K. and Wyant, R. E.: "Propping Fractures with Aluminum Particles", *J. Pet. Tech.* (June, 1961) 583-589.

Kern, L. R., Perkins, T. K. and Wyant, R. E.: "The Mechanics of Sand Movement in Fracturing", *Trans.*, AIME (1959) **216,** 403-405.

Kirsch, G.: "Die Theorie der Elastizitat und die Bedurfnisse der Festigkeitslehre", *Zeitschr. des Vereines Deutscher Ingenieure* (1898) **XLII**, No. 29, 797.

Knox, John A., Lasater, R. M. and Dill, W. R.: "A New Concept in Acidizing Utilizing Chemical Retardation", paper SPE 975 presented at SPE 39th Annual Fall Meeting, Houston, Tex., Oct. 11-14, 1964.

Lagrone, K. W. and Rasmussen, J. W.: "A New Development in Completion Methods — The Limited Entry Technique", *J. Pet. Tech.* (July, 1963) 695-702.

Lamb, Horace: *Hydrodynamics,* 6th ed., Dover Publications, New York (1945).

Lamé and Clapeyron: "Memoire sur l'equilibre interieur des corps solides homogenes: *Memoirs presents par divers savans*" (1833) **4.**

Landrum, B. L. and Crawford, P. B.: "Horizontal Fractures Do Affect Ultimate Recovery", *Pet. Eng.* (June, 1961) B-80.

Lefkovits, H. C. and Matthews, C. S.: "Application of Decline Curves to Gravity-Drainage Reservoirs in the Stripper Stage", *Trans.,* AIME (1958) **213,** 275-280.

Lowe, D. K. and Huitt, J. L.: "Propping Agent Transport in Horizontal Fractures", *J. Pet. Tech.* (June, 1966) 753-764.

Lowe, D. K., McGlothlin, B. B., Jr., and Huitt, J. L.: "A Computer Study of Horizontal Fracture Treatment Design", *J. Pet. Tech.* (April, 1967) 559-569.

Mallinger, M. A., Rixe, F. H. and Howard, G. C.: "Development and Use of Propping Agent Spacers to Increase Well Productivity", *Drill. and Prod. Prac.,* API (1964) 88.

Maly, Joe W. and Morton, Tom E.: "Selection and Evaluation of Wells for Hydrafrac Treatment", *Oil and Gas J.* (May 3, 1951) No. 52, 126.

Matthews, C. S.: "Analysis of Pressure Build-up and Flow Test Data", *J. Pet. Tech.* (Sept., 1961) 862-870.

Matthews, C. S., Brons, F. and Hazebroek, P.: "A Method for Determination of Average Pressure in a Bounded Reservoir", *Trans.,* AIME (1954) **201**, 182-191.

McGlothlin, B. B., Jr., and Huitt, J. L.: "Relation of Formation Rock Strength to Propping Agent in Hydraulic Fracturing", *J. Pet. Tech.* (March, 1966) 377-384.

McGuire, W. J., Harrison, E. and Kieschnick, W. F.: "The Mechanics of Fracture Induction and Extension", *Trans.,* AIME (1954) **201**, 252-263.

McGuire, W. J. and Sikora, V. J.: "The Effect of Vertical Fractures on Well Productivity", *Trans.,* AIME (1960) **219,** 401-403.

McLamore, R. and Gray, K. E.: "The Mechanical Behavoir of Anisotropic Sedimentary Rock", paper 66-Pet-2 presented at the Petroleum Mechanical Engineering Conference, ASME, New Orleans, La., Sept. 18-21, 1966.

McLeod, H. O. and Coulter, A. W.: "The Use of Alcohol in Gas Well Stimulation", paper SPE 1633 presented at SPE Third Annual Eastern Regional Meeting, Columbus, Ohio, Nov. 10-11, 1966.

Mead, H.: "Another Concept for Final Buildup Pressure", paper 1111-G presented at SPE 33rd Annual Fall Meeting, Houston, Tex., Oct. 5-8, 1958.

Miles, A. J. and Topping, A. D.: "Stresses Around a Deep Well", *Trans.,* AIME (1949) **179,** 186-191.

Miller, C. C., Dyes, A. B. and Hutchinson, C. A., Jr.: "Estimation of Permeability and Reservoir Pressure from Bottom Hole Pressure Build-Up Characteristics", *Trans.,* AIME (1950) **189,** 91-104.

Monaghan, P. H., Salathiel, R. A., Morgan, B. E. and Kaiser, A. P., Jr.: "Laboratory Studies of Formation Damage in Sands Containing Clays", *Trans.,* AIME (1959) **216,** 209-215.

Morrisson, T. E. and Henderson, J. H.: "Gravity Drainage of Oil Into Large Horizontal Fractures", *Trans.,* AIME (1960) **219,** 7-15.

Murphy, W. B. and Juch, A. H.: "Pin-Point Sandfracturing — A Method of Simultaneous Injection Into Selected Sands", *J. Pet. Tech.* (Nov., 1960) 21-24.

Muskat, M.: *Flow of Homogeneous Fluids Through Porous Media,* McGraw-Hill Book Co., Inc., New York (1937) 145.

Muskat, M.: *Physical Principles of Oil Production,* McGraw-Hill Book Co., Inc., New York (1949) 242.

Nadai, A.: *Theory of Flow and Fracture of Solids,* 2nd ed., McGraw-Hill Book Co., Inc., New York (1950) **I,** 208, 340-343.

Nathans, M. W. and Holzer, A.: "Shock-Induced Transitions—Sample History and Sample Recovery", Report PNE-112F Part 1, AEC-UCRL (June, 1963).

Neill, G. H., Brown, R. W. and Simmons, C. M.: "An Inexpensive Method of Multiple Fracturing", *Drill. and Prod. Prac.,* API (1957) 27.

Neill, G. H., Dobbs, J. B., Pruitt, G. T. and Crawford, H. R.: "Field and Laboratory Results of Carbon Dioxide and Nitrogen in Well Stimulation", *J. Pet. Tech.* (March, 1964) 243-248.

Nikuradse, J.: *Forsch, Gebiete Ingenieurw Forschungsheft* (Sept.-Oct., 1932); reprinted in *Pet. Eng.* (1940) **XI,** Nos. 6, 8, 9, 11 and 12.

Nordyke, M. D.: "Cratering Experience with Chemical and Nuclear Explosives", *Engineering with Nuclear Explosives,* USAEC TID 7695 (1964) 51-73.

"Oil's Interest in A-Bomb Grows, But . . .", *Pet. Week* (Sept. 26, 1958) 13.

Orowan, E.: "Fatigue and Fracturing of Metals", paper presented at M.I.T. Symposium, Boston, Mass., June, 1950.

Ousterhout, R. S. and Hall, C. D., Jr.: "Reduction of Friction Loss in Fracturing Operations", *J. Pet. Tech.* (March, 1961) 217-222.

Perkins, T. K. and Kern, L. R.: "Widths of Hydraulic Fractures", *J. Pet. Tech.* (Sept., 1961) 937-949.

Perry, J. H.: *Chemical Engineers Handbook,* McGraw-Hill Book Co., Inc., New York (1950) 1238-1240.

Petroleum Production Handbook, Thomas C. Frick and R. William Taylor, Eds., McGraw-Hill Book Co., Inc., New York (1962) **2**, 47.

Phansalkar, A. K., Roebuck, A. H. and Scott, J. B.: U. S. Patent No. 3,046,222 (July 24, 1962).

Pinson, J. M., Simmons, J., Landrum, B. L. and Crawford, P. B.: "Effect of Large Elliptical Fractures on Sweep Efficiencies in Water Flooding or Fluid Injection Programs", *Prod. Monthly* (Nov., 1963) 20.

Platt, George: "A Study of Transient Pressure Behavior in Porous Media", unpublished MS thesis, U. of Tulsa, Tulsa, Okla. (1952).

Powell, J. P. and Johnston, K. H.: "Waterflood Fracturing Pays Off", *IPAA Monthly* (Oct., 1960) 22.

Prats, M.: "Effect of Vertical Fractures on Reservoir Behavoir — Incompressible Fluid Case", *Soc. Pet. Eng. J.* (June, 1961) 105-118.

"Project Gasbuggy—Feasibility Study", El Paso Natural Gas Co., USAEC and USBM (May 14, 1965).

"Project Rulison—New Try at Nuclear Stimulation", *World Oil* (Sept., 1969) 67-71.

Rabb, D. D.: "Large Borehole Drilling Cost", Lawrence Radiation Laboratory, Livermore, Calif. (to be published).

Rawson, D. E.: "Review and Summary of Some Project Gnome Results", Report UCRL 7166, Lawrence Radiation Laboratory, Livermore, Calif. (Dec., 1962).

Rawson, D. E.: "Review and Summary of Some Project Gnome Results", *Trans., AGU* (March, 1963).

Rawson, D. E., Boardman, C. R. and Jaffe, N.: "The Environment Created by a Nuclear Detonation in Salt", Report PNE-107F AEC (to be published).

Reistle, C. E.: U. S. Patent No. 2,368,424 (Jan. 30, 1945).

Reynolds, J. J., Popham, J. L., Scott, J. B. and Coffer, H. F.: "Hydraulic Fracture — Field Test to Determine Areal Extent and Orientation", *J. Pet. Tech.* (April, 1961) 371-376.

Rixe, F. H.: "Review of Oil Well Fracturing Theories", unpublished report, Pan American Petroleum Corp. (March 16, 1967).

Rixe, F. H., Fast, C. R. and Howard, G. C.: "Selection of Propping Agents for Hydraulic Fracturing", *Drill. and Prod. Prac.*, API (1963) 138.

Roberts, George, Jr.: "A Review of Hydraulic Fracturing and Its Effect on Exploration", paper presented at AAPG Meeting, Wichita, Kans., Oct. 2, 1953.

Root, R. L.: U. S. Patent No. 3,254,719 (June 7, 1966).

"Rulison Goes off Without Hitch after Needless Delay", *Oil and Gas J.* (Sept. 15, 1969) 40.

"Rulison Nuclear Device Is Finally Detonated", *World Oil* (Oct., 1969).

Sallee, W. L. and Rugg, F. E.: "Artificial Formation Fracturing in Southern Oklahoma and North Central Texas", *Pet. Eng.* (Feb., 1954) B-75; and *Bull.*, AAPG (Nov., 1953) **37,** No. 11, 2539-2550.

Scott, P. P., Jr., Bearden, W. G. and Howard, G. C.: "Rock Rupture as Affected by Fluid Properties", *Trans.*, AIME (1953) **198,** 111-124.

Seely, F. B.: *Resistance of Materials,* 3rd ed., John Wiley and Sons, Inc., New York (1947) 272.

Smith, C. F., Pavlich, J. P. and Slovinsky, R. L.: "Potassium, Calcium Treatments Inhibit Clay Swelling", *Oil and Gas J.* (Nov. 30, 1964) 80.

Smith, J. E.: "Design of Hydraulic Fracture Treatments", paper SPE 1286 presented at SPE 40th Annual Fall Meeting, Denver, Colo., Oct. 3-6, 1965.

Stead, F. W.: "Distribution in Groundwater of Radionuclides from Underground Nuclear Explosions", *Engineering with Nuclear Explosions*, USAEC TID 7695 (1964) 127-138.

Stekoll, Marion H.: "New Light on Fracturing Through Perforations", *Oil and Gas J.* (Oct. 29, 1956) 95.

Stewart, J. B. and Coulter, A. W.: "Increased Fracturing Efficiency by Fluid Loss Control", *Pet. Eng.* (June, 1959) B-43.

Teplitz, A. J. and Hassebroek, W. E.: "An Investigation of Oil Well Cementing", *Drill. and Prod. Prac.*, API (1946) 76.

Timoshenko, S.: *Strength of Materials,* 2nd ed., D. van Nostrand Co., Inc., New York (1941) Part I, 281 and 305; Part II, 474-476, 480.

Torrey, P. D.: "Progress in Squeeze Cementing Application and Technique", *Oil Weekly* (July 29, 1940).

van Everdingen, A. F.: "The Skin Effect and Its Influence on the Productive Capacity of a Well", *Trans.*, AIME, (1953) **198,** 171-176.

van Everdingen, A. F. and Hurst, W.: "The Application of the Laplace Transformation to Flow Problems in Reservoirs", *Trans.,* AIME (1949) **186,** 305-324.

van Poollen, H. K.: "Do Fracture Fluids Damage Productivity?", *Oil and Gas J.* (May 27, 1957) 120.

van Poollen, H. K.: "Theories of Hydraulic Fracturing", *Quarterly,* Colorado School of Mines, Golden, Colo. (July, 1957) **52**, No. 3, 113-131.

van Poollen, H. K., Tinsley, John M. and Saunders, Calvin D.: "Hydraulic Fracturing: Fracture Flow Capacity vs Well Productivity", *Trans.*, AIME (1958) **213**, 91-95.

Wahl, H. A.: "Fracture Design in Liquid Saturated Reservoirs", *J. Pet. Tech.* (April, 1963) 437-442.

Wahl, H. A.: "Horizontal Fracture Design Based on Propped Fracture Area", *J. Pet. Tech.* (June, 1965) 723-731.

Wahl, H. A. and Campbell, J. M.: "Sand Movement in Horizontal Fractures", paper SPE 564 presented at SPE Production Research Symposium, Norman, Okla., April 29-30, 1963.

Walsh, J. B. and Brace, W. F.: "A Fracture Criterion for Brittle Anisotropic Rocks", *J. Geophys. Research* (1964) **LXIX,** No. 16, 3449.

Wasson, J. A.: "The Application of Hydraulic Fracturing in the Recovery of Oil by Waterflooding: A Summary", IC 8175, USBM (1963).

Waters, A. B. and Tinsley, John M.: "Gas Well Deliverability Improved by Planned Fracturing Treatments", paper SPE 1275 presented at SPE 40th Annual Fall Meeting, Denver, Colo., Oct. 3-6, 1965.

Westergaard, H. M.: "Plastic State of Stress Around a Deep Well", *J. Boston Soc. Civil Eng.* (Jan., 1940).

Wilsey, L. E. and Bearden, W. G.: "Reservoir Fracturing — A Method of Oil Recovery from Extremely Low Permeability Formations", *Trans.*, AIME (1954) **201,** 169-175.

Wyllie, M. R. J. and Gregory, A. R.: "Fluid Flow Through Unconsolidated Porous Aggregates", *Ind. and Eng. Chem.* (1955) **XLVII**, 1379.

Yuster, S. T. and Calhoun, J. C., Jr.: "Pressure Parting of Formations in Water Flood Operations — Part I", *Oil Weekly* (March 12, 1945) 34.

Yuster, S. T. and Calhoun, J. C., Jr.: "Pressure Parting of Formations in Water Flood Operations — Part II", *Oil Weekly* (March 19, 1945) 34.

Author Index

(List of authors referred to in the Monograph text, references and the bibliography.)

Subject Index

G

H

I

R

S